TAKING SIDES

Clashing Views on Controversial

Environmental Issues

TENTH EDITION

TAKING SIDES

Clashing Views on Controversial

Environmental Issues

TENTH EDITION

Selected, Edited, and with Introductions by

Thomas A. Easton
Thomas College

and

Theodore D. Goldfarb

McGraw-Hill/Dushkin
A Division of The McGraw-Hill Companies

This book is dedicated to my children and grandchildren as well as all other children for whom the successful resolution of these issues is of great urgency (T. D. G.)

Photo Acknowledgment
Cover image: © 2003 by PhotoDisc, Inc.

Cover Art Acknowledgment
Charles Vitelli

Manufactured in the United States of America

Tenth Edition

123456789BAHBAH6543

Library of Congress Cataloging-in-Publication Data
Main entry under title:
Taking sides: clashing views on controversial environmental issues/selected, edited, and with introductions by Thomas A. Easton and Theodore D. Goldfarb.—10th ed.
Includes bibliographical references and index.
1. Environmental policy. 2. Environmental protection. I. Easton, Thomas A., *comp.* II. Goldfarb, Theodore D., *comp.*
363.7
0-07-285531-2
1091-8825

Printed on Recycled Paper

Preface

Theodore D. Goldfarb, who was a professor of chemistry at the State University of New York at Stony Brook, ably edited *Taking Sides: Clashing Views on Controversial Environmental Issues* through its first nine editions. In the spring of 2002, Ted succumbed after a long battle with cancer. I have since been asked to assume the editorship of this book.

I have already edited *Taking Sides: Clashing Views on Controversial Issues in Science, Technology, and Society* through five editions and will continue as editor of that title. As a professor of science at Thomas College, I have taught ecology, environmental science, and environmentalism for many years, so I am very pleased by the many new opportunities presented by editing *Taking Sides: Environmental Issues.*

As Ted noted in the preface to the ninth edition of this book, "Faculty are divided about whether or not it is appropriate to use a classroom to advocate a particular position on a controversial issue.... No matter whether the goal is to attempt an objective presentation or to encourage advocacy, it is necessary to present both sides of any argument. To be a successful proponent of any position, it is essential to understand your opponents' arguments."

Which answer to the issue question—yes or no—is the correct answer? Perhaps neither. Perhaps both. Students should read, think about, and discuss the readings and then come to their own conclusions without letting my or their instructor's opinions (which are likely to show at least some of the time!) dictate theirs. The additional readings mentioned in the introductions and postscripts should prove helpful.

It is worth stressing that the issues covered in this book are all live issues; that is, the debates they represent are active and ongoing. Some have been active for years; others are new. All are controversial, and I have chosen essays that show the opposing viewpoints on these issues as clearly and as understandably (nontechnically) as possible.

This edition of *Taking Sides: Environmental Issues* contains 38 readings arranged in pro and con pairs to form 19 issues. For each issue, an *introduction* provides historical background and a brief description of the debate. The *postscript* after each pair of readings offers more recent contributions to the debate, additional references, and sometimes a hint of future directions. Each part is preceded by an *On the Internet* page that lists several links that are appropriate for further pursuing the issues in that part.

Changes to this edition About two-thirds of this book consists of new material. The book's volume introduction is new. Also, there are six completely new issues: *Is Biodiversity Overprotected?* (Issue 2); *Are Environmental Regulations Too Restrictive?* (Issue 3); *Do Environmentalists Overstate Their Case?* (Issue 6); *Should the Arctic National Wildlife Refuge Be Opened to Oil Drilling?* (Issue 7);

Should DDT Be Banned Worldwide? (Issue 8); and *Do Human Activities Threaten to Change the Global Climate?* (Issue 12). In addition, for nine of the issues retained from the previous edition, one or both of the readings have been replaced. In all, 24 of the 38 readings are new to this edition.

A word to the instructor   An *Instructor's Manual With Test Questions* (multiple-choice and essay) is available through the publisher for the instructor using *Taking Sides* in the classroom. Also available is a general guidebook, *Using Taking Sides in the Classroom,* which offers suggestions for adapting the pro-con approach in any classroom setting. An online version of *Using Taking Sides in the Classroom* and a correspondence service for Taking Sides adopters can be found at http://www.dushkin.com/usingts/.

Taking Sides: Clashing Views on Controversial Environmental Issues is only one title in the Taking Sides series. If you are interested in seeing the table of contents for any of the other titles, please visit the Taking Sides Web site at http://www.dushkin.com/takingsides/.

Thomas A. Easton
Thomas College

Contents In Brief

iv

Contents

Janet N. Abramovitz, a senior researcher at the Worldwatch Institute, argues that if we fail to attach economic value to supposedly free services provided by nature, we are more likely to misuse and destroy the ecosystems that provide those services. Professors of applied ecology Marino Gatto and Giulio A. De Leo contend that the pricing approach to valuing nature's services is misleading because it falsely implies that only economic values matter.

Professor of economics David N. Laband argues that the public demands excessive amounts of biodiversity largely because decision makers and voters do not have to bear the costs of producing it. In an interview with science writer Kris Christen, biologist E. O. Wilson argues that biodiversity is crucial to human survival and that efforts need to be increased to protect it. He maintains that the loss of species reduces the productivity and stability of natural ecosystems and that with each species lost, potential drugs and other valuable resources are also lost.

Peter W. Huber, a senior fellow at the Manhattan Institute, argues that the environment is best protected by traditional conservation, which puts

PART 2 ENVIRONMENTAL IMPACTS 117

Professor of economics Dwight R. Lee argues that the economic and other benefits of Arctic National Wildlife Refuge (ANWR) oil are so great that even environmentalists should agree to permit drilling—and they probably would if they stood to benefit directly. Physicist Amory B. Lovins and lawyer L. Hunter Lovins assert that recovering ANWR oil is too costly and too vulnerable to disruption. They hold that alternatives such as developing greater fuel efficiency are wiser choices for meeting future energy needs.

Anne Platt McGinn, a senior researcher at the Worldwatch Institute, argues that although DDT is still used to fight malaria, there are other, more effective and less environmentally harmful methods. She maintains that DDT should be banned or reserved for emergency use. Roger Bate, director of Africa Fighting Malaria, asserts that DDT is the cheapest and most effective way to combat malaria and that it should remain available for use.

The national academies of science of the United Kingdom, the United States, Brazil, China, India, Mexico, and the Third World argue that genetically modified crops hold the potential to feed the world during the twenty-first century while also protecting the environment. Brian Halweil, a researcher at the Worldwatch Institute, argues that the genetic modification of crops threatens to produce pesticide-resistant insect pests and herbicide-resistant weeds, will victimize poor farmers, and is unlikely to feed the world.

Professor of urban and environmental policy Sheldon Krimsky summarizes the evidence indicating that many chemicals released to the environment affect the endocrine systems of animals and humans and may threaten human health with cancers, reproductive anomalies, and neurological effects. Toxicologist Stephen H. Safe argues that the suggestion that industrial estrogenic compounds contribute to increased cancer incidence and reproductive problems in humans is not plausible.

Carol M. Browner, administrator for the Environmental Protection Agency (EPA), summarizes the evidence and arguments that were the basis for the EPA's proposal for more stringent standards for ozone and particulates. Daniel B. Menzel, a professor of environmental medicine and a researcher on air pollution toxicology, argues that adequate research has not been done to demonstrate that the new standards will result in the additional public health benefits that would justify the difficulty and expense associated with their implementation.

The Intergovernmental Panel on Climate Change states that global warming appears to be real, with strong effects on sea level, ice cover, and rainfall patterns to come, and that human activities—particularly emissions of carbon dioxide—are to blame. Neuroscience researcher Kevin A. Shapiro argues that past global warming predictions have been wrong and that the data do not support calls for immediate action to reduce emissions of carbon dioxide.

PART 3 DISPOSING OF WASTES 257

DuPont corporate counsel Bernard J. Reilly argues that the Superfund legislation has led to unfair standards and waste cleanup cost delegation. *Audubon* contributing editor Ted Williams warns against turning Superfund into a public welfare program for polluters.

Environmental Defense Fund scientist Richard A. Denison and economic analyst John F. Ruston rebut a series of myths that they say have been promoted by industrial opponents in an effort to undermine the environmentally valuable and successful recycling movement. Engineering and economics researchers Chris Hendrickson, Lester Lave, and Francis McMichael assert that ambitious recycling programs are often too costly and are of dubious environmental value.

Secretary of Energy Spencer Abraham argues that the Yucca Mountain, Nevada, nuclear waste disposal site is suitable technically and scientifically and that its development serves the U.S. national interest in numerous ways. Science writer Jon Christensen argues that it is impossible to forecast with confidence that nuclear waste entombed in Yucca Mountain will not threaten the environment over the next 10,000 (or more) years.

Professor of management Dinah M. Payne and professor of accounting Cecily A. Raiborn argue that environmental responsibility and sustainable development are essential parts of modern business ethics and that only through them can both business and humans thrive. Professor of economics Jacqueline R. Kasun argues that sustainable development poses threats to human freedom, dignity, and material welfare.

Introduction

Environmental Issues:
The Never-Ending Debate

Thomas A. Easton

One of the courses I teach is called "Environmentalism: Philosophy, Ethics, and History." I begin this course by explaining the roots of the word *ecology,* from the Greek word *oikos* (house or household), and then assigning the students to write a brief paper about their own households. How much, I ask them, do you need to know about the place where you live? And why?

The answers vary. Some of the resulting papers focus on people—roommates if the "household" is a dorm room, spouse and children if the student is older, parents and siblings if the student lives at home—and the need to cooperate and get along, and sometimes the need not to overcrowd. Some pay attention to houseplants, pets, and occasionally even bugs and mice. Some focus on economics—possessions, services, and their costs; where the checkbook is kept; where the bills accumulate; the importance of paying those bills; and, of course, the importance of earning money to pay those bills. Some focus on maintenance—cleaning, cleaning supplies, repairs, and whom to call if something major breaks. For some, the emphasis is on operation—running the garbage disposal, grocery shopping, working the lights and the stove, and so on. A very few recognize the presence of toxic chemicals under the sink and in the medicine cabinet and the need for precautions in their handling.

Not surprisingly, some students initially object that this exercise seems trivial. "What does this have to do with environmentalism?" they ask. Yet the course is rarely very old before most are saying, "Ah! I get it!" That nice, homey microcosm has a great many of the features of the macrocosmic environment, and the multiple ways that people can look at the microcosm mirror the ways that people look at the macrocosm. It is all there, as is the question of priorities: Which is most important—people, fellow creatures, economics, maintenance, operation, waste disposal, food supply, or toxics control? Or are they all equally important?

And how does one decide? I try to illuminate this question by describing a parent trying to teach a teenager not to sit on a woodstove. In July, the kid asks, "Why?" and continues to perch. In August, likewise. And still in September. But in October or November, the kid yells "Ouch!" and jumps off in a hurry.

That is, people seem to learn best when they get burned.

This is surely true in our homely *oikos,* where we may not realize that our fellow creatures deserve attention until the houseplants die of neglect or cockroaches invade the cupboards. Similarly, economics comes to the fore when the phone gets cut off, repairs when a pipe ruptures, air quality when the air conditioner breaks or strange fumes rise from the basement, and garbage disposal when the bags pile up and begin to stink. Toxics control suddenly matters when a child or pet gets into the rat poison.

In the larger *oikos* of environmentalism, such events are analogous to the loss of a species or an infestation by another, to floods and droughts, to lakes being turned into cesspits by raw sewage, to air being fouled by industrial smokestacks, to the contamination of groundwater by toxic chemicals, to the death of industries and the loss of jobs, and to famine and plague and even war.

If nothing is wrong in our households, we are not very likely to realize that there is something we should be paying attention to. And this, too, has its parallel in the larger world. Indeed, the history of environmentalism is, in part, a history of people carrying on with business as usual until something goes obviously awry. Then, if they can agree on the nature of the problem (Did the floor cave in because the joists were rotten or because there were too many people at the party?), they might learn something about how to prevent its recurrence.

The Question of Priorities

It is a truism that agreement is difficult. In environmental matters, people argue endlessly over whether or not anything is actually wrong, what a problem's eventual impact will be, what (if anything) can be done to repair any resulting damage, and how to prevent recurrence—not to mention who is to blame and who should take responsibility for fixing the problem! Part of the reason is simple: Different things matter to different people. For example, individual citizens might want clean air and water, or cheap food, or a convenient commute. Politicians might favor sovereignty over international cooperation. Economists and industrialists might consider a few coughs (or worse) a cheap price to pay for wealth or jobs.

No one now seems to think that protecting the environment is not important. But different groups—even different environmentalists—have different ideas of what "environmental responsibility" means. To a paper company that cuts trees for pulp, it might mean leaving a screen of trees (a "beauty strip") beside the road and minimizing erosion. To hikers following trails through or within view of the same tract of land, that might not be enough; they might want the trees left alone. The hikers might also object to people using trail bikes and all-terrain vehicles on the trails. They might even object to hunters and anglers, whose activities they see as diminishing the wilderness experience. They might push for protecting the land as limited-access wilderness. The hunters and anglers would object to that, of course, because they want to be able to use their vehicles to bring their game home or to bring their boats to their favorite rivers and lakes. They could also argue, with some justification, that their license fees support a great deal of environmental protection work.

To a corporation, dumping industrial waste into a river might make perfect sense because alternative ways of disposing of waste are likely to cost more and diminish profits. Of course, the waste renders the water less useful to wildlife or humans living downstream, who might well object. Yet preventing the corporation from dumping might be seen as depriving it of property. A similar problem arises when regulations prevent people and corporations from using land—and making money—as they had planned. Conservatives have argued that environmental regulations thus violate the Fifth Amendment to the U.S. Constitution, which says, "No person shall ... be deprived of ... property, without due process of law; nor shall private property be taken for public use, without just compensation."

One might think that the dangers of dumping industrial waste into rivers are obvious. But scientists can and do disagree about the consequences of such activities, even given the same evidence. For instance, a chemical in waste may clearly cause cancer in laboratory animals. Is it therefore a danger to humans? A scientist working for a company that is dumping that chemical in a river might maintain that no such danger has been proven. Yet a scientist working for an environmental group such as Greenpeace might argue that the danger is obvious because carcinogens generally affect more than one species.

Scientists are human. They have values rooted in political ideology and religion. They might feel that the individual matters more than corporations or society, or vice versa. They might also favor short-term benefits over long-term benefits, or vice versa. And scientists, citizens, corporations, and government all reflect prevailing social attitudes. When America was expanding westward, the focus was on building industries, farms, and towns. When problems arose, there was vacant land waiting to be moved to. But when the expansion was done, problems became more visible and less avoidable. People could see that there were trade-offs involved in human activity: more industry meant more jobs and more wealth, but there was a price in air and water pollution and adverse effects on human health (among other things).

Nowhere, perhaps, are these trade-offs more obvious than in the former Soviet Union, which was infamous for refusing to admit that industrial activity was anything but desirable. Any citizen who spoke up about environmental problems risked being jailed. The result, which became visible to Western nations after the fall of the Iron Curtain in 1990, was industrial zones in which rivers had no fish, children were sickly, and life expectancies were reduced. The fate of the Aral Sea, a vast inland body of water once home to a thriving fishery and a major regional transportation route, is emblematic: Because the Soviet Union wanted to increase its cotton production, it diverted the rivers that delivered most of the Aral Sea's freshwater supply for irrigation. The sea then began to lose more water to evaporation than it gained, and it rapidly shrank, exposing a sea bottom that is so contaminated by industrial wastes and pesticides that wind-borne dust is now responsible for a great deal of human illness. The fisheries are dead, and freighters lie rusting on bare ground where waves once lapped.

The Environmental Movement

The twentieth century saw immense changes in the conditions of human life and in the environment that surrounds and supports human life. According to historian J. R. McNeill, in *Something New Under the Sun: An Environmental History of the Twentieth-Century World* (W. W. Norton, 2000), the environmental impacts that resulted from the interactions of burgeoning population, technological development, shifts in energy use, politics, and economics during that period are unprecedented in both degree and kind. Yet a worse impact might be that people have come to accept as "normal" a very temporary situation that "is an extreme deviation from any of the durable, more 'normal,' states of the world over the span of human history, indeed over the span of earth history." Thus, people are not prepared for the inevitable and perhaps drastic changes ahead.

Environmental factors cannot be denied their role in human affairs. Nor can human affairs be denied their place in any effort to understand environmental change. As McNeill says, "Both history and ecology are, as fields of knowledge go, supremely integrative. They merely need to integrate with each other."

The environmental movement, which grew during the twentieth century in response to increasing awareness of human impacts, is a step in that direction. Yet environmental awareness was evident long before the modern environmental movement. When he was young, John James Audubon (1785–1851), famous for his bird paintings, was an enthusiastic slaughterer of birds (some of which he used as models for his paintings). Later in life, he came to appreciate that birds were diminishing in numbers, as was the American bison, and he called for conservation measures. His was a minority voice, however. It was not until later in the century that John Muir (1838–1914), founder of the Sierra Club, began to call for the preservation of natural wilderness, untouched by human activities. In 1890 Gifford Pinchot (1865–1946) found "the nation... obsessed by a fury of development. The American Colossus was fiercely intent on appropriating and exploiting the riches of the richest of all continents." Under President Theodore Roosevelt, he became the first head of the U.S. Forest Service and a strong voice for conservation (not to be confused with preservation; Pinchot's conservation meant using nature in such a way that it was not destroyed; his aim was "the greatest good of the greatest number in the long run"). In the 1930s Aldo Leopold (1887–1948), best known for his concept of the "land ethic" and his book *A Sand County Almanac, and Sketches Here and There* (Oxford University Press, 1949), argued that people had a responsibility not only to maintain the environment but also to repair damage that was done in the past.

The modern environmental movement was kick-started by the publication of Rachel Carson's *Silent Spring* (Houghton Mifflin, 1962). In the 1950s Carson realized that the use of pesticides was having unintended consequences —the death of nonpest insects, food chain accumulation of poisons and the consequent loss of birds, and even human illness—and meticulously documented the case. When her book was published, she and it were immediately vilified by pesticide proponents in government, academia, and industry (most notably, the pesticides industry). There was no problem, the critics said; the negative effects, if any, were outweighed by the benefits; and she—a *woman* and a nonscientist —could not possibly know what she was talking about. But the facts won out. A decade later, DDT was banned, and other pesticides were regulated in ways unheard of before Carson spoke out.

Other issues have followed or are following a similar course.

The situation before Rachel Carson and *Silent Spring* is nicely captured by Judge Richard Cudahy, who, in "Coming of Age in the Environment," *Environmental Law* (Winter 2000), writes, "It doesn't seem possible that before 1960 there was no 'environment'—or at least no environmentalism. I can even remember the Thirties, when we all heedlessly threw our trash out of car windows, burned coal in the home furnace (if we could afford to buy any), and used a lot of lead for everything from fishing sinkers and paint to no-knock gasoline. Those were the days when belching black smoke meant a welcome end to the Depression and little else."

Historically, humans have felt that their own well-being mattered more than anything else. The environment existed to be used. Unused, it was only wilderness or wasteland, awaiting the human hand to "improve" it and make it valuable. This is not surprising, for the natural tendency of the human mind is to appraise all things in relation to the self, the family, and the tribe. An important aspect of human progress has lain in enlarging our sense of "tribe" to encompass nations and groups of nations. Some now take it as far as the human species. Some include other animals. Some embrace plants, bacteria, and even landscapes, as well.

The more limited standard of value remains common. Add to that a sense that wealth is not just desirable but a sign of virtue (the Puritans brought an explicit version of this with them when they colonized North America; see Lynn White, Jr., "The Historical Roots of Our Ecological Crisis," *Science* [March 10, 1967]), and it is hardly surprising that humans have used and still use the environment intensely. People also tend to resist any suggestion that they should restrain their use out of regard for other living things. Human needs, many insist, come first.

The unfortunate consequences include the loss of other species. For example, lions vanished from Europe about 2000 years ago. The dodo of Mauritius was extinguished in the 1600s (see the American Museum of Natural History's account at http://www.amnh.org/exhibitions/expeditions/treasure_fossil/Treasures/Dodo/dodo.html?dinos). And the last of North America's passenger pigeons died in a Cincinnati zoo in 1914 (see http://www.amnh.org/exhibitions/expeditions/treasure_fossil/Treasures/Passenger_Pigeons/pigeons.html?dinos). Concern for species was at first limited to those of obvious value to humans. In 1871 the U.S. Commission on Fish and Fisheries was created and charged with finding solutions to the decline in food fishes and with promoting aquaculture; the first federal legislation designed to protect game animals was the Lacey Act of 1900. It was not until 1973 that the U.S. Endangered Species Act was adopted to shield all species from human impacts.

Other unfortunate consequences of human activities include dramatic erosion, air and water pollution, oil spills, accumulations of hazardous (including nuclear) waste, famine, and disease. Among the many "hot stove" incidents that have caught public attention are the following:

- The Dust Bowl—in 1934 wind blew soil from drought-stricken farms in Oklahoma all the way to Washington, D.C.;
- Cleveland's Cuyahoga River caught fire in the 1960s;
- The Donora, Pennsylvania, smog crisis—in one week in October 1948, 20 people died and over 7,000 were sickened;
- The London smog crisis in December 1952—4,000 dead;
- The *Torrey Canyon* and *Exxon Valdez* oil spills, which fouled shores and killed seabirds, seals, and fish;
- Love Canal, where industrial wastes seeped from their burial site into homes and contaminated ground water;
- Union Carbide's toxics release at Bhopal, India—3,800 dead and up to 100,000 ill, according to Union Carbide; others claim a higher toll;
- The Three Mile Island and Chernobyl nuclear accidents;
- The decimation of elephants and rhinoceroses to satisfy a market for tusks and horns;
- The loss of forests—in 1997 fires set to clear Southeast Asian forest lands produced so much smoke that regional airports had to close;
- Ebola, a virus that kills nine-tenths of those it infects, apparently first struck humans because growing populations reached into its native habitat;
- West Nile Fever, a mosquito-borne virus with a much less deadly record than Ebola, was brought to North America by travelers or immigrants from Egypt;
- Acid rain, global climate change, and ozone depletion, all caused by substances released into the air by human activities.

The alarms have been raised by many people in addition to Rachel Carson. For instance, in 1968 (when world population was only a little over half of what it is today) Paul Ehrlich described the ecological threats of a rapidly growing population in *The Population Bomb* (Ballantine Books), and Garrett Hardin described the consequences of using self-interest alone to guide the exploitation of publicly owned resources (such as air and water) in his influential essay "The Tragedy of the Commons," *Science* (December 13, 1968). (In 1974 Hardin introduced the unpleasant concept of "lifeboat ethics," which says that if there are not enough resources to go around, some people must do without.) In 1972 a group of economists, scientists, and business leaders calling themselves "The Club of Rome" published *The Limits to Growth* (Universe Books), an analysis of population, resource use, and pollution trends that predicted difficult times within a century. The study was redone in 1992 using more powerful computer models, and the researchers came to very similar conclusions (see *Beyond the Limits: Confronting Global Collapse, Envisioning a Sustainable Future* [Chelsea Green, 1992]).

The following list of selected U.S. and UN laws, treaties, conferences, and reports illustrates the national and international responses to these alarms:

1967 The U.S. Air Quality Act set standards for air pollution.

1968 The UN Biosphere Conference discussed global environmental problems.

1969 The U.S. Congress passed the National Environmental Policy Act, which (among other things) required federal agencies to prepare environmental impact statements for their projects.

1970 The first Earth Day demonstrated so much public concern that the Environmental Protection Agency (EPA) was created; the Endangered Species Act, Clean Air Act, and Safe Drinking Water Act soon followed.

1971 The U.S. Environmental Pesticide Control Act gave the EPA the authority to regulate pesticides.

1972 The UN Conference on the Human Environment, held in Stockholm, Sweden, recommended government action and led to the UN Environment Programme.

1973 The Convention on International Trade in Endangered Species of Wild Fauna and Flora (CITES) restricted trade in threatened species; because enforcement was weak, however, a black market flourished.

1976 The U.S. Resource Conservation and Recovery Act and the Toxic Substances Control Act established control over hazardous wastes and other toxic substances.

1979 The Convention on Long-Range Transboundary Air Pollution addressed problems such as acid rain (recognized as crossing national borders in 1972).

1982 The Law of the Sea addressed marine pollution and conservation.

1982 The second UN Conference on the Human Environment (the Stockholm +10 Conference) renewed concerns and set up a commission to prepare a "global agenda for change," leading to the 1987 Brundtland report (*Our Common Future*).

1983 The U.S. Environmental Protection Agency and the U.S. National Academy of Science issued reports calling attention to the prospect of global warming as a consequence of the release of greenhouse gases such as carbon dioxide.

1987 The Montreal Protocol (strengthened in 1992) required nations to phase out the use of chlorofluorocarbons (CFCs), the chemicals responsible for stratospheric ozone depletion.

1987 The Basel Convention controlled cross-border movement of hazardous wastes.

1988 The UN assembled the Intergovernmental Panel on Climate Change, which would report in 1995, 1998, and 2001 that the dangers of global warming were real, large, and increasingly ominous.

1992 The UN Convention on Biological Diversity required nations to act to protect species diversity.

1992 The UN Conference on Environment and Development (also known as the Earth Summit), held in Rio de Janeiro, Brazil, issued a broad call for environmental protections.

1992 The UN Convention on Climate Change urged restrictions on carbon dioxide release to avoid climate change.

1994 The UN Conference on Population and Development, held in Cairo, Egypt, called for the stabilization and reduction of global population growth, largely by improving women's access to education and health care.

1997 The Kyoto Protocol attempted to strengthen the 1992 Convention on Climate Change by requiring reductions in carbon dioxide emissions, but U.S. resistance limited its success.

2000 The Treaty on Persistent Organic Pollutants required nations to phase out the use of many pesticides and other chemicals.

2002 The UN World Summit on Sustainable Development, held in Johannesburg, South Africa, brought together representatives of governments, nongovernmental organizations, businesses, and other groups to examine "difficult challenges, including improving people's lives and conserving our natural resources in a world that is growing in population, with ever-increasing demands for food, water, shelter, sanitation, energy, health services and economic security."

Rachel Carson would surely have been pleased by these responses, for they suggest both concern over the problems identified and determination to solve those problems. But she would just as surely have been frustrated, for a simple listing of laws, treaties, and reports does nothing to reveal the endless wrangling and the way political and business forces try to block progress whenever it is seen as interfering with their interests. Agreement on banning chlorofluorocarbons was relatively easy to achieve because CFCs were not seen as essential to civilization and because substitutes were available. Restraining greenhouse gas emissions is harder because fossil fuels are considered essential. Also, although fuel substitutes do exist, they are more expensive.

The Globalization of the Environment

Years ago environmental problems were largely seen as local. A smokestack belched smoke and made the air foul. A city sulked beneath a layer of smog. Bison or passenger pigeons declined in numbers and even vanished. Rats flourished in a dump where burning garbage produced clouds of smoke and runoff contaminated streams and groundwater and made wells unusable. Sewage, chemical wastes, and oil killed the fish in streams, lakes, rivers, and harbors. And toxic chemicals such as lead and mercury entered the food chain and affected the health of both wildlife and people.

By the 1960s it was becoming clear that environmental problems did not respect borders. Smoke blows with the wind, carrying one locality's contamination to others. Water flows to the sea, carrying sewage and other wastes with it. Birds migrate, carrying with them whatever toxins they have absorbed with their food. In 1972 researchers reported that most of the acid rain falling on Sweden came from other countries. Other researchers have shown that the rise and fall of the Roman Empire can be tracked in Greenland, where glaciers preserve lead-containing dust deposited over the millennia—the amount rises as Rome flourished, falls with the Dark Ages, and rises again with the Renaissance and Industrial Revolution. Today it is common knowledge that pesticides and other chemicals can show up in places where they have never been used (such as the Arctic), even years after their use has been discontinued. The 1979 Convention on Long-Range Transboundary Air Pollution has been strengthened several times with amendments to address persistent organic pollutants, heavy metals, and other pollutants.

There are also new environmental problems that exist only in a global sense. Ozone depletion, first identified in the stratosphere over Antarctica, threatens to increase the amount of ultraviolet light reaching the ground, thereby increasing the incidence of skin cancer and cataracts, among other things. The cause is the use of chlorofluorocarbons in refrigeration, air conditioning, aerosol cans, and electronics (for cleaning grease off circuit boards) by the industrialized world. The effect is global. Worse yet, the cause is rooted in northern lands such as the United States and Europe, but the worst effects may be felt where the sun shines brightest—in the tropics, which are dominated by developing nations. A serious issue of justice or equity is therefore involved.

A similar problem arises with global warming, which is also rooted in the industrialized world and its use of fossil fuels. The expected climate effects will hurt mostly the poorer nations of the tropics, perhaps worst of all those on low-lying South Pacific islands, which are expected to be wholly inundated by rising seas.

Both the developed and the developing world are aware of the difficulties posed by environmental issues. In Europe, "green" political parties play a growing part in government. In Japan, some environmental regulations are more demanding than those of the United States. Developing nations understandably place dealing with their growing populations high on their list of priorities, but they also play an important role in UN conferences on environmental issues, often demanding more responsible behavior from developed nations such as the United States (which often resists these demands; it has refused to ratify international agreements such as the Kyoto Protocol, for example).

Western scholars have been known to suggest that developing nations should forgo industrial development because if their huge populations ever attain the same per-capita environmental impact as the populations of wealthier lands, the world will be laid waste. It is not hard to understand why the developing nations object to such suggestions; they too want a better standard of living. Nor do they think it fair that they should suffer for the environmental sins of others.

Are global environmental problems so threatening that nations must surrender their sovereignty to international bodies? Should the United States or Europe have to change energy supplies to protect South Pacific nations? Should developing nations be obliged to reduce birth rates or forgo development because their population growth is seen as exacerbating pollution or threatening biodiversity?

Questions such as these play an important part in global debates today. They are not easy to answer, but their very existence says something important about the general field of environmental studies. This field is based in the science of ecology, a word whose root is that same *oikos* with which I began. Ecology focuses on living things and their interactions with each other and their surroundings. It deals with resources, limits, and coexistence. It can see problems, their causes, and even potential solutions. And it can turn its attention to human beings as easily as it can to deer mice.

Yet human beings are not mice. We have economies and political systems, vested interests, and conflicting priorities and values. Ecology is only one part of environmental studies. Other sciences—chemistry, physics, climatology, epidemiology, geology, and more—are involved. So are economics, history, law, and politics. Even religion can play a part.

Unfortunately, no one field sees enough of the whole to predict problems (the chemists who developed CFCs could hardly have been expected to realize what would happen when these chemicals reached the stratosphere). Environmental studies is a field for teams. That is, it is a holistic, multidisciplinary field.

This gives us an important basic principle to use when evaluating arguments on either side of any environmental issue: Arguments that fail to recognize the complexity of the issue are necessarily suspect. On the other hand, arguments that endeavor to convey the full complexity of an issue may be impossible to understand. A middle ground is essential for clarity, but any reader or student must realize that something important might be left out.

Current Environmental Issues

In 2001 the National Research Council's Committee on Grand Challenges in Environmental Sciences published *Grand Challenges in Environmental Sciences* (National Academy Press) in an effort to reach "a judgment regarding the most important environmental research challenges of the next generation—the areas most likely to yield results of major scientific and practical importance if pursued vigorously now." These areas include the following:

- Biogeochemical cycles (the cycling of plant nutrients; the ways human activities affect them; and the consequences for ecosystem functioning, atmospheric chemistry, and human activities)
- Biological diversity
- Climate variability
- Hydrologic forecasting (groundwater, droughts, floods, etc.)
- Infectious diseases
- Resource use
- Land use
- Reinventing the use of materials (e.g., recycling)

Some of these "grand challenges" are covered in this book. There are, of course, a great many other environmental issues—many more than can be covered in any one book such as this one. I have not tried to deal here with invasive species, the Endangered Species Act, the removal of dams to restore populations of anadromous fishes such as salmon, the depletion of aquifers, floodplain development, urban planning, and many others. My sample of the variety of available issues begins with the more philosophical ones. For instance, as I have already said, many people believed (and still believe) that nature has value only when it is turned to human benefit. One consequence of this belief is that it may be easier to convince people that nature is worth protecting if one can somehow calculate a cash value for nature "in the raw." Some environmentalists object to attempts to do this because they believe that economic value is not the only value or even one that should matter at all (see Issue 1).

What other values might be considered? Perhaps nature has a value all its own or a right to exist unmolested. Perhaps human property rights should take precedence (see Issue 3). Perhaps the aim should be social justice (Issue 4).

There is also considerable debate over the "precautionary principle," which says in essence that even if we are not sure that our actions will have unfortunate consequences, we should take precautions just in case (see Issue 5). This principle plays an important part in many environmental debates, from those over the value of preserving biodiversity (Issue 2), to the wisdom of opening the Arctic National Wildlife Refuge to oil drilling (Issue 7), to the folly (or wisdom) of burying nuclear waste under Yucca Mountain in Nevada (Issue 15).

Should we be concerned about the environmental impacts of specific human actions or products? Here, too, we can consider the wisdom of opening the Arctic National Wildlife Refuge to oil drilling as well as the conflict between the value of DDT for preventing malaria and its impact on ecosystems (Issue 8), the hormone-like effects of some pesticides and other chemicals on both wildlife and humans (Issue 10), and the hazards of air pollutants (Issue 11) and global warming (Issue 12). Another concern is genetic engineering: although it promises to do wonders for food production, some worry about its potential effects on ecosystems (Issue 9).

Waste disposal is a problem area all its own. It encompasses not only nuclear waste (Issue 15) but also hazardous waste (Issue 13) and municipal waste (Issue 14). A new angle on hazardous waste comes from the popularity of the personal computer—or more specifically, from the huge numbers of computers that are discarded each year.

What solutions are available? Some are specific to particular issues, as recycling is to waste handling (Issue 14). Some are more general, as evidenced by the issue on whether or not population growth is a primary cause of environmental problems (Issue 16).

Some analysts argue that whatever solutions are needed, government does not need to impose them all. Some maintain that private industry can be trusted to find and implement enough solutions voluntarily to reduce the need for regulations (Issue 18). Such voluntary action would perhaps be more likely if government could find a way to motivate industry, such as with tradable pollution rights (Issue 17).

The overall aim, of course, is to avoid disaster and enable human life and civilization to continue prosperously into the future. The term for this is *sustainable development* (see Issue 19), and it was the chief concern of the UN World Summit on Sustainable Development, held in Johannesburg, South Africa, in August 2002. On the other hand, there are people who think that this is a non-issue because people today are better off than ever before and because, in their eyes, "environmentalism" might more honestly be called "exaggerationism" (see Issue 6).

On the Internet ...

Leadership for Environment and Development International

Leadership for Environment and Development (LEAD), set up in 1991 by the Rockefeller Foundation, is an international network of professionals committed to sustainable development. Among other things, it provides this Web site, where people can calculate their "ecological footprint"—a rough estimate of how much of the earth's land, water, and other resources must be used to meet the needs of their personal lifestyle.

http://www.lead.org/leadnet/footprint/

The Convention on International Trade in Endangered Species of Wild Fauna and Flora

The Convention on International Trade in Endangered Species of Wild Fauna and Flora (CITES) is an international agreement between governments that aims to ensure that international trade in specimens of wild animals and plants does not threaten their survival.

http://www.cites.org

Environmental Justice

The U.S. Environmental Protection Agency pursues environmental justice under the Office of Enforcement and Compliance Assurance as part of its "firm commitment to the issue of environmental justice and its integration into all programs, policies, and activities, consistent with existing environmental laws and their implementing regulations."

http://www.epa.gov/compliance/environmentaljustice/

National Wildlife Federation

The National Wildlife Federation is the nation's largest member-supported conservation group, dedicated to protecting wildlife, wilderness, and the environment.

http://www.nwf.org

Natural Resources Defense Council

The Natural Resources Defense Council is one of the most active environmental research and advocacy organizations. Among its many concerns are clean air and water, energy, global warming, toxic chemicals, and nuclear waste.

http://www.nrdc.org

PART 1

Philosophy and Politics

*E*nvironmental debates are rooted in questions of values and are inevitably political in nature. It is worth stressing that people who consider themselves to be environmentalists can be found on both sides of most of the issues in this book. They differ in what they see as their own self-interests and even in what they see as humanity's long-term interests.

To understand the general issues raised in this section is to be prepared for the more specific issues raised later in this book.

- Should a Price Be Put on the Goods and Services Provided by the World's Ecosystems?

- Is Biodiversity Overprotected?

- Are Environmental Regulations Too Restrictive?

- Should Environmental Policy Attempt to Cure Environmental Racism?

- Is the Precautionary Principle a Sound Basis for International Policy?

- Do Environmentalists Overstate Their Case?

1

ISSUE 1

Should a Price Be Put on the Goods and Services Provided by the World's Ecosystems?

YES: Janet N. Abramovitz, from "Putting a Value on Nature's 'Free' Services," *World Watch* (January/February 1998)

NO: Marino Gatto and Giulio A. De Leo, from "Pricing Biodiversity and Ecosystem Services: The Never-Ending Story," *BioScience* (April 2000)

ISSUE SUMMARY

YES: Janet N. Abramovitz, a senior researcher at the Worldwatch Institute, argues that if we fail to attach economic value to supposedly free services provided by nature, we are more likely to misuse and destroy the ecosystems that provide those services.

NO: Professors of applied ecology Marino Gatto and Giulio A. De Leo contend that the pricing approach to valuing nature's services is misleading because it falsely implies that only economic values matter.

Human activities frequently involve trading a swamp or forest or mountainside for a parking lot or housing development or farm. People generally agree that the parking lot, housing development, or farm is a worthwhile project, for it has obvious benefits—expressible in economic terms—to human beings. But are there costs as well? Construction costs, labor costs, and material costs can easily be calculated, but what about the swamp? The forest? The species living there?

How much is a species worth? One approach to answering this question is to ask people how much they would be willing to pay to keep a species alive. If the question is asked when there are a million species in existence, few people will likely be willing to pay much. But if the species is the last one remaining, they might be willing to pay a great deal. Most people would agree that both answers fail to get at the true value of a species, for nature is not expressible solely in terms of cash values. Yet some way must be found to weigh the effects of human activities on nature against the benefits gained from those activities. If it is not, we will continue to degrade the world's ecosystems and threaten our own continued well-being.

Traditional economics views nature as a "free good." That is, forests generate oxygen and wood, clouds bring rain, and the sun provides warmth, all without charge to the humans who benefit. At the same time, nature has provided ways for people to dispose of wastes—such as dumping raw sewage into rivers or emitting smoke into the air—without paying for the privilege. This "free" waste disposal has turned out to have hidden costs in the form of the health effects of pollution (among other things), but it has been up to individuals and governments to bear the costs associated with those effects. The costs are real, but in general, they have not been borne by the businesses and other organizations that produced them. They have thus come to be known as "external" costs.

Environmental economists have recognized the problem of external costs, and government regulators have devised a number of ways to make those who are responsible accept the bill, such as instituting requirements for pollution control and fining those who exceed permitted emissions. Yet some would say that this approach does not help enough.

The *ecosystem services* approach recognizes that undisturbed ecosystems do many things that benefit us. A forest, for instance, slows the movement of rain and snowmelt into streams and rivers; if the forest is removed, floods may follow (a connection that recently forced China to deemphasize forest exploitation). Swamps filter the water that seeps through them. Food chains cycle nutrients necessary for the production of wood and fish and other harvests. Bees pollinate crops and make food production possible. These services are valuable—even essential—to us, and anything that interferes with them must be seen as imposing costs just as significant as the illnesses associated with pollution.

How can those costs be assessed? In 1997 Robert Costanza and his colleagues published an influential paper entitled "The Value of the World's Ecosystem Services and Natural Capital" in the May 15 issue of the journal *Nature.* In it, the authors listed a variety of ecosystem services and attempted to estimate what it would cost to replace those services if they were somehow lost (such as by building a sewage treatment plant to replace a swamp). The total bill for the entire biosphere came to $33 trillion (the middle of a $16–54 trillion range), compared to a global gross national product of $25 trillion. Costanza et al. stated that this was surely an underestimate.

What good is such an estimate? Perhaps it could motivate governments to greater efforts to protect the environment in order to avoid colossal financial (or other) future consequences. This point is made in the following selection by Janet N. Abramovitz, who argues that nature's services are responsible for the vast bulk of the value in the world's economy and that attaching economic value to those services may encourage their protection. In the second selection, Marino Gatto and Giulio A. De Leo argue that the pricing approach to valuing nature's services is misleading because it ignores equally important "nonmarket" values.

Janet N. Abramovitz

 **YES**

Putting a Value on Nature's "Free" Services

During the last half of 1997, massive fires swept through the forests of Sumatra, Borneo, and Irian Jaya, which together form a stretch of the Indonesian archipelago as wide as all of Europe. By November, almost 2 million hectares had burned, leaving the region shrouded in haze and more than 20 million of its people breathing hazardous air. Tens of thousands of people had been treated for respiratory ailments. Hundreds had died from illness, accidents and starvation. The fires, though by then out of control, had been set deliberately and systematically—not by small farmers, and not by El Niño, but by commercial outfits operating with implicit government approval. Strange as this immolation of some of the world's most valuable natural assets may seem, it was not unique. The same year, a large part of the Amazon Basin in Brazil was blanketed by smoke for similar reasons. The fires in the Amazon have been set annually, but in 1997 they destroyed over 50 percent more forest than the year before, which in turn had recorded five times as many fires (some 19,115 fires during a single six-week period) as in 1995.

For the timber and plantation barons of Indonesia, as for the cattle ranchers and frontier farmers of Amazonia, setting fires to clear forests has become standard practice. To them, the natural rainforests are an obstruction that must be sold or burned to make way for their profitable pulp and palm oil plantations. Yet, these are the same forests that for many others serve as both homes and livelihoods. For the hundreds of millions who live in Indonesia and in the neighboring nations of Malaysia, Singapore, Brunei, southern Thailand and the Philippines, it is becoming painfully apparent that without healthy forests, it is difficult to remain healthy people.

As this issue of WORLD WATCH went to press [December 1997], the fires in Southeast Asia were still generating enough smoke to be visible from space. Some relief was expected with the arrival of the seasonal rains, but those rains were past due—in part because of an unusually strong El Niño effect. Along with the trees, the region's large underground peat deposits have caught on fire, and such fires are perniciously difficult to put out; they can continue smoldering for years.

From Janet N. Abramovitz, "Putting a Value on Nature's 'Free' Services," *World Watch* (January/February 1998). Copyright © 1998 by The Worldwatch Institute. Reprinted by permission. http://www.worldwatch.org.

When the smoke does finally clear, Southeast Asia—and the world—will attempt to tally the costs. There are the costs of impaired health and sometimes death, from both lung diseases and accidents caused by poor visibility. There is the productivity that was lost as factories, schools, roads, docks, and airports were shut down (over 1,000 flights in and out of Malaysia were cancelled in September alone); there are the crop yields that fell as haze kept the region in day-long twilight, and the harvests of forest products that were wiped out. Timber (some of the most valuable species in the world) and wildlife (some of the most endangered in the world) are still being consumed by flames. Over three-fourths of the world's remaining wild orangutans live on the fire-ravaged provinces of Sumatra and Kalimantan. Some of them, caught fleeing the flames, have become part of the illegal trade. Because of their location, the Indonesian fires, like those in the Amazon, have dealt a heavy blow to the biodiversity of the earth as a whole.

As the smoke billowed dramatically from Southeast Asia, a much less visible—but similarly costly—ecological loss was taking place in a very different kind of location. While the Indonesian haze was being photographed from satellites, this other loss might not be noticed by a person standing within an arm's length of the evidence—yet, in its implications for the human future, it is a close cousin of the Asian catastrophe. In the United States, more than 50 percent of all honeybee colonies have disappeared in the last 50 years, with half of that loss occurring in just the last 5 years. Similar losses have been observed in Europe. Thirteen of the 19 native bumblebee species in the United Kingdom are now extinct. These bees are just two of the many kinds of pollinators and their decline is costing farmers, fruit growers, and beekeepers hundreds of millions of dollars in losses each year.

What the ravaged Indonesian forests and disappearing bees have in common is that they are both examples of "free services" that are provided by nature and consumed by the human economy—services that have immense economic value, but that go largely unrecognized and uncounted until they have been lost. Many of those services are indispensable to the people who exploit them, yet are not counted as real benefits, or as a part of GNP.

Though widely taken for granted, the "free" services provided by the natural world form the invisible foundation that supports all societies and economies. We rely on the oceans to provide abundant fish, on forests for wood and new medicines, on insects and other creatures to pollinate our crops, on birds and frogs to keep pests in check, and on forests and rivers to supply clean water. We take it for granted that when we need timber we can cut trees, or that when we need water we can find a spring or drill a well. We assume that clean air will blow the smog out of our cities, that the climate will be stable and predictable, and that the mounting quantity of waste we generate will continue to disappear, if we can just get it out of sight. Nature's services have always been there, free for the taking, and our expectations—and economies—are based on the premise that they always will be. A timber magnate or farmer may have to pay a price for the land, but assumes that what happens naturally on the land—the growing of trees, or pollinating of crops by wild bees, or filtering of fresh water—usually happens for free. We are like young children who think that food comes from the refrigerator, and who do not yet understand that what now seems free is not.

Ironically, by undervaluing natural services, economies unwittingly provide incentives to misuse and destroy the very systems that produce those services; rather than protecting their assets, they squander them. Nature, in turn becomes increasingly less able to supply the prolific range services that the earth's expanding population and economy demand.... It is no exaggeration to suggest that the continued erosion of natural systems threatens not only the continuing viability of today's human enterprise, but ultimately the prospects for our continued existence.

Underpinning the steady stream of services nature provides to us, there is a more fundamental service these systems provide—a kind of self-regulating process by which ecosystems and the biosphere are kept relatively stable and resilient. The ability to withstand disturbances like fires, floods, diseases, and droughts, and to rebound from the shocks these events inflict, is essential to keeping the life-support system operating. As systems are simplified by monoculture or cut up by roads, and the webs that link systems become disconnected, they become more brittle and vulnerable to catastrophic, irreversible decline. We are being confronted by ample evidence, now—from the breakdown of the ozone layer to the increasingly severity of fires, floods and droughts, to the diminished productivity of fruit and seed sets in wild and agricultural plants —that the biosphere is becoming less resilient.

Unfortunately, much of the human economy is based on practices that convert natural systems into something simpler, either for ease of management (it's easier to harvest straight rows of trees that are all the same age than to harvest carefully from complex forests) or to maximize the production of a desired commodity (like corn). But simplified systems lack the resilience that allows them to survive short-term shocks such as outbreaks of diseases or pests, or forest fires, or even longer-term stresses such as that of global warming. One reason is that the conditions within these simplified systems are not hospitable to all of the numerous organisms and processes needed to keep such systems running. A tree plantation or fish farm may provide some of the products we need, but it cannot supply the array of services that natural diverse systems do—and must do—in order to survive over a range of conditions. To keep our own economies sustainable, then, we need to use natural systems in ways that capitalize on, rather than destroy, their regenerative capacity. For humans to be healthy and resilient, nature must be too.

Resiliency is destroyed by fragmentation, as well as by simplification. Fires in healthy rainforests are very rare. By nature, they are too wet to burn. But as they are opened up and fragmented by roads and logging and pasture, they become drier and more prone to fire. When fire strikes forests that are not adapted to fire (as is the case in the rainforests of both Brazil and Indonesia), it is exceptionally destructive and tends to kill a majority of the trees. The fires in Southeast Asia's peat swamp rainforests bring further disruption, by releasing long-sequestered carbon into the atmosphere.

The fires in Indonesia are not being started by poor slash-and-burn peasants, but by "slash-and-burn industrialists"—owners of rubber, palm oil, rice, and timber plantations who have been taking advantage of a dry year to clear as much natural forest as they can. Though it issued a recent law forbidding the burning, the government in Indonesia is in fact pushing for higher production levels from these export sectors. In both the rainforests and the peat swamps, it has given the plantation owners large concessions to encourage continued "conversion" to one-crop commodities. And the government continues to push costly agricultural settlements into peat forests ill-suited to rice. After the fires became a serious regional problem (and international embarrassment), the government revoked the permits of 29 companies, but such actions were too little, too late.

The current fires are not the first to ravage parts of Southeast Asia; extensive logging in Indonesia and Malaysia led to a major conflagration in 1983 that burned over 3 million hectares and wiped out $5 billion worth of standing timber in Indonesia alone. After 1983, fires that had once been rare became a common occurrence. The 1997 fire will likely turn out to have been the most costly yet. Unless policies change, the fires will be reignited this year.

What Forests Do

Around the world, the degradation, fragmentation, and simplification—or "conversion"—of ecosystems is progressing rapidly. Today, only 1 to 5 percent of the original forest cover of the United States and Europe remains. One-third of Asia's forest has been lost since 1960, and half of what remains is threatened by the same industrial forest activities responsible for the Indonesian fires. In the Amazon, 13 percent of the natural cover has already been cleared, mostly for cattle pasture. In many countries, including some of the largest, more than half of the land has been converted from natural habitat to other uses that are less resilient. In countries that stayed relatively undisturbed until the 1980s, significant portions of remaining ecosystems have been lost in the last decade. These trends have been accelerating everywhere. As the natural ecosystems disappear, so do many of the goods and services they provide.

That may seem to contradict the premise that people want those goods and services and would not deliberately destroy them. But there's a logical explanation: governments and business owners typically perceive that the way they can make the most profit from an ecosystem is to maximize its production of a single commodity, such as timber from a forest. For the community (or society) as a whole, however, that is often the least profitable or sustainable use. The economic values of other uses, and the number of people who benefit, added up, can be enormous. A forest, if not cut down to make space for a one-commodity plantation, can produce a rich variety of nontimber forest products (NTFPs) on one hand, while providing essential watershed protection and climate regulation, on the other. These uses not only have more immediate economic value but can also be sustained over a longer term and benefit more people.

In 1992, alternative management strategies were reviewed for the mangrove forests of Bintuni Bay in Indonesia. When nontimber uses such as fish, locally used products, and erosion control were included in the calculations, the researchers found that the most economically profitable strategy was to keep the forest standing with only a modest amount of timber cutting—yielding $4,800 per hectare. If the forest was managed only for timber-cutting, it would yield only $3,600 per hectare. Over the longer term, it was calculated that keeping the forest intact would ensure continued local uses of the area worth $10 million a year (providing 70 percent of local income) and protect fisheries worth $25 million a year—values that would be lost if the forest were cut.

The variety and value of goods produced and collected from forests, and their importance to local livelihoods and national economies, is an economic reality worldwide. For instance, rattan—a vine that grows naturally in tropical forests—is widely used to make furniture. Global trade in rattan is worth $2.7 billion in exports each year, and in Asia it employs a half-million people. In Thailand, the value of rattan exports is equal to 80 percent of the legal timber exports. In India, such "minor" products account for three fourths of the net export earnings from forest produce, and provide more than half of the formal employment in the forestry sector. And in Indonesia, hundreds of thousands of people make their livelihoods collecting and processing NTFPs for export, a trade worth at least $25 million a year. Many of these forests were destroyed in the fire.

Even so, non-timber commodities are only part of what is lost when a forest is converted to a one-commodity industry. There is a nexus between the two catastrophes of the Indonesian fires and the North American and European bee declines, for example, since forests provide habitat for bees and other pollinators. They also provide habitat for birds that control disease-carrying and agricultural pests. Their canopies break the force of the winds and reduce rainfall's impact on the ground, which lessens soil erosion. Their roots hold soil in place, further stemming erosion. In purely monetary terms, a forest's capacity to protect a watershed alone can exceed the value of its timber. Forests also act as effective water-pumping and recycling machinery, helping to stabilize local climate. And, through photosynthesis, they generate enough of the planet's oxygen, while absorbing and storing so much of its carbon (in living trees and plants), that they are essential to the stability of climate worldwide.

Beyond these general functions, there are services that are specific to particular kinds of forests. Mangrove forests and coastal wetlands, notably, play critical roles in linking land and sea. They buffer coasts from storms and erosion, cycle nutrients, serve as nurseries for coastal and marine fisheries, and supply critical resources to local communities. For flood control alone, the value of mangroves has been calculated at $300,000 per kilometer of coastline in Malaysia—the cost of the rock walls that would be needed to replace them. Protecting coasts from storms will be especially important as climate change makes storms more violent and unpredictable. One force driving the accelerated loss of these mangroves in the last two decades has been the explosive growth of intensive commercial aquaculture, especially for shrimp export. Another has been the excess diversion of inland rivers and streams, which reduces downstream flow and allows the coastal waters to become too salty to support the coastal forests.

The planet's water moves in a continuous cycle, falling as precipitation and moving slowly across the landscape to streams and rivers and ultimately to the sea, being absorbed and recycled by plants along the way. Yet, human actions have changed even that most fundamental force of nature by removing natural plant cover, draining swamps and wetlands, separating rivers from their floodplains, and paving over land. The slow natural movement of water across the landscape is also vital for refilling nature's underground reservoirs, or aquifers, from which we draw much of our water. In many places, water now races across the landscape much too quickly, causing flooding and droughts, while failing to adequately recharge aquifers.

The value of a forested watershed comes from its capacity to absorb and cleanse water, recycle excess nutrients, hold soil in place, and prevent flooding. When plant cover is removed or disturbed, water and wind not only race across the land, but carry valuable topsoil with them. According to David Pimentel, an agricultural ecologist at Cornell University, exposed soil is eroded at several thousand times the natural rate. Under normal conditions, each hectare of land loses somewhere between 0.004 and 0.05 tons of soil to erosion each year— far less than what is replaced by natural soil building processes. On lands that have been logged or converted to crops and grazing, however, erosion typically takes away 17 tons in a year in the United States or Europe, and 30 to 40 tons in Asia, Africa, or South America. On severely degraded land, the hemorrhage can rise to 100 tons in a year. The eroded soil carries nutrients, sediments, and chemicals valuable to the system it leaves, but often harmful to the ultimate destination.

One way to estimate the economic value of an ostensibly free service like that of a forested watershed is to estimate what it would cost society if that service had to be replaced. New York City, for example, has always relied on the natural filtering capacity of its rural watersheds to cleanse the water that serves 10 million people each day. In 1996, experts estimated that it would cost $7 billion to build water treatment facilities adequate to meet the city's future needs. Instead, the city chose a strategy that will cost it only one-tenth that amount: simply helping upstream counties to protect the watersheds around its drinking water reservoirs.

Even an estimate like that tends to greatly understate the real value, however, because it covers the replacement cost of only one of the many services the ecosystem provides. A watershed, for example, also contributes to the regulation of the local climate. After forest cover is removed, an area can become hotter and drier, because water is no longer cycled and recycled by plants (it has been estimated that a single rainforest tree pumps 2.5 million gallons of water into the atmosphere during its lifetime.) Ancient Greece and turn-of-the-century Ethiopia, for example, were moister, wooded regions before extensive deforestation, cultivation, and the soil erosion that followed transformed them into the hot, rocky countries they are today. The global spread of desertification offers brutal evidence of the toll of lost ecosystem services.

The cumulative effects of local land use changes have global implications. One of the planet's first ecosystem services was the production of oxygen over billions of years of photosynthetic activity, which allowed oxygen-breathing organisms—such as ourselves—to evolve. Humans have begun to unbalance the global climate regulation system, however, by generating too much carbon dioxide and reducing the capacity of ecosystems to absorb it. Burning forests and peat deposits only makes the problem worse. The fires in Asia sent about as much carbon into the atmosphere last year as did all of the factories, power plants, and vehicles in the United Kingdom. For carbon sequestration alone, economists have been able to estimate the value of intact forests at anywhere from several hundred to several thousand dollars per hectare. As the climate changes the value of being able to regulate local and global climates will only increase.

What Bees Do

If we are often blind to the value of the free products we take from nature, it is even easier to overlook the value of those products we don't harvest directly—but without which our economies could not function. Among these less conspicuous assets are the innumerable creatures that keep potentially harmful organisms in check, build and maintain soils, and decompose dead matter so it can be used to build new life, as well as those that pollinate crops. These various birds, insects, worms, and microorganisms demonstrate that small things can have hugely disproportionate value. Unfortunately, their services are in increasingly short supply because pesticides, pollutants, disease, hunting, and habitat fragmentation or destruction have drastically reduced their numbers and ability to function. As Stephen Buchmann and Gary Paul Nabhan put it in a recent book on pollinators, "nature's most productive workers [are] slowly being put out of business."

Pollinators, for example, are of enormous value to agriculture and the functioning of natural ecosystems. Without them, plants cannot produce the seeds that ensure their survival—and ours. Unlike animals, plants cannot roam around looking for mates. To accomplish sexual reproduction and ensure genetic mixing, plants have evolved strategies for moving genetic material from one plant to the next, sometimes over great distances. Some rely on wind or water to carry pollen to a receptive female, and some can self-pollinate. The most highly evolved are those that use flowers, scents, oils, pollens, and nectars to attract and reward animals to do the job. In fact, more than 90 percent of the world's quarter-million flowering plant species are animal-pollinated. When animals pick up the flower's reward, they also pick up its pollen on various body parts—faces, legs, torsos. Laden with sticky yellow cargo, they can appear comical as they veer through the air—but their evolutionary adaptations are uncannily potent.

Developing a mutually beneficial relationship with a pollinator is a highly effective way for a plant to ensure reproductive success, especially when individuals are isolated from each other. Spending energy producing nectars and extra pollen is a small price to pay to guarantee reproduction. Performing this matchmaking service are between 120,000 and 200,000 animal species, including bees, beetles, butterflies, moths, ants, and flies, along with more than 1,000 species of vertebrates such as birds, bats, possums, lemurs, and even geckos. New evidence shows that many more of these pollinator species than previously believed are threatened with extinction.

Eighty percent of the world's 1,330 cultivated crop species (including fruits, vegetables, beans and legumes, coffee and tea, cocoa, and spices) are pollinated by wild and semi-wild pollinators. One-third of U.S. agricultural output is from insect-pollinated plants (the remainder is from wind-pollinated grain plants such as wheat, rice, and corn). In dollars, honeybee pollination services are 60 to 100 times more valuable than the honey they produce. The value of wild blueberry bees is so great, with each bee pollinating 15 to 19 liters (about 40 pints) of blueberries in its life, that they are viewed by farmers as "flying $50 bills."

Without pollinator services, crops would yield less, and wild plants would produce few seeds—with large economic and ecological consequences. In Europe, the contribution of honey bee pollination to agriculture was estimated to be worth $100 billion in 1989. In the Piedmont region of Italy, poor pollination of apple and apricot orchards cost growers $124 million in 1996. The most pervasive threats to pollinators include habitat fragmentation and disturbance, loss of nesting and over-wintering sites, intense exposure of pollinators to pesticides and of nectar plants to herbicides, breakdown of "nectar corridors" that provide food sources to pollinators during migration, new diseases, competition from exotic species, and excessive hunting. The rapid spread of two parasitic mites in the United States and Europe has wiped out substantial numbers of honeybee colonies. A "forgotten pollinators" campaign was recently launched by the Arizona Sonoran Desert Museum and others, to raise awareness of the importance and plight of these service providers.

Ironically, many modern agricultural practices actually limit the productivity of crops by reducing pollination. According to one estimate, for example, the high levels of pesticides used on cotton reduce annual yields by 20 percent (worth $400 million) in the United States alone by killing bees and other insect pollinators. One-fifth of all honeybee losses involve pesticide exposure, and honeybee poisonings may cost agriculture hundreds of millions of dollars each year. Wild pollinators are particularly vulnerable to chemical poisoning because their colonies cannot be picked up and moved in advance of spraying the way domesticated hives can. Herbicides can kill the plants that pollinators need to sustain themselves during the "off-season" when they are not at work pollinating crops. Plowing to the edges of fields to maximize planting area can reduce yields by disturbing pollinator nesting sites. Just one hectare of unplowed land, for example, provides nesting habitat for enough wild alkali bees to pollinate 100 hectares of alfalfa.

Domesticated honeybees cannot be expected to fill the gap left when wild pollinators are lost. Of the world's major crops, only 15 percent are pollinated by domesticated and feral honeybees, while at least 80 percent are serviced by wild pollinators. Honeybees do not "fit" every type of flower that needs pollination. And because honeybees visit so many different plant species, they are not very "efficient"—that is, there is no guarantee that the pollen will be carried to a potential mate of the same species and not deposited on a different species.

Many plants have developed interdependencies with particular species of pollinators. In peninsular Malaysia, the bat *Eonycteris spelea* is thought to be the exclusive pollinator of the durian, a large spiny fruit that is highly valued in Southeast Asia. The bats' primary food supply is a coastal mangrove that flowers continuously throughout the year. The bats routinely fly tens of kilometers from their roost sites to the mangrove stands, pollinating durian trees along the way. However, mangrove stands in Malaysia and elsewhere are under siege, as are the inland forests. Without both, the bats are unlikely to survive.

Pollinators that migrate long distances, such as bats, monarch butterflies, and hummingbirds, need to follow routes that offer a reliable supply of nectar-providing plants for the full journey. Today, however, such nectar corridors are being stretched increasingly thin and are breaking. When the travelers cannot rest and "refuel" every day, they may not survive the journey.

The migratory route followed by long-nosed bats from their summer breeding colonies in the desert regions of the U.S. Southwest to winter roosts in central Mexico illustrates the problems faced by many service providers. To fuel trips of up to 150 kilometers a night, these bats rely on the sequential flowering of at least 16 plant species—particularly century plants and columnar cacti. Along much of the migratory route, the nectar corridor is being fragmented. On both U.S. and Mexican rangelands, ranchers are converting native vegetation into exotic pasture grasses for grazing cattle. In the Mexican state of Sonora, an estimated 376,000 hectares have been stripped of nectar source plants. In parts of the Sierra Madre, the bat-pollinators are threatened by competition from human bootleggers, who have been over-harvesting century plants to make the alcoholic beverage mescal. And the latest threat comes from dynamiting and burning of bat roosts by Mexican ranchers attempting to eliminate vampire bats that feed on cattle and spread livestock diseases. The World Conservation Union estimates that worldwide, 26 percent of bat species are threatened with extinction.

Many of the disturbances that have harmed pollinators are also hurting creatures that provide other beneficial services, such as biological control of pests and disease. Much of the wild and semi-wild habitat inhabited by beneficial predators such as birds has been wiped out. The "pest control services" that nature provides are incalculable, and do not have the fundamental flaws of chemical pesticides (which kill beneficial insects along with the pests and harm people). Individual bat colonies in Texas can eat 250 tons of insects each night. Without birds, leaf-eating insects are more abundant and can slow the growth of trees or damage crops. Biologists Paul and Anne Ehrlich speculate that without birds, insects would have become so dominant that humans might never have been able to achieve the agricultural revolution that set the stage for the rise of civilization.

It is not too late to provide essential protections to the providers of such essential services—by using no-till farming to reduce soil erosion and allow nature's underground economy to flourish, by cutting back on the use of toxic agricultural chemicals, and by protecting migratory routes and nectar corridors to ensure the survival of wild pollinators and pest control agents.

Buffer areas of native vegetation and trees can have numerous beneficial effects. They can serve as havens for resident and migratory insects and animals that pollinate crops and control pests. They can also help to reduce wind erosion, and to absorb nutrient pollution that leaks from agricultural fields. Such zones have been eliminated from many agricultural areas that are modernized to accommodate new equipment or larger field sizes. The "sacred groves" in South Asian and African villages—natural areas intentionally left undeveloped —still provide such havens. Where such buffers have been removed, they can be reestablished; they can be added not only around farmers' fields, but along highways and river banks, links between parks, and in people's back yards.

People can also encourage pollinators by providing nesting sites, such as hollow logs, or by ensuring that pollinators have the native plants they need during the "off-season" when they are not working on the agricultural crops. Changing some prevalent cultural or industrial practices, too, can help. There is the practice, for example, of growing tidy rows of cocoa trees. These may make for a handsome plantation. But midges, the only known pollinator of cultivated cacao (the source of chocolate), prefer an abundance of leaf litter and trees in a more natural array. Plantations that encourage midges can have ten times the yield of those that don't.

Scientists have begun to ratchet up their study of wild pollinators and to domesticate more of them. The bumblebee, for example, was domesticated ten years ago and is now a pollinator of valuable greenhouse grown crops.

The Other Service Economy

Natural services have been so undervalued because, for so long, we have viewed the natural world as an inexhaustible resource and sink. Human impact has been seen as insignificant or beneficial. The tools used to gauge the economic health and progress of a nation have tended to reinforce and encourage these attitudes. The gross domestic product (GDP), for example, supposedly measures the value of the goods and services produced in a nation. But the most valuable goods and services—the ones provided by nature, on which all else rests—are measured poorly or not at all. . . . The unhealthy dynamic is compounded by the fact that activities that pollute or deplete natural capital are counted as contributions to economic wellbeing. As ecologist Norman Myers puts it, "Our tools of economic analysis are far from able to apprehend, let alone comprehend, the entire range of values implicit in forests."

When economies and societies use misleading signals about what is valuable, people are encouraged to make decisions that run counter to their own long-range interests—and those of society and future generations. Economic calculations grossly underestimate the current and future value of nature. While a fraction of nature's goods are counted when they enter the marketplace, many of them are not. And nature's services—the life-support systems—are not counted at all. When the goods are considered free and therefore valued at zero, the market sends signals that they are only economically valuable when converted into something else. For example, the profit from deforesting land is counted as a plus on a nation's ledger sheet, because the trees have been converted to saleable lumber or pulp, but the depletions of the timber stock, watershed, and fisheries are not subtracted.

Last year, an international team of researchers led by Robert Costanza of the University of Maryland's Institute for Ecological Economics, published a landmark study on the importance of nature's services in supporting human economies. The study provides, for the first time, a quantification of the current economic value of the world's ecosystem services and natural capital. The researchers synthesized the findings of over 100 studies to compute the average per hectare value for each of the 17 services that world's ecosystems provide. They concluded that the current economic value of the world's ecosystem services is in the neighborhood of $33 trillion per year, exceeding the global GNP of $25 trillion.

Placing a monetary value on nature in this way has been criticized by those who believe that it commoditizes and cheapens nature's infinite value. But in practice, we all regularly assign value to nature through the choices we make. The problem is that in normal practice, many of us don't assign such value to nature until it is converted to something man-made—forests to timber, or swimming fish to a restaurant meal. With a zero value, it's easy to see why nature has almost always been the loser in standard economic equations. As the authors of the Costanza study note, " . . . the decisions we make about ecosystems imply valuations (although not necessarily expressed in monetary terms). We can choose to make these valuations explicit or not . . . but as long as we are forced to make choices, we are going through the process of valuation." The study is also raising a powerful new challenge to those traditional economists who are accustomed to keeping environmental costs and benefits "external" to their calculations.

While some skeptics will doubtless argue that the global valuation reported by Costanza and his colleagues overestimates the current value of nature's services, if anything it is actually a very conservative estimate. As the authors point out, values for some biomes (such as mountains, arctic tundra, deserts, urban parks) were not included. Further, they note that as ecosystem services become scarcer, their economic value will only increase.

Clearly, failure to value nature's services is not the only reason why these services are misused. Too often, illogical and inequitable resource use continues —even in the face of evidence that it is ecologically, economically, and socially unsustainable—because powerful interests are able to shape policies by legal or illegal means. Frequently, some individuals or entities get the financial benefits from a resource while the losses are distributed across society. Economists call this "socializing costs." Stated simply, the people who get the benefits are not the ones who pay the costs. Thus, there is little economic incentive for those exploiting a resource to use it judiciously or in a manner that maximizes public good. Where laws are lax or are ignored, and where people do not have an opportunity for meaningful participation in decision-making, such abuses will continue.

The liquidation of 90 percent of the Philippines' forest during the 1970s and 1980s under the Ferdinand Marcos dictatorship, for example, made a few hundred families over $42 billion richer. But 18 million forest dwellers became much poorer. The nation as a whole went from being the world's second largest log exporter to a net importer. Likewise, in Indonesia today, the "benefits" from burning the forest will enrich a relatively few well-connected individuals and companies but tens of millions of others are bearing the costs. Even in wealthy nations, such as Canada, the forest industry wields heavy influence over how the forests are managed, and for whose benefit.

We have already seen that the loss of ecosystem services can have severe economic, social, and ecological costs even though we can only measure a fraction of them. The loss of timber and lives in the Indonesian fires, and the lower production of fruits and vegetables from inadequate pollination, are but the tip of the iceberg. The other consequences for nature are often unforeseen and unpredictable. The loss of individual species and habitat, and the degradation and simplification of ecosystems, impair nature's ability to provide the services we need. Many of these changes are irreversible, and much of what is lost is simply irreplaceable.

By reducing the number of species and the size and integrity of ecosystems, we are also reducing nature's capacity to evolve and create new life. Almost half of the forests that once covered the Earth are now gone, and much of what remains is in fragmented patches. In just a few centuries we have gone from living off nature's interest to spending down the capital that has accumulated over millions of years of evolution. At the same time we are diminishing the capacity of nature to create new capital. Humans are only one part of the evolutionary product. Yet we have taken on a major role in shaping its future production course and potential. We are pulling out the threads of nature's safety net even as we depend on it to support the world's expanding human population and economy.

In that expanding economy, consumers now need to recognize that it is possible to reduce and reverse the destructive impact of our activities by consuming less and by placing fewer demands on those services we have so mistakenly regarded as free. We can, for example, reduce the high levels of waste and overconsumption of timber and paper. We can also increase the efficiency of water and energy use. In agricultural fields we can leave hedgerows and unplowed areas that serve as nesting and feeding sites for pollinators. We can sharply reduce reliance on agricultural chemicals, and improve the timing of their application to avoid killing pollinators.

Maintaining nature's services requires looking beyond the needs of the present generation, with the goal of ensuring sustainability for many generations to come. We have no honest choice but to act under the assumption that future generations will need at least the same level of nature's services as we have today. We can neither practically nor ethically decide what future generations will need and what they can survive without.

Marino Gatto and Giulio A. De Leo  **NO**

Pricing Biodiversity and Ecosystem Services: The Never-Ending Story

In 1844, the French engineer Jules Juvénal Dupuit introduced cost–benefit analysis to evaluate investment projects.... The application of cost–benefit analysis to ecological issues fell out of favor three decades ago, and it was gradually replaced by multicriteria analysis in the decision-making process for projects that have an impact on the environment. Although multicriteria analysis is currently used for environmental impact assessments [EIA] in many nations, [recently] the concept of cost–benefit analysis has again become fashionable, along with the various pricing techniques associated with it, such as contingent valuation methods, hedonic prices, and costs of replacement of ecological services.... Economists have generated a wealth of virtuosic variations on the theme of assessing the societal value of biodiversity, but most of these techniques are invariably based on price—that is, on a single scale of values, that of goods currently traded on world markets.

Perhaps the most famous recent study on the issue of pricing biodiversity and ecological services is that by Costanza et al., who argued that if the importance of nature's free benefits could be adequately quantified in economic terms, then policy decisions would better reflect the value of ecosystem services and natural capital. Drawing on earlier studies aimed at estimating the value of a wide variety of ecosystem goods and services, Costanza et al. estimated the current economic value of the entire biosphere at $16–54 trillion per year, with an average value of approximately $33 trillion per year. By contrast, the gross national product of the United States totals approximately $18 trillion per year. The paper, as its authors intended, stimulated much discussion, media attention, and debate. A special issue of *Ecological Economics* (April 1998) was devoted to commentaries on the paper, which, with few exceptions, were laudatory. Some economists have questioned the actual numbers, but many scientists have praised the attempt to value biodiversity and ecosystem functions.

Although Costanza et al. acknowledged that their estimates were crude and imperfect, they also pointed the way to improved assessments. In particular, they noted the need to develop comprehensive ecological economic models that could adequately incorporate the complex interdependencies between ecosystems and economic systems, as well as the complex individual dynamics of both types of systems. Despite the authors' caveats and the fact that many economists have been circumspect in applying their own tools to decisions regarding natural systems, the monetary approach is perceived by scientists, policymakers, and the general public as extremely appealing; a number of biologists are also of the opinion that attaching economic values to ecological services is of paramount importance for preserving the biosphere and for effective decision-making in all cases where the environment is concerned.

In this article, we espouse a contrary view, stressing that, for most of the values that humans attach to biodiversity and ecosystem services, the pricing approach is inadequate—if not misleading and obsolete—because it implies erroneously that complex decisions with important environmental impacts can be based on a single scale of values. We contend that the use of cost–benefit analysis as the exclusive tool for decision-making about environmental policy represents a setback relative to the existing legislation of the United States, Canada, the European Union, and Australia on environmental impact assessment, which explicitly incorporates multiple criteria (technical, economic, environmental, and social) in the process of evaluating different alternatives. We show that there are sound methodologies, mainly developed in business and administration schools by regional economists and by urban planners, that can assist decision-makers in evaluating projects and drafting policies while accounting for the nonmarket values of environmental services.

The Limitations of Cost–Benefit Analysis and Contingent Valuation Methods

Historically, the first important implementation of cost–benefit analysis at the political level came in 1936, with passage of the US Flood Control Act. This legislation stated that a public project can be given a green light if the benefits, to whomsoever they accrue, are in excess of estimated costs. This concept implies that all benefits and costs are to be considered, not just actual cash flows from and to government coffers. However, public agencies (e.g., the US Army Corps of Engineers) quickly ran into a problem: They were not able to give a monetary value to many environmental effects, even those that were predictable in quantitative terms. For instance, engineers could calculate the reduction of downstream water flow resulting from construction of a dam, and biologists could predict the river species most likely to become extinct as a consequence of this flow reduction. However, public agencies were not able to calculate the cost of each lost species. Therefore, many ingenious techniques for the monetary valuation of environmental goods and services have been devised since the 1940s. These techniques fall into four basic categories.

- **Conventional market approaches.** These approaches, such as the replacement cost technique, use market prices for the environmental service that is affected. For example, degradation of vegetation in developing countries leads to a decrease in available fuelwood. Consequently, animal dung has to be used as a fuel instead of a fertilizer, and farmers must therefore replace dung with chemical fertilizers. By computing the cost of these chemical fertilizers, a monetary value for the degradation of vegetation can then be calculated.

- **Household production functions.** These approaches, such as the travel cost method, use expenditures on commodities that are substitutes or complements for the environmental service that is affected. The travel cost method was first proposed in 1947 by the economist Harold Hotelling, who, in a letter to the director of the US National Park Service, suggested that the actual traveling costs incurred by visitors could be used to develop a measure of the recreation value of the sites visited.
- **Hedonic pricing.** This form of pricing occurs when a price is imputed for an environmental good by examining the effect that its presence has on a relevant market-priced good. For instance, the cost of air and noise pollution is reflected in the price of plots of land that are characterized by different levels of pollution, because people are willing to pay more to build their houses in places with good air quality and little noise....
- **Experimental methods.** These methods include contingent valuation methods, which were devised by the resource economist Siegfried V. Ciriacy-Wantrup. Contingent valuation methods require that individuals express their preferences for some environmental resources by answering questions about hypothetical choices. In particular, respondents to a contingent valuation methods questionnaire will be asked how much they would be willing to pay to ensure a welfare gain from a change in the provision of a nonmarket environmental commodity, or how much they would be willing to accept in compensation to endure a welfare loss from a reduced provision of the commodity.

Among these pricing techniques, the contingent valuation methods approach is the only one that is capable of providing an estimate of existence values, in which biologists have a special interest. Existence value was first defined by Krutilla as the value that individuals may attach to the mere knowledge that rare and diverse species, unique natural environments, or other "goods" exist, even if these individuals do not contemplate ever making active use of or benefiting in a more direct way from them. The name "contingent valuation" comes from the fact that the procedure is contingent on a constructed or simulated market, in which people are asked to manifest, through questionnaires and interviews, their demand function for a certain environmental good (i.e., the price they would pay for one extra unit of the good versus the availability of the good)....

The limits of cost–benefit analysis were discussed in the 1960s, after more than two decades of experimentation. In particular, many authors pointed out that cost–benefit analysis encouraged policymakers to focus on things that can be measured and quantified, especially in cash terms, and to disregard problems that are too large to be assessed easily. Therefore, the associated price might not reflect the "true" value of social equity, environmental services, natural capital, or human health. In particular, economists themselves recognize that the increasingly popular contingent valuation methods are undermined by several conceptual problems, such as free-riding, overbidding, and preference reversal.

When it comes to monetary valuation of the goods and services provided by natural ecosystems and landscapes specifically, a number of additional problems undermine the effectiveness of pricing techniques and cost–benefit analysis. These problems include the very definition of "existence" value, the dependence of pricing techniques on the composition of the reference group, and the significance of the simulated market used in contingent valuation.

The definition of "existence" value   A classic example of contingent valuation methods is to ask for the amount of money individuals are willing to pay to ensure the continued existence of a species such as the blue whale. However, the existence value of whales does not take into account potential indirect services and benefits provided by these mammals. It is just the value of the existence of whales for humans, that is, the satisfaction that the existence of blue whales provides to people who want them to continue to exist. Therefore, there is a real risk that species with very low or no aesthetic appeal or whose biological role has not been properly advertised will be given a low value, even if they play a fundamental ecological function. Without adequate information, most people do not understand the extent, importance, and gravity of most environmental problems. As a consequence, people may react emotionally and either underestimate or overestimate risks and effects.

Therefore, it is not surprising that five of the seven guidelines issued by the National Oceanic and Atmospheric Administration [NOAA] about how to conduct contingent valuation discuss how to properly inform and question respondents to produce reliable estimates (e.g., in-person interviews are preferred to telephone surveys to elicit values). Of course, acquisition of reliable and complete information is always possible in theory, but in practice strict adherence to NOAA guidelines makes contingent valuation methods expensive and time consuming.

Difficulties with the reference group for pricing   Pricing techniques such as contingent valuation methods provide information about individual willingness to pay or willingness to accept, which must be summed up in the final balance of cost–benefit analysis. Therefore, the outcome of cost–benefit analysis depends strongly on the group of people that is taken as a reference for valuation—particularly on their income. Van der Straaten noted that the Exxon *Valdez* oil spill in 1989 provides a good example of this dependence. The population of the United States was used as a reference group to calculate the damage to the existence value of the affected species and ecosystems using contingent valuation methods. Exxon was ultimately ordered to pay $5 billion to compensate the people of Alaska for their losses. This huge figure was a consequence of the high income of the US population. If the same accident had occurred in Siberia, where salaries are lower, the outcome would certainly have been different.

This example shows that contingent valuation methods simply provide information about the preferences of a particular group of people but do not necessarily reflect the ecological importance of ecosystem goods and services. Moreover, the outcome of cost–benefit analysis depends on which individual willingness to pay or willingness to accept are included in the cost–benefit analysis. If the quality of the Mississippi River is at issue, should the analysis be restricted to US citizens living close to the river, or should the willingness to pay of Californians and New Yorkers be included too? According to Krutilla's definition of existence value, for many environmental goods and ecological services that may ultimately affect ecosystem integrity at the global level, the preferences of the entire human population should potentially be considered in the analysis. Because practical reasons obviously preclude doing so, contingent valuation methods will inevitably only provide information about the preferences of specific groups of people. For many of the ecological services that may be considered the heritage of humanity, contingent valuation methods analyses performed locally in a particular economic situation should be extrapolated only with great caution to other areas. The process of placing a monetary value on biodiversity and ecosystem functioning through nonuser willingness to pay is performed in the same way as for user willingness to pay, but the identification of people who do not use an environmental good directly and still have a legitimate interest in its preservation is problematic.

Significance of the simulated market   Contingent valuation methods are contingent on a market that is constructed or simulated, not real. It is difficult to believe in the efficiency of what Adam Smith called the "invisible hand" of the market for a process that is the artificial production of economic advisors and does not possess the dynamic feedback that characterizes real competitive markets. Is it even possible to simulate a market where units of biodiversity are bought and sold? As Friend stated, "these contingency evaluation methods (CVM) tend to create an illusion of choice based on psychology (willingness) and ideology (the need to pay) which is supposed, somewhat mysteriously, to reflect an equilibrium between the consumer demand for and producer supply of environmental goods and services."

Many additional criticisms of pricing ecological services are more familiar to biologists. For many ecological services, there is simply no possibility of technological substitution. Moreover, the precise contribution of many species is not known, and it may not be known until the species is close to extinction. . . . In addition, specific ecosystem services, as evaluated by Costanza et al., should not be separated from one another and valued individually because the importance of any piece of biodiversity cannot be determined without considering the value of biodiversity in the aggregate. And finally, the use of marginal value theory may be invalidated by the erratic and catastrophic behavior of many ecological systems, resulting in potentially detrimental effects on the health of humans, the productivity of renewable resources, and the vitality and stability of societies themselves.

Despite the efforts of many economists, we believe that some goods and services, especially those related to ecosystems, cannot reasonably be given a monetary value, although they are of great value to humans. Economists coined the term "intangibles" to define these goods. Cost–benefit analysis cannot easily deal with intangibles. As Nijkamp wrote, more than 20 years ago, "the only reasonable way to take account of intangibles in the traditional cost–benefit analysis seems to be the use of a balance with a debit and a credit side in which all intangible project effects (both positive and negative) are represented in their own (qualitative or quantitative) dimensions" as secondary information. In other words, the result of cost–benefit analysis is primarily a single number, the net monetary benefit that comprises all the effects that can be sensibly converted into monetary returns and costs.

Commensurability of Different Objectives and Multicriteria Analysis

Cost–benefit analysis includes intangibles in the decision-making process only as ancillary information, with the main focus being on those effects that can be converted to monetary value. This approach is not a balanced solution to the problem of making political decisions that are acceptable to a wide number of social groups with a range of legitimate interests....

However, even if the attempt to put a price on everything is abandoned, it is not necessary to give up the attempt to reconcile economic issues with social and environmental ones. Social scientists long ago developed multicriteria techniques to reach a decision in the face of multiple different and structurally incommensurable goals. The most important concept in multicriteria analysis was actually conceived by an Italian economist, Vilfredo Pareto, at the end of the nineteenth century. It is best explained by a simple example. Suppose that a natural area hosting several rare species is a target for the development of a mining activity. Alternative mining projects can have different effects in terms of profits from mining (measured in dollars) and in terms of sustained biodiversity (measured in suitable units, for instance, through the Shannon index). Profit from mining can be corrected using welfare economics to include those environmental and social effects that can be priced (e.g., the benefit of providing jobs to otherwise unemployed people, the cost of treating lung disease of miners, and the cost of the loss of the tourists who used to visit the natural area)....

The methods of multicriteria analysis are intended to assist the decision-maker in choosing among ... alternatives ... (a task that is particularly difficult when there are several incommensurable objectives, not just two). Nevertheless, the initial step of determining [these] alternatives is of enormous importance, for three reasons. First, [doing so] makes perfect sense even if there is no way of pricing a certain environmental good because each objective can be expressed in its own proper units without reduction to a common scale. Second, the determination of all the feasible alternatives ... requires the joint effort of a multidisciplinary team that includes, for example, economists, engineers, and biologists and that must predict the effects of alternative decisions on all of the different environmental and social components to which humans are sensitive and which, therefore, deserve consideration. Third, the determination of [feasible alternatives] allows the objective elimination of inadequate alternatives because [they are] independent of the subjective perception of welfare ... [and] in essence describe the tradeoff between the various incommensurable objectives when every effort is made to achieve the best results in all respects; the attention of the authority that must make the final decision is thus directed toward genuine potential solutions because nonoptimal decisions have already been discarded.

It should be noted that a cost–benefit analysis does not elicit tradeoffs between incommensurable goods because it also gives a green light to projects ..., provided that the benefits that can be converted into a monetary scale exceed the costs.... Cost–benefit analysis, however, is not useful for eliciting the tradeoffs between two incommensurable goods, neither of which is monetary. For instance, there might be a conflict between the goals of preserving wildlife within a populated area and minimizing the risk that wild animals are vectors of dangerous diseases. A multicriteria analysis can describe this tradeoff, whereas a cost–benefit analysis cannot.

Another philosophical point concerning the issue of commensurability is the question of implicit pricing. Economists often argue that to make a decision is to put an implicit price on such intangibles as human life or aesthetics and, therefore, to reduce their value to a common scale (as pointed out also by Costanza et al.)....

Environmental Impact Assessment and Multiattribute Decision-Making

Because of the flaws of cost–benefit analysis, many countries have taken a different approach to decision-making through the use of environmental impact assessment legislation (e.g., the United States in 1970, with the signing of the National Environmental Policy Act, NEPA; France in 1976, with the act 76/629; the European Union in 1985, with the directive 85/337). Environmental impact assessment procedures, if properly carried out, represent a wiser approach than setting an a priori value of biodiversity and ecosystem services because these procedures explicitly recognize that each situation, and every regulatory decision, responds to different ethical, economic, political, historical, and other conditions and that the final decision must be reached by giving appropriate consideration to several different objectives. As Canter noted, all projects, plans, and policies that are expected to have a significant environmental impact would ideally be subject to environmental impact assessment.

The breadth of goals embraced by environmental impact assessment is much wider than that of cost–benefit analysis. Environmental impact assessment provides a conceptual framework and formal procedures for comparing different alternatives to a proposed project (including the possibilities of not development a site, employing different management rules, or using mitigation measures); for fostering interdisciplinary team formation to investigate all possible environmental, social, and economic consequences of a proposed activity; for enhancing administrative review procedures and coordination among the agencies involved in the process; for producing the necessary documentation to enhance transparency in the decision-making process and the possibility of reviewing all the objective and subjective steps that resulted in a given conclusion; for encouraging broad public participation and the input of different interest groups; and for including monitoring and feedback procedures. Classical multiattribute analysis can be used to rank different alternatives.... Ranking usually requires the use of value functions to transform environmental and other indicators (e.g., biological oxygen demand or animal density) to levels of satisfaction on a normalized scale, and the weighting of factors to combine value functions and to rank the alternatives. These weights explicitly reflect the relative importance of the different environmental, social, and economic compartments and indicators.

A wide range of software packages for decision support can assist experts in organizing the collected information; in documenting the various phases of EIA; in guiding the assignment of importance weights; in scaling, rating, and ranking alternatives; and in conducting sensitivity analysis for the overall decision-making process. This last step, of testing the robustness and consistency of multiattribute analysis results, is especially important because it shows how sensitive the final ranking is to small or large changes in the set of weights and value functions, which often reflect different and subjective perspectives. It is important to stress that, although the majority of environmental impact assessments have been conducted on specific projects, such as road construction or the location of chemical plants, there is no conceptual barrier to extending the procedure to evaluation of plans, programs, policies, and regulations. In fact, according to NEPA, the procedure is mandatory for any federal action with an important impact on the environment. The extension of environmental impact assessment to a level higher than a single project is termed "strategic environmental assessment" and has received considerable attention.

Conclusions

An impressive literature is available on environmental impact assessment and multiattribute analysis that documents the experience gained through 30 years of study and application. Nevertheless, these studies seem to be confined to the area of urban planning and are almost completely ignored by present-day economists as well as by many ecologists. Somewhere between the assignment of a zero value to biodiversity (the old-fashioned but still used practice, in which environmental impacts are viewed as externalities to be discarded from the balance sheet) and the assignment of an infinite value (as advocated by some radical environmentalists), lie more sensible methods to assign value to biodiversity than the price tag techniques suggested by the new wave of environmental economists. Rather than collapsing every measure of social and environmental value onto a monetary axis, environmental impact assessment and multiattribute analysis allow for explicit consideration of intangible nonmonetary values along with classical economic assessment, which, of course, remains important. It is, in fact, possible to assess ecosystem values and the ecological impact of human activity without using prices. Concepts such as

Odum's eMergy [the available energy of one kind previously required to be used up directly and indirectly to make the product or service] and Rees' ecological footprint [the area of land and water required to support a defined economy or population at a specified standard of living], although perceived by some as naive, may aid both ecologists and economists in addressing this important need.

To summarize our viewpoint, economists should recognize that cost–benefit analysis is only part of the decision-making process and that it lies at the same level as other considerations. Ecologists should accept that monetary valuation of biodiversity and ecosystem services is possible (and even helpful) for part of its value, typically its use value. We contend that the realistic substitute for markets, when they fail, is a transparent decision-making process, not old-style cost–benefit analysis. The idea that, if one could get the price right, the best and most effective decisions at both the individual and public levels would automatically follow is, for many scientists, a sort of Panglossian obsession. In reality, there is no simple solution to complex problems. We fear that putting an a priori monetary value on biodiversity and ecosystem services will prevent humans from valuing the environment other than as a commodity to be exploited, thus reinvigoraing the old economic paradigm that assumes a perfect substitution between natural and human-made capital. As Rees wrote, "for all its theoretical attractiveness, ascribing money values to nature's services is only a partial solution to the present dilemma and, if relied on exclusively, may actually be counterproductive."

POSTSCRIPT

Should a Price Be Put on the Goods and Services Provided by the World's Ecosystems?

In "Can We Put a Price on Nature's Services?" *Report From the Institute for Philosophy and Public Policy* (Summer 1997), Mark Sagoff objects that trying to attach a price to ecosystem services is futile because it legitimizes the accepted cost-benefit approach and thereby undermines efforts to protect the environment from exploitation. The March 1998 issue of *Environment* contains environmental economics professor David Pearce's detailed critique of the 1997 Costanza et al. study. Pearce objects chiefly to the methodology, not the overall goal of attaching economic value to ecosystem services. Costanza et al. reply to Pearce's objections in the same issue. Pearce and Edward B. Barbier have published *Blueprint for a Sustainable Economy* (Earthscan, 2000), in which they discuss how governments worldwide are now applying economics to environmental policy.

Despite the controversy over the worth of assigning economic values to various aspects of nature, researchers continue the effort. Gretchen C. Daily et al., in "The Value of Nature and the Nature of Value," *Science* (July 21, 2000), discuss valuation as an essential step in all decision making and argue that efforts "to capture the value of ecosystem assets . . . can lead to profoundly favorable effects." Daily and Katherine Ellison continue the theme in *The New Economy of Nature: The Quest to Make Conservation Profitable* (Island Press, 2002). In "What Price Biodiversity?" *Ecos* (January 2000), Steve Davidson describes an ambitious program funded by the Commonwealth Scientific and Industrial Research Organization (CSIRO) and the Myer Foundation that is aimed at developing principles and methods for objectively valuing "ecosystem services—the conditions and processes by which natural ecosystems sustain and fulfil human life—and which we too often take for granted. These include such services as flood and erosion control, purification of air and water, pest control, nutrient cycling, climate regulation, pollination, and waste disposal."

For a recent textbook in environmental economics, see Barry C. Field and Martha K. Field, *Environmental Economics* (Irwin/McGraw-Hill, 2001). Jeff Gersh, in "Bigger, Badder—But Not Better: A New Breed of Economists Exposes the Myth of Unlimited Growth," *The Amicus Journal* (Winter 1999), contends that few economists recognize that continued growth or development is both undesirable and impossible, although many environmental thinkers have come to that conclusion. Finally, E. O. Wilson argues that economic development and environmental protection can coexist in *The Future of Life* (Alfred A. Knopf, 2002).

ISSUE 2

Is Biodiversity Overprotected?

YES: David N. Laband, from "Regulating Biodiversity: Tragedy in the Political Commons," *Ideas on Liberty* (September 2001)

NO: E. O. Wilson, from "Why Biodiversity Matters," interview by Kris Christen, *OECD Observer* (Summer 2001)

ISSUE SUMMARY

YES: Professor of economics David N. Laband argues that the public demands excessive amounts of biodiversity largely because decision makers and voters do not have to bear the costs of producing it.

NO: In an interview with science writer Kris Christen, biologist E. O. Wilson argues that biodiversity is crucial to human survival and that efforts need to be increased to protect it. He maintains that the loss of species reduces the productivity and stability of natural ecosystems and that with each species lost, potential drugs and other valuable resources are also lost.

Extinction is normal. Indeed, 99.9 percent of all the species that have ever lived are extinct, according to some estimates. But the process is normally spread out over time, with the formation of new species by mutation and selection balancing the loss of old ones to disease, new predators, climate change, habitat loss, and other factors. Today, human activities are an important cause of species loss mostly because humans destroy or alter habitat but also because of hunting, the introduction of competitors, and the introduction of diseases. Current estimates put the loss of mammal and bird species at one per year. As many as one-fifth of all species may disappear from the earth during the next 30 years.

Awareness of the problem has been growing. In 1973 the United States adopted the Endangered Species Act to protect species that were so reduced in numbers or restricted in habitat that a single untoward event could wipe them out. The act barred construction projects that would further threaten endangered species. In one famous case, construction on the Tellico Dam on the Little Tennessee River in Loudon County, Tennessee, was halted because it threatened the snail darter, a small fish. Another case involved the spotted owl, which was threatened by logging in the Northwest. Those in favor of the dam or the timber industry felt that the endangered species was trivial compared to the human benefits at stake. Those in favor of the act argued that the loss of a single species might not matter to the world, but where one species went, others would follow. Protecting one species also protects others. However, the number of threatened and endangered species has not diminished. In fact, that number has increased more than sevenfold, from 174 in 1976 to 1,244 as of November 2000.

Internationally, species protection is covered by the Convention on International Trade in Endangered Species of Wild Fauna and Flora (CITES). This agreement has banned trade in such natural products as elephant ivory to prevent the continued slaughter of elephants. Less successfully, it has also tried to protect rhinoceroses (killed for their horns) and about 5,000 other species of animals and 25,000 species of plants, including some whole groups, such as primates, cetaceans (whales, dolphins, and porpoises), sea turtles, parrots, corals, cacti, and orchids.

Is it enough to stop construction projects and ban trade? Some argue that efforts should be made to undo some of the damage that has already been done. For example, there is a movement to tear down dams that block the path of migratory fish, such as salmon and shad, so that they may once more breed and multiply (see http://www.amrivers.org). For another example, urbanization and agricultural development have greatly altered the Everglades in Florida: rivers have been straightened, water has been diverted, and land has been drained for farms. This activity has resulted in low water tables, increased fire danger, smaller bird populations, and the decline of the Florida panther, among other negative impacts. Currently, the Army Corps of Engineers is planning to undo some of the changes to the Everglades' water flow in order to restore the natural habitat as much as possible. The Comprehensive Everglades Restoration Plan (CERP) was approved by Congress in 2000, will cost almost $8 billion, and will take more than 30 years. Construction is scheduled to start in 2003 or 2004. See Phyllis McIntosh, "Reviving the Everglades," *National Parks* (January 2002).

Is too much being done to protect the species with which we share the earth? In the following selection, David N. Laband argues that it is, largely because the people who set environmental policy do not need to pay for protection efforts themselves. Instead, the costs of protecting biodiversity are unfairly laid upon landowners. In the second selection, Kris Christen interviews E. O. Wilson, who supports the protection of biodiversity as crucial to human survival. He argues that the loss of species reduces the productivity and stability of natural ecosystems and that with each species lost, potential drugs and other valuable resources are also lost.

David N. Laband    **YES**

Regulating Biodiversity: Tragedy in the Political Commons

Last summer, lightning struck and killed an enormous pine tree on one side of my backyard. At about the same time, voracious pine bark beetles girdled and killed an equally impressive pine tree on the other side. Now bereft of needles, these two arboreal giants pose a potential threat to my house: if they were to fall at just the right angle, the damage could be substantial. In the interest of safety, my wife wants to have the trees removed; for the sake of promoting biodiversity on my two-acre lot, I do not.

Our personal dilemma mirrors a much larger struggle that quietly threatens to destroy the rights of private timberland owners across the United States —the desire of urban dwellers to have their cake and eat it too. They demand houses made of wood, wood furniture, paper and paper products, and so on, while also demanding environmental amenities such as aesthetically pleasing landscape views, biodiversity, and animal habitat. At a personal level this can't be done. If the trees are removed, my wife has peace of mind, but the many animals that depend on dead pine trees for their existence, either directly or indirectly, will vanish. If the trees stay, we will be promoting the ecological diversity of our property, but my wife will worry about our house with every gust of wind. We can't have it both ways. Similarly, at a macro level, there is a trade-off between production/consumption of timber and production/consumption of related environmental amenities.

The Role of Intensively Managed Forests

The problem of how to grow and harvest increasing amounts of timber while simultaneously producing a steadily increasing array and level of environmental amenities associated with forested land has resulted in an industry-wide discussion of how to simultaneously achieve both objectives. There is a growing appreciation within the forestry community for the prospect that intensively managed forests may yield increasing amounts of wood while minimizing the total acreage from which wood is harvested. This maximizes the amount of acreage available to meet other demands—such as agricultural production, animal habitat, and other environmental amenities associated with natural forests.

From David N. Laband, "Regulating Biodiversity: Tragedy in the Political Commons," *Ideas on Liberty* (September 2001). Copyright © 2001 by The Foundation for Economic Education. Reprinted by permission of *Ideas on Liberty* and the author.

However, intensively managed forests have come under heavy fire from self-proclaimed environmentalists. In these so-called plantation forests, man, not nature, regenerates the trees, which accordingly grow in even-aged stands. Their well-being is affected by the application of herbicides and pesticides, as well as by occasional thinning and fire management. In contrast to naturally (re)generated timberland, plantation timberland has been described as an "ecological desert," with the stated or implied conclusion that the nature and extent of biological diversity associated with natural forests is both greater and therefore more desirable than that associated with plantation forests.[1]

The Threat to Private Landowners and Social Welfare

Such pejorative rhetoric is both misleading and counterproductive. The unfortunate but nonetheless compelling truth is that we can't have our cake and eat it too. We must make responsible choices about what to produce and how to produce it. A serious threat to private landowners develops when citizens living in urban areas demand that private owners of timberland (definitionally located in rural areas) produce environmental amenities such as aesthetically pleasing views, biodiversity, animal habitat, and the like, *provided the urbanites don't have to pay for it.*

Further, they seek to enforce their demands by using the political process to pass regulations that require landowners disproportionately to bear the cost of producing these environmental amenities. For example, Oregon law requires private timberland owners to replant within two years areas from which they cut trees. Other regulations forbid clearcutting of timberland. Federal regulations pertaining to endangered species are incredibly restrictive and intrusive with respect to an individual's property rights. The pursuit of environmental amenities that we are told are vital to some vaguely defined public interest through policies that impose virtually all the costs on relatively small numbers of private landowners generates what might be termed a "tragedy of the political commons."

Garrett Hardin introduced us to the tragedy of the commons.[2] Hardin developed a stylized example of a communal pasture open to all comers. There are no private property rights to the pasture, or rules, customs, or norms for shared use. In this setting, each shepherd, seeking to maximize the value of his holdings, keeps adding sheep to his flock as long as doing so adds an increment of gain. Further, the shepherds graze their sheep on the commons as long as the pasture provides any sustenance. Ignorant of the effects of their individual actions on the others, the shepherds collectively (and innocently) destroy the pasture. As Hardin concludes: "Therein is the tragedy. Each man is locked into a system that compels him to increase his herd without limit—in a world that is limited. Ruin is the destination toward which all men rush, each pursuing his own best interest in a society that believes in freedom of the commons."

Man's exploitation of the political commons is analogous to his exploitation of natural-resource commons. Our majority-rule voting process, which permits a majority of citizens to impose differential costs on the minority, encourages overprotection of endangered species, and overproduction of biodiversity, animal habitat, and landscape views. This occurs because each individual who bears a negligible portion of the costs of providing environmental amenities has a private incentive to keep demanding additional environmental protections as long as there is *any* perceived marginal benefit. As with the overgrazed pasture, the result of overprotecting Bambi is, as has become apparent all over the eastern United States, disastrous. Moreover, and not surprisingly, we are starting to hear real concern voiced about the recent proliferation of other animal species such as black bears, mountain lions, and coyotes. We are creating social tragedies that result from the political commons.

The tragedy is compounded by the incentives generated for private landowners by the heavy hand of command-and-control policies. When government abrogates property rights without compensation, landowners have strong incentives to mitigate their expected losses. They can do so by changing their land use from timber production to housing or commercial development. There is no incentive to promote habitat for endangered species; doing so means only that use of one's land will be seriously compromised by the highly restrictive provisions of the Endangered Species Act. Instead, a landowner who finds a member of an endangered species on his property has a well-understood incentive to "shoot, shovel, and shut up." Such behaviors are not likely to further environmental objectives.

Other People's Costs

It is relatively easy to demonstrate that because private timberland owners bear the cost of producing biodiversity, nonland-owners demand excessive amounts of it. The first point to be made in this regard is that urbanites do not in fact place a high value on biodiversity. One need look no further than the readily observable behavior of urbanites for proof of this claim. Urbanites have the ability and prerogative to produce biodiversity on their own residential property. That is, they could let their residential lots grow wild with natural flora and fauna. This would, without question, promote ecological diversity. In practice, virtually no residential property owners, living anywhere in the United States, do this. Instead, they invest (implicitly through their time and explicitly by purchase) hundreds, if not thousands, of dollars annually in the care and maintenance of their lawns and grounds in a decidedly unnatural state. Like owners of intensively managed timberland, owners of residential property chemically treat and harvest the growth on their property. In so doing, they create a landscape with relatively little floral or faunal diversity. What this behavior reveals, of course, is that urban dwellers place a higher value on having their own aesthetically pleasing ecological deserts than on personally promoting local biodiversity, even when the latter would save them hundreds, perhaps thousands, of dollars each year. The clear implication is that urbanites simply do not attach much importance to biodiversity.

This leads directly to a second point: notwithstanding that biodiversity is of little importance to them personally, urbanites may favor local, state, and federal statutes that ostensibly enhance biodiversity, provided such statutes impose the cost burden on rural landowners. The feel-good benefit of such regulation may be small, but with no personal costs to worry about, urbanites can be convinced to vote for them. However, if there were even a moderate cost to urban dwellers, we can be reasonably certain that restrictive regulations would not be passed. This explains why, for example, Oregon's replanting regulations are not imposed on owners of residential properties who cut down trees.

Earth's limited resources cannot provide all things to all people simultaneously. For that matter, the earth cannot provide all things just to self-proclaimed environmentalists. Consequently, responsible choices about the use of resources must be made. It is irresponsible to enact environmental policies that impose costs disproportionately on private timberland owners. Such policies lead to overproduction of environmental protection because urban voters who place little value on environmental amenities support regulations that impose little or no cost on themselves personally. Further, these policies create incentives for private timberland owners to minimize, not maximize, their production of environmental amenities. This problem of incompatible incentives makes it less likely that public policy will actually attain its stated objectives.

Notes

1. National Audubon Society, www.audubon.org/campaign/fh/chipmills.htm, no date.
2. Garrett Hardin, "The Tragedy of the Commons," *Science* (162), 1968, pp. 1243–48; see www.dieoff.org/page95.htm.

Why Biodiversity Matters

Kris Christen: You have written that "the loss of genetic and species diversity... is the folly our descendants are least likely to forgive us." What will our descendants see that will lead them to place this happenstance highest on their list of unforgiven ills?

E.O. Wilson: We still haven't woken to the fact that while all the changes in the environment having to do with pollution, ozone depletion and global warming are vitally important, they can be reversed—while on the other hand species extinction, the loss of biodiversity, cannot be reversed. We are not deliberately trying to wipe out the Creation, but we are, by general agreement among experts on biodiversity, heading toward extinction of as many as 20% of species in the next 30 years.

Why do I and other ecologists consider that unforgivable? Because each species is a masterpiece of evolution and, depending on the species, has been evolving into its present state for some thousands to tens of millions of years. The average life span of a species before humanity came along was between half a million years in mammals and, in some groups like the insects, 10 million years. To wipe out species at the rate we are now inflicting has been to increase the extinction rate by between a hundred and a thousand times.

By impoverishing the planet of life forms, we also reduce the productivity and stability of natural ecosystems.

What we lose in terms of natural products through the extinction of the species that uniquely produced them is extraordinary. A wondrous example of this is a substance discovered... at Harvard from a small tree collected in Sarawak, at the northern end of Borneo. Random screening of *Calophyllum lanigerum* revealed a substance that is completely effective against the AIDS virus. Upon closer study, the substance proved to be an inhibitor of reverse transcriptase, which stops reproduction of the AIDS virus in its tracks. When collectors were sent back to get more samples, the tree was gone; the forest around it had been cut over, and it took a long search to find other specimens of this rather rare tree. This substance never would have been discovered if the species had been extinguished completely, and it easily could have been extinguished before anyone did that survey.

From E. O. Wilson, "Why Biodiversity Matters," interview by Kris Christen, *OECD Observer* (Summer 2001). Excerpted from "Biodiversity at the Crossroads," *Environmental Science and Technology*, vol. 34, no. 5 (March 1, 2000), pp. 123A–128A. Copyright © 2000 by The American Chemical Society. Reprinted by permission.

Christen: What do you see as the most difficult environmental problems we face as we head into the new millennium?

Wilson: That's easy. Land degradation and the loss of irreplaceable, non-renewable natural resources, including the natural environment, by the combination of continued population growth and the drive of people everywhere to increase their consumption and, with it, their quality of life.

At the present time, the ecological footprint—the amount of productive land used per capita for food production, water and waste management, habitation, transportation, and other necessities—for the United States is about 12 acres. In developing countries, its about 1 acre. So, with 80% of the world's population in the developing countries and virtually all of the projected population growth over the next few decades occurring there, the pressures upon the earth's resources and its flora and fauna are going to be enormous because these people are understandably anxious to increase their ecological footprint.

We need to identify the world's conservation hot spots and go all out to save them. Hot spots are those natural environments that have the largest number of plants and animals found nowhere else and are themselves endangered. Hawaii is one of the hottest spots in the world, with the highest rates of extinction as well as the greatest endangerment of plant and animal species. Other notorious hot spots include Madagascar, Ecuador's mountain forests, Brazil's Atlantic Forest, the Western Ghats of India, the forest on the southern slopes of the Himalayas, and now, increasingly, coral reefs. The magnitude of the catastrophe has been measured, and there's nothing to be gained anymore by just wringing one's hands. We have to devise a strategy for the next century that can pull us through the bottleneck and that will depend on the full engagement of the best we have to offer in science and technology. Its also going to have to involve a shift of world opinion away from purely econometric measures of success and progress, toward an environmental ethic that says what really counts is quality of life for all generations to come.

Christen: Do you have faith that the application of science and technology will prove an adequate remedy?

Wilson: Yes. Its got to. The stakes are too high. Can we do it? There's a certain level of urgency, when one says "Yes, because we have no other choice", and that is truly our current circumstance.

Christen: What positive movement have you seen in biodiversity and in educating people as a solution to turning around species extinction?

Wilson: Countries are now setting a few areas of the world aside—where forests and other habitats were being wiped out and large numbers of species along with them—particularly in the developing world where it's most needed. These countries signed on to the 1992 Earth Summit Convention on Biological Diversity. As a result of government intervention, the destruction is beginning to slow down—for example, in Brazil's Atlantic Forest. Also, in Brazil recently, a substantial amount of the Pantenal, the South American equivalent of the Everglades with a magnificently rich diversity of organisms, was purchased and set aside in a reserve. So bits and pieces around the world are being saved, but it's still far below what's needed—even below the 3% or 4% of the surface of the land composing the most important hot spots. As for education, its still entirely inadequate worldwide, even in this country, where awareness of the biodiversity crisis has been slowly spreading.

Christen: Concerning preservation of biodiversity, is gene patenting part of the solution or part of the problem?

Wilson: It's a potential problem, but it's certainly also, when handled right, a big part of the solution. Pharmaceutical companies have been slow to screen and make use of natural products because of fear of not being able to secure patent rights, because theoretically, or I should say legally, a natural product is not patentable. It takes an enormous amount of money to find, perfect, test, and market a new pharmaceutical. It's a risky venture economically. It would be unfortunate to depend solely on the synthesis of new pharmaceuticals from the ground up, because millions of organisms have been in an arms race with one another for hundreds of millions of years, slugging it out with bacteria and developing all sorts of anticancer materials and other defensive substances by Darwinian trial and error. Each one is a potential pharmacopoeia of materials waiting to be discovered. For example, insects, which are the dominant and most diverse creatures on the land around the world, are surely loaded with substances of this kind, yet they have scarcely been looked at by government or commercial laboratories.

Christen: What's your opinion about biotechnological applications in the area of plant engineering as to their effect on biodiversity?

Wilson: Here, I may run into trouble with some of my colleagues, but I'm all for biotechnology. We need all the science and technology we can get to sustain both humanity and the natural world. We need to push bioengineering to the limit in creating more productive crops, particularly crops that can live in already devastated, soil-impoverished environments, low-diversity saline environments, and other biologically marginal land where productivity can be increased to the maximum allowed by photosynthetic potential and with a minimal effect on biodiversity. We absolutely must increase the world food supply.

Christen: But is this technology something we might live to regret?

Wilson: The risk involved, of course, is that new life forms can be created that penetrate and endanger natural ecosystems, or they may transfer their genes to natural species in ways that help lead to their extinction. But so far, there has been very little sign that this is a general risk. Intact, healthy, natural ecosystems are hard to penetrate, even by natural species introduced as invasive forms from other countries . . . The side effects are a risk, but I don't see them as a very large risk at this point. In any case, its one worth taking, given the benefits and given our ability (and responsibility) to monitor each situation and regulate it.

Christen: What ethical considerations should we be pondering as this technology moves forward?

Wilson: I think there's a very strong moral argument to be made for getting the maximum gains possible through bioengineering. But there is an equally strong moral argument to be made for protecting all the biodiversity we can. It seems to me that the research laboratories... should have as part of their standard practice a strong environmental programme. They should not just be satisfied with ensuring new biological strains are safe, but also take a proactive role in helping to preserve the natural environment... Corporate leaders, who will be judged accordingly, will want to play a prominent role.

Christen: How big, in your opinion, do land and marine reserves need to be to preserve all the world's biodiversity?

Wilson: As big as we can manage to make them. I'll tell you why. As you reduce the size of a reserve, or any habitat, you automatically reduce the number of species that can live sustainably on that reserve. The amount of reduction is roughly the following: A 90% reduction in area eventually results in a 50% decrease in the number of species. Although it may take a number of years, it still happens very rapidly in ecological time... Tragically, we'll always lose species. It's part of the mission of conservation biology to figure out the designs and exchange of fauna and flora, recovery and enlargement of natural reserves, to lose as few as possible, and to keep as much biodiversity as possible.

Christen: In what ways might governments and citizens be persuaded to do more to protect biodiversity?

Wilson: Education, education, education. To that end, what we need are more public philosophers, government advisors, and media people with a scientific background, or at least enough knowledge of biodiversity, so it comes into focus for them, and therefore for their audiences. This has been a major failure of the media in this country, particularly in presenting key scientific issues. It is scandalous in the case of biodiversity... How can you interest the public? Granted there's nothing duller than river pollution. But it's different with vanishing ecosystems and species. We need to dramatise the world's real biodiversity issues better because it's not getting across to Joe Sixpack as to why any of this matters.

Christen: In your book *Consilience*, you propose unifying all the major branches of knowledge. What does your ideal curriculum look like for students in the 21st century?

Wilson: Let me say right away that I'm not exactly urging unification; I'm reporting that it is happening and urging that it be speeded up... The social sciences, by general agreement, are far weaker than they should be for the problems that they're supposed to solve. In my opinion, the major reason for that inadequacy is that they lack a foundation of the kind that biology has in chemistry and chemistry has in physics . . . The social sciences, in particular, are going to live in a dreamworld as far as the environment is concerned until they become linked more solidly to biology—not just the biology of the mind and a realistic view of the human condition, but biology that includes studies of the environment. In other words, the real world.

POSTSCRIPT

Is Biodiversity Overprotected?

There is debate over whether the best way to protect biodiversity is to protect individual endangered species or to protect habitat, which must also shield all the species that share that habitat. In 1998 a bill that would have substituted habitat conservation plans for the protection of individual species was proposed in the U.S. Senate but never came to a vote. It was opposed by environmentalists and House representatives, who argued that habitat conservation does not adequately protect endangered species, and by conservatives who wanted the bill to compensate property owners who lost property value or income opportunities because of habitat protection restrictions.

Mark L. Shaffer, J. Michael Scott, and Frank Casey, in "Noah's Options: Initial Cost Estimates of a National System of Habitat Conservation Areas in the United States," *Bioscience* (May 2002), state, "Solving the habitat portion of the endangered species and biodiversity conservation problems is neither trivial nor overwhelming. A national system of habitat conservation areas in the United States could be secured for an initial annual investment between $5 billion and $8 billion, sustained over 30 years, or roughly one-fourth to one-third the cost of maintaining our national highway system over the same period."

The sixth International Conference on Biological Diversity was held in The Hague in April 2002. Sponsored by the United Nations Environment Programme, it emphasized efforts to protect genetic resources of the sort mentioned by Wilson in order to "give biodiversity-rich countries additional incentives to conserve and sustainably use their resources." In addition, the conference addressed the hazards posed to biodiversity by exotic (nonnative) species, a topic that is itself controversial. Peter Warshall, in "Green Nazis?" *Whole Earth* (March 2001), says that conservatives are attacking attempts to control or eradicate nonnative species as motivated by a kind of environmental racism.

Although the Endangered Species Act was due for reauthorization in 1993, the necessary legislation has not yet been enacted. Useful articles in favor of reauthorization include John Volkman's "Making Room in the Ark," *Environment* (May 1992) and T. H. Watkins's "What's Wrong With the Endangered Species Act?" *Audubon* (January/February 1996). Bonnie B. Burgess discusses the history and future of the Endangered Species Act in *Fate of the Wild: The Endangered Species Act and the Future of Biodiversity* (University of Georgia Press, 2001). Burgess feels that the act is itself endangered because of attacks from conservatives and obstacles erected by the government. Finally, Omar N. White addresses the question of the act's constitutionality in "The Endangered Species Act's Precarious Perch: A Constitutional Analysis Under the Commerce Clause and the Treaty Power," *Ecology Law Quarterly* (February 2000).

ISSUE 3

Are Environmental Regulations Too Restrictive?

YES: Peter W. Huber, from "Saving the Environment From the Environmentalists," *Commentary* (April 1998)

NO: Paul R. Ehrlich and Anne H. Ehrlich, from "Brownlash: The New Environmental Anti-Science," *The Humanist* (November/December 1996)

ISSUE SUMMARY

YES: Peter W. Huber, a senior fellow at the Manhattan Institute, argues that the environment is best protected by traditional conservation, which puts human concerns first.

NO: Environmental scientists Paul R. Ehrlich and Anne H. Ehrlich argue that many objections to environmental protections are self-serving and based in bad or misused science.

Concern for the environment in America is not much more than a century old. In 1785 Thomas Jefferson invented the idea (if not the wording) of NIMBY ("Not In My Back Yard") when he wrote, "Let our workshops remain in Europe." He thought that an American factory system would have undesirable social, moral, and aesthetic effects. Clearly, he was alone in that thought, for America developed its industrial base very quickly. The workshop builders flourished, and the effects that concerned Jefferson did indeed come to pass. The first national park, Yosemite, resulted from legislation signed by President Abraham Lincoln in 1864. Yellowstone was approved in 1872. Both were responses to an awareness that if the areas' unique features were not protected, they would be destroyed by ranchers, miners, loggers, and market hunters, as had already happened elsewhere.

By the 1960s people were beginning to realize that other activities, such as the use of pesticides, also threatened treasured features of the environment, such as songbirds (see Rachel Carson, *Silent Spring* [Houghton Mifflin, 1962]), as well as human health. The result was government regulation of pesticides, air pollution, water pollution, and much, much more. Lead has been removed from gasoline and paint, chlorofluorocarbons from aerosol deodorants and refrigerants, and phosphates from laundry detergents. Developers have been told they cannot fill in swamps and other wetlands. Loggers have been forbidden to log in many areas. And commercial fishing seasons have been limited or eliminated entirely.

The economic impact of environmental regulation has not been as great as it might have been if Jefferson had had his way in 1785, but in each case someone's economic benefit has been interfered with. In other cases—such as when users of off-road vehicles have been barred from driving on the nesting grounds of rare shorebirds—the freedom to do as one wishes has been interfered with. In nations such as China, which instituted a "one child per couple" population control policy in 1979, the freedom in question is the freedom to have as many children as one wishes. As the environmental regulations have proliferated, so has the interference with freedoms that people once took for granted. And so have the objections. Conservative politicians and lobbyists for industry, recreation, and home-owner groups struggle to block or weaken every new environmental regulation and to repeal old regulations, often in the name of individual freedom and property rights. Environmentalists counter that freedom must be tempered by responsibility; individual freedom and property rights must have limits, or we will destroy what lets us and our children live on earth.

The issue is not just America's; it is the world's. Environmentalists are active everywhere, identifying problems, promoting a sense of crisis, and saying what must be done, what behaviors must be controlled, and what freedoms must be limited. They have been successful enough to rouse fears among some political conservatives of a liberal-environmentalist conspiracy to take over the world and impose an antifreedom world government. Such conservatives welcome an approach that is designed to weaken environmental regulation in favor of the economy.

In the following selections, Peter W. Huber argues that the environment is best protected by traditional conservation. This puts human concerns before those of what he considers to be the insignificant and unattractive portions of the world that are favored by environmentalists and their "pervasive, manipulative, and intrusive bureaucracy." Paul R. Ehrlich and Anne H. Ehrlich maintain that many objections to environmental protections are essentially self-serving. They assert that antienvironmentalists deny the facts in favor of religious, economic, and political ideologies.

Saving the Environment From the Environmentalists

As a political movement, environmentalism was invented by a conservative Republican. He loved wild animals. He particularly loved to shoot them.

In the spring of 1908, with time running out on his second term, President Theodore Roosevelt held a hugely successful conference on conservation. The report that emerged, T.R. would declare, was "one of the most fundamentally important documents ever laid before the American people." He promptly called a hemispheric conference on the same theme, and was working on a global one when he left office in March 1909.

He had learned his conservation the hard way. After Grover Cleveland defeated the Republicans in 1884, T.R. returned to his Chimney Butte ranch in the Dakota Territory with plans to increase his cattle herd fivefold. Armed neighbors came by to complain. As H. W. Brands recounts in his recent *T.R.: The Last Romantic,* "the potential for overstocking the range weighed constantly on the minds of the ranchers of the plains."

Although Roosevelt faced down his angry neighbors, he also set about finding a political solution to the problem that concerned them, forming and becoming president of the Little Missouri Stockmen's Association. He would only regret not starting earlier. The Dakota pastures were badly overgrazed in the summer of 1886, and many herds, T.R.'s among them, were destroyed in the dreadfully harsh winter that followed.

Occupying the White House two decades later, T.R. and his chief forester, Gifford Pinchot, would be the first to apply the word "conservation" to describe environmental policy. By then, Roosevelt had come to view the misuse of natural resources as "the fundamental problem which underlies almost every other problem of our national life."

⋯⊙⋯

The administration of Theodore Roosevelt was certainly not the first to show such concern. Congress had proclaimed Yellowstone a national park in 1872. Yosemite, Sequoia, and General Grant national parks were established in 1890. The first U.S. forest reserve, forerunner of the national forests, was proclaimed in the area around Yellowstone National Park in 1891. Presidents Harrison, Cleveland, and McKinley transferred some 50 million acres of timberland into the reserve system.

T.R.'s distinction was to give conservation its name and, more importantly, to transform it into an enduringly popular political movement. On the way to adding 150 million more acres to the country's forest reserves, he would persuade the great mass of ordinary Americans that conservation was in their own best interests.

What with two world wars and a depression intervening, it would take another six decades to complete a federal legal framework for conservation. In the meantime, much occurred to affect conventional notions of the environment. The radioactive aftermath of Hiroshima taught a first, ghastly lesson about insidious environmental poison. There followed popularized accounts of industrial equivalents of Hiroshima—fallout without the bomb. Rachel Carson defined the new genre in 1962, with the publication of *The Silent Spring,* about the dangers of pesticides.

All this became reflected in law. The Clean Air, Clean Water, and Resource Conservation and Recovery Acts of the 1960's, like the Endangered Species Act passed unanimously by the Senate in 1973, *seemed* to be cut from the same old conservationist cloth woven by T.R. (though they concerned smoke, sewage, and landfills rather than parks and mountains). But even as they completed and somewhat extended the framework for traditional conservation, these laws also quietly launched a new era—the era of environmentalism.

Regulating multifarious forms of pollution—the purpose of the clean-air, clean-water, and landfill acts—required a more elaborate regulatory structure than regulating parks and reserves. President Nixon had to establish a new cabinet-level body, the Environmental Protection Agency (EPA), to take charge. More significantly, each of the laws also included something quite new: an open-ended "toxics" provision, a general invitation to monitor the micro-environment for poisons and regulate them as needed. Even the Endangered Species Act, though written mainly with the likes of cougars in mind, was drafted broadly enough to protect unpleasant rodents like the kangaroo rat, and would soon be amended to prevent not only hunting but also "harming," which a federal court then construed to cover "habitat modification."

A mere statutory afterthought in the 1960's, the micro-environment was getting entire acts of its own a decade later. The Toxic Substances Control Act was promulgated in 1976. Then, in 1980, came Superfund. And thus, somewhere between Vietnam and the discovery of alarming concentrations of chemicals in the soil and groundwater at a town in upstate New York called Love Canal, a legal infrastructure for the new environmentalism slipped into place. Conservation was not abandoned. But politically it was overtaken, subsumed into something bigger. Bigger precisely because it concerned the very small.

⁂

Over time, the distinctions between conservation and environmentalism have been obscured. But they really are two different schools.

Conservation happens in places we can see, and draw on a map. Yellowstone starts here and ends there. Bison, eagles, and rivers are only somewhat harder to track.

T.R. had no trouble seeing the things that made him a conservationist. Forests were being leveled, ranges overgrazed, and game depleted. Hunters and hikers, cattlemen, farmers, and bird-watchers could easily grasp all this, too. The political choices T.R. was urging were based on these considerations. Americans would want to preserve Yellowstone for the same reason they might some day wish to climb Everest: because it was there, because they knew it was there, and because they desired to keep it there.

If conservation happens in places we can see, micro-environmentalism happens everywhere. The microcosm is so populous, the forces of dispersion so inexorable, that in every breath we take we inhale many of the very molecules once breathed by Moses and Caesar. At that level of things, everything gets polluted, even though no one can see it, and it is all too easy to suggest causes and effects. Fish die, frogs are deformed, breast cancers proliferate, immune systems collapse, sperm counts plummet, learning disabilities multiply: every time, invisible toxics are assumed to be the culprit.

To believe wholeheartedly in micro-environmentalism one must either be a savant or put a great deal of trust in savants. In particular, one must put one's trust in computer models. The model is everything. Only the model can say just where the dioxin came from, or how it may affect our cellular protein. Only the model will tell us whether our backyard barbecues (collectively, of course) are going to alter rainfall in Rwanda. Only the model can explain why a relentless pursuit of the invisible—halogenated hydrocarbons, heavy metals, or pesticides —will save birds or cut cancer rates. The cry of the loon gives way to the hum of the computer. T.R. trades in his double-barreled shotgun for a spectrometer.

But precisely because it involves things so very small, the microcosm requires management that is very large. Old-style conservationists maintained reasonably clean lines between private and public space. They may have debated how many Winnebagos to accommodate in Yellowstone, how much logging, hunting, fishing, or drilling for oil to tolerate on federal reserves, but the debates were confined by well-demarcated boundaries. Everyone knew where public authority began and ended. Yellowstone required management of a place, not a populace. Municipal sewer pipes and factory smokestacks may have required more management, but still of a conventional kind. The new models are completely different, so different that they are tended by a new oligarchy, a priesthood of scientists, regulators, and lawyers.

With detectors and computers that claim to count everything everywhere, micro-environmentalism never has to stop. With the right models in hand, it is easy to conclude that your light bulb, flush toilet, and hair spray, your washing machine and refrigerator and compost heap, are all of legitimate interest to the authorities. Nothing is too small, too personal, too close to home to drop beneath the new environmental radar. It is not Yellowstone that has to be fenced, but humanity itself. That requires a missionary spirit, a zealous willingness to work door to door. It requires propagandists at the EPA, lesson plans in public schools, and sermons from the modern pulpit. Children are taught to enlighten —perhaps even to denounce—their backsliding parents.

⋗⟨◉⟩⋗

At this point, environmental discourse often degenerates into a fractious quarrel about underlying facts. One side insists that tetraethyl lead, pseudo-estrogen, and low-frequency electromagnetic radiation seriously harm human health. The other side says they do not. One side says these things will hurt birds, frogs, and forests, and have already done so. The other side says they have not and will not.

One might suppose that science would settle such disputes. But it cannot. In a classic essay from 1972, the nuclear physicist Alvin Weinberg explained why. He coined a term, "trans-science," to describe the study of problems too large, diffuse, rare, or long-term to be resolved by scientific means. It would, for instance, take eight billion mice to perform a statistically significant test of the health effects of radiation at exposure levels the EPA deems to be "safe." The model used to set that threshold may be right, or it may be way off; the only certainty is that no eight-billion-mouse experiment is going to happen.

The same goes for any model of very-low-probability accidents—an earthquake precipitating the collapse of the Hoover dam, say, leading to the inundation of the Imperial Valley of California. Statistical models can be built, and have been, but their critical, constituent parts cannot be tested. And similarly with all the most far-reaching models of micro-environmentalism, a realm of huge populations (molecules, particles) paired with very weak or slow effects. Whether we are talking about global warming, ozone depletion, species extinction, radiation, halogens, or heavy metals, whether the concern is for humans or frogs, redwoods or sandworts, the time frames are too long, the effects too diffuse, the confounding variables too numerous.

You may doubt this if you get your environmental trans-science the way most people do, for the mass media always convey a greater sense of certitude. There is no news in reporting "Dog May or May Not Bite Man; Scientists Waffle." Instead, *Newsweek* gives us: "Meteorologists disagree about the cause and extent of the cooling trend. But they are almost unanimous in the view that the trend will reduce agricultural productivity for the rest of the century." That was in 1975. They were still almost unanimous in 1992, according to Vice President Al Gore; but about what? "Scientists have concluded—almost unanimously—that global warming is real and the time to act is now." (I owe this juxtaposition to the *Economist,* December 20, 1997.) If the papers give you the various sides of the trans-scientific debate at all, they give it in different editions; sometimes, the editions are published twenty years apart.

It is a fair bet that now and again a model will predict things exactly right. It is a fairer bet that much of the time it will not. Indeed, if overall statistics confirm anything, it is that environmental toxins of human origin are not the main cause of anything much. The more industrialized we become, the longer we live and the healthier we grow. There is a model—quite a credible one, in fact —that purports to prove that a steady dose of low-level radiation, like the one you get living in a high-altitude locale like Denver, or at some suitable distance from Chernobyl, actually improves your health.

Nor are these the only problems. Suspect toxins vastly outnumber modelers. The list of things we might reasonably worry about grows faster than new rules can be published in the *Federal Register*. But the axiology of science, its priorities of investigation and research, the criteria for what to study and what not to, are matters of taste, budget, values—everything but science itself. Scientific priorities, Weinberg notes, are themselves trans-scientific. So are all the engineering issues, the practical fixes that regulators prescribe. Science will never tell us just how much scrubber or converter to stick on a tailpipe or smokestack, how much sand and gravel at the end of a sewer pipe, how much plastic and clay around the sides of a dump.

So, in the end, the micro-environmentalist just names his favorite poison, and gets on with making sure that nobody drinks it. The process is arrayed in the sumptuary of science, but the key calls are political. Micro-environmentalism ends up as a pursuit of politics by other means.

⋅◈⋅

There is nothing wrong with politics, of course—T.R. reveled in them. But here too there is an essential difference between the old conservationism and the new environmentalism.

All the choices old-style conservationists make are conventionally political. The Clinton administration recently designated as a national monument a vast stretch of land in Utah, from Bryce Canyon to the Colorado River, and from Boulder to the Arizona state line. It was a controversial call: the area includes the Kaiparowits plateau, where a Dutch-owned concern was slated to begin mining massive coal formations. T.R. would certainly have understood the controversy over the Kaiparowits plateau, and would likely have approved the decision to conserve.

In the new environmentalism, by contrast, conventional political process decides little. The clauses about toxics that were inserted as an afterthought in the clean-air and clean-water acts, and as the central thought in Superfund, are just a stew of words. They articulate no standard, set no budget, establish no limits. In T.R.'s day they would not even have passed constitutional muster. The Supreme Court would have cited the "nondelegation doctrine," which, then at least, forbade Congress to delegate responsibilities wholesale to the executive branch.

Today the delegation goes a lot further. Though nominally in the hands of the President and overseen by Congress, political authority for micro-environmental matters is now centered in the new trans-scientific oligarchy. The key calls are still stroke-of-the-pen political, at bottom, but no ordinary observer can see to the bottom. The only thing ordinary Americans may dimly realize is that somewhere deep in the EPA it has been deemed wise to spend more money digging up an industrial park in New Jersey than ever was spent conserving a forest in the Adirondacks.

Politicians know how to reward friends and punish enemies, but democratic politics tends, as a whole, to be pretty even-handed. When the old conservationists took your land, they paid you for it, and the money came from taxes and user fees. That was about as fair as the income tax—not very, but fair enough. In the new environmentalism, most of the taxing occurs off the public books. There is a great deal of creeping, uncompensated expropriation, and a freakish rain of ruin on those unlucky enough to discover the wrong rodent, marsh, or buried chemical on their land. Any amount of public environmental good, however small, can entail any private financial burden, however large.

We have likewise lost all pragmatic sense of when enough is enough. Conservation, driven as it must be through normal political channels, can be pushed only so far. The Clinton administration had to trade political chips for the Kaiparowits plateau; nobody feared it would soon seize the rest of Utah. Conservation works, politically, because the boundaries are reasonably well defined and because it targets real estate, not molecules. By contrast, most of the Northeast could be placed in regulatory receivership for its countless microenvironmental derelictions. Whereas hikers and hunters occupy a seat or two at the political table, synthetic estrogens and carbon dioxide have somehow escaped from the coils of politics, and the priesthood can pursue them without restraint.

The "remedial" efforts that emerge from this pursuit end up repelling even the intended beneficiaries. Contact with Superfund has become socially poisonous. The very arrival of the EPA in a community shatters property values, repels new industrial investment, and throws a region's entire future into doubt. Environmental regulation has in effect become a mirror image of the problems it is supposed to solve, leaking into society cancerous plumes of lawyers, administrators, and consultants, the brokers of ignorance, speculation, and uncertainty.

Theodore Roosevelt was no Ralph Waldo Emerson, Henry David Thoreau, or John Muir. These "preservationists" revered wilderness for its own sake. Muir, founder of the Sierra Club, adamantly opposed building the Tuolumne River dam in Yosemite to supply water to San Francisco. T.R. supported it, consistent with his "wise-use" philosophy of conservation. For T.R., the whole point of conserving nature was to continue using it—forests for lumber, ranges for grazing, rivers for electrical power. Hunters, cattlemen, ranchers were to be involved in conservation because it was in their own self-interest. "Despite occasional moments of doubt," writes H. W. Brands, T.R. "passionately believed in the capacity of the ordinary people of America to act in the public welfare, once they were alerted to the true nature of that welfare."

That was the faith that defined the first century of conservationism. Congress had established Yellowstone National Park as a "pleasuring ground" for people. The national parks would include forests, seashores, lakeshores, and scenic trails but also monuments, historical sites, and battlefields—man's creation alongside nature's. T.R.'s distant cousin Franklin, too, was an ardent conservationist, and during his presidency he established his own share of national parks and forests; but he also built roads, bridges, tunnels, airports, and skyscrapers. Like T.R., he believed there was room enough in nature for man.

Today, the preservationist vision is back on top. The quasi-pagan nature worship of the late 19th century has been reworked as the trans-scientific demonology of the late 20th. Those who believe in the new methods and models do not even credit the distinction between conservation and preservation. The computer models can link any human activity, however small, to any environmental consequence, however large—it is just a matter of tracing out small effects through space and time, down the rivers, up the food chains, and into the roots, the egg shells, or the fatty tissue of the breast. This is what chaos theorists call the "butterfly effect," traced out by computer. If you believe in the computer, you must believe that the only way really to "conserve" is not to touch at all.

◈

Is it possible to change course, and if so, how? The answer comes in two parts, the philosophical and the practical-political.

There was never much high-church philosophy to T.R.'s conservationism. It was inspired by an abiding appreciation for the beauty of nature—that is, by aesthetics. And it was disciplined by a real sense (this may seem a curious thing to say of a man like T.R.) of humility. Not much philosophy there, but enough.

A sense of aesthetics would get us a long way in reforming environmental discourse. It would, to begin with, help us cut through the scientism, the fussy bureaucratic detail. It would let us ignore the priesthood and dispense with its soaring intellectual cathedrals. It would save us the enormous expense and inconvenience of digging up New Jersey and conserving our own trash. It would allow us to spend our energy and dollars on places that are simply beautiful, and oppose things for no fancier reason than that they are ugly.

The aesthetic approach does not mean ignoring the micro-environment completely, still less rejecting every commandment ever prescribed by the priesthood. Priests and propagandists have every right to help shape our aesthetic preferences, for better or worse; they just should not be allowed to palm off their art as science. Purity is beautiful, and industrial byproducts in our drinking water are ugly, even if invisible and harmless. (Fluoride and chlorine in the water are sort of ugly, too, even if they give us healthier teeth and guts.)

There is also an aesthetic case to be made for frugality: we are not going to run out of space for dumps, but garbage is not beautiful, and making do with less often is. By the same token, however, profligate excess in the digging up of dumps is as ugly as profligate excess in the original dumping. T.R.-style conservationists would devote far more energy to parks and forests, to sewage treatment and cleaner smokestacks, and far less to part-per-billion traces of dioxin. Whatever impact pesticides may have, setting aside 100 million acres of forest will likely protect more birds than trying to bankrupt the DuPont corporation through the Superfund. The most beautiful way to purify water is probably the most effective way, too: maintain unspoiled watersheds. While an "almost unanimous" priesthood forecast cooling in 1975, and warming in 1992, the conservationist just went on planting trees, the most pleasant and practical way to suck carbon out of the air, however it may (or may not) affect global climate.

As for a sense of humility, it might usefully take the form of a wariness of grand public works. T.R. endorsed his share of them; FDR endorsed many more. In retrospect, it seems clear that more of the megalithic government projects of those days should have been opposed. They certainly should be as we go forward. Yesterday the federal dollar erected huge dams and drained swamps; today federal money is used to unleash those same rivers, and convert sugar plantations back into swamp. (The swamp programs are doubly expensive because the government also props up the price of sugar.) A consistent conservationism might have blocked more of the before, and thus saved us from having to do much of the after.

A consistent philosophy of moderation and caution could also do much to blunt the vindictive, punitive impulses of the modern environmentalist—and thereby help make things greener. In the aftermath of the Exxon Valdez spill, the multi-billion dollar steam-cleaning of rocks in Prince William Sound did far more harm than good, stripping away the organic seeds of rebirth along with the oil. In places where the cleanup was left to the wind and the waves, "nature," Scientific American would conclude, "fared better on its own." But the frenzied demands that Exxon be made to pay and pay overwhelmed every other impulse, to the point where increasing the damage to the oil company became much more important than abating damage to the Sound.

⟞⟐⟝

So much for philosophy. Politically, the most important principle is that whereas the environmentalist mission is exclusionary, the conservationist mission is populist and inclusionary, welcoming humankind as an integral and legitimate part of nature's landscape. Conservationism does not see man as a tapeworm in the bowel of nature. Symbiosis is possible. And when a choice has to be made, as it sometimes must, people come first.

The old conservationists were reluctant collectivists; the new environmentalists, eager ones. Having successfully conflated eagles with snail darters, halogenated hydrocarbons with the mountain peaks of Yosemite, the new environmentalists claim to speak for them all. This is an agenda that fits easily into a left-wing shoe. Running the whole environment—literally, "that which surrounds"—is an opportunity the Left gladly welcomes. The micro-environment is the best part of all, requiring as it does a pervasive, manipulative, and intrusive bureaucracy—for the Left, political ambrosia.

In reply, the Right has nothing better to offer than a long tradition of creating parks, husbanding wildlife, and venerating natural heritages of every kind. Politically speaking, however, that should be enough. It is the old conservation, not the new, that welcomes the family in the camper. It is the old that dispenses with oligarchy and caters to the common tastes of the common man.

It is the old that is the legacy of T.R., a man who so loved to shoot wild animals that he resolved to conserve the vast open spaces in which they live.

Besides, too-eager collectivists never end up conserving anything; only the reluctant ones do. (Behold the land once called East Germany: Love Canal, border to border, perfected by Communists.) The old conservationism, of parks and forests and Winnebagos, advances the green cause *because* of the Winnebago. The man in the Winnebago is enlisted in the cause precisely by an appeal to his own private sense of what is beautiful, and therefore to what he wants for himself and his family.

What is wrong with that?

NO    Paul R. Ehrlich and Anne H. Ehrlich

Brownlash: The New Environmental Anti-Science

Humanity is now facing a sort of slow-motion environmental Dunkirk. It remains to be seen whether civilization can avoid the perilous trap it has set for itself. Unlike the troops crowding the beach at Dunkirk, civilization's fate is in its own hands; no miraculous last-minute rescue is in the cards. Although progress has certainly been made in addressing the human predicament, far more is needed. Even if humanity manages to extricate itself, it is likely that environmental events will be defining ones for our grandchildren's generation —and those events could dwarf World War II in magnitude.

Sadly, much of the progress that has been made in defining, understanding, and seeking solutions to the human predicament over the past 30 years is now being undermined by an environmental backlash. We call these attempts to minimize the seriousness of environmental problems the *brownlash* because they help to fuel a backlash against "green" policies. While it assumes a variety of forms, the brownlash appears most clearly as an outpouring of seemingly authoritative opinions in books, articles, and media appearances that greatly distort what is or isn't known by environmental scientists. Taken together, despite the variety of its forms, sources, and issues addressed, the brownlash has produced what amounts to a body of anti-science—a twisting of the findings of empirical science—to bolster a predetermined worldview and to support a political agenda. By virtue of relentless repetition, this flood of anti-environmental sentiment has acquired an unfortunate aura of credibility.

It should be noted that the brownlash is not by any means a coordinated effort. Rather, it seems to be generated by a diversity of individuals and organizations. Some of its promoters have links to right-wing ideology and political groups. And some are well-intentioned individuals, including writers and public figures, who for one reason or another have bought into the notion that environmental regulation has become oppressive and needs to be severely weakened. But the most extreme—and most dangerous—elements are those who, while claiming to represent a scientific viewpoint, misstate scientific findings to support their view that the U.S. government has gone overboard with regulation, especially (but not exclusively) for environmental protection, and that subtle, long-term problems like global warming are nothing to worry about.

The words and sentiments of the brownlash are profoundly troubling to us and many of our colleagues. Not only are the underlying agendas seldom revealed but, more important, the confusion and distraction created among the public and policymakers by brownlash pronouncements interfere with and prolong the already difficult search for realistic and equitable solutions to the human predicament.

Anti-science as promoted by the brownlash is not a unique phenomenon in our society; the largely successful efforts of creationists to keep Americans ignorant of evolution is another example, which is perhaps not entirely unrelated. Both feature a denial of facts and circumstances that don't fit religious or other traditional beliefs; policies built on either could lead our society into serious trouble.

Fortunately, in the case of environmental science, most of the public is fairly well informed about environmental problems and remains committed to environmental protection. When polled, 65 percent of Americans today say they are willing to pay good money for environmental quality. But support for environmental quality is sometimes said to be superficial; while almost everyone is in favor of a sound environment—clean air, clean water, toxic site cleanups, national parks, and so on—many don't feel that environmental deterioration, especially on a regional or global level, is a crucial issue in their own lives. In part this is testimony to the success of environmental protection in the United States. But it is also the case that most people lack an appreciation of the deeper but generally less visible, slowly developing global problems. Thus they don't perceive population growth, global warming, the loss of biodiversity, depletion of groundwater, or exposure to chemicals in plastics and pesticides as a personal threat at the same level as crime in their neighborhood, loss of a job, or a substantial rise in taxes.

So anti-science rhetoric has been particularly effective in promoting a series of erroneous notions, including:

- Environmental scientists ignore the abundant good news about the environment.
- Population growth does not cause environmental damage and may even be beneficial.
- Humanity is on the verge of abolishing hunger; food scarcity is a local or regional problem and not indicative of overpopulation.
- Natural resources are superabundant, if not infinite.
- There is no extinction crisis, and so most efforts to preserve species are both uneconomic and unnecessary.
- Global warming and acid rain are not serious threats to humanity.
- Stratospheric ozone depletion is a hoax.
- The risks posed by toxic substances are vastly exaggerated.
- Environmental regulation is wrecking the economy.

How has the brownlash managed to persuade a significant segment of the public that the state of the environment and the directions and rates in which it is changing are not causes for great concern? Even many individuals who are sensitive to local environmental problems have found brownlash distortions of global issues convincing. Part of the answer lies in the overall lack of scientific knowledge among United States citizens. Most Americans readily grasp the issues surrounding something familiar and tangible like a local dump site, but they have considerably more difficulty with issues involving genetic variation or the dynamics of the atmosphere. Thus it is relatively easy to rally support against a proposed landfill and infinitely more difficult to impose a carbon tax that might help offset global warming.

Also, individuals not trained to recognize the hallmarks of change have difficulty perceiving and appreciating the gradual deterioration of civilization's life-support systems. This is why record-breaking temperatures and violent storms receive so much attention while a gradual increase in annual global temperatures—measured in fractions of a degree over decades—is not considered newsworthy. Threatened pandas are featured on television, while the constant and critical losses of insect populations, which are key elements of our life-support systems, pass unnoticed. People who have no meaningful way to grasp regional and global environmental problems cannot easily tell what information is distorted, when, and to what degree.

Decision-makers, too, have a tendency to focus mostly on the more obvious and immediate environmental problems—usually described as "pollution" —rather than on the deterioration of natural ecosystems upon whose continued functioning global civilization depends. Indeed, most people still don't realize that humanity has become a truly global force, interfering in a very real and direct way in many of the planet's natural cycles.

For example, human activity puts ten times as much oil into the oceans as comes from natural seeps, has multiplied the natural flow of cadmium into the atmosphere eightfold, has doubled the rate of nitrogen fixation, and is responsible for about half the concentration of methane (a potent greenhouse gas) and more than a quarter of the carbon dioxide (also a greenhouse gas) in the atmosphere today—all added since the industrial revolution, most notably in the past half-century. Human beings now use or co-opt some 40 percent of the food available to all land animals and about 45 percent of the available freshwater flows.

Another factor that plays into brownlash thinking is the not uncommon belief that environmental quality is improving, not declining. In some ways it is, but the claim of uniform improvement simply does not stand up to close scientific scrutiny. Nor does the claim that the human condition in general is improving everywhere. The degradation of ecosystem services (the conditions and processes through which natural ecosystems support and fulfill human life) is a crucial issue that is largely ignored by the brownlash. Unfortunately, the superficial progress achieved to date has made it easy to label ecologists doomsayers for continuing to press for change. At the same time, the public often seems unaware of the success of actions taken at the instigation of the environmental movement. People can easily see the disadvantages of environmental regulations but not the despoliation that would exist without them. Especially resentful are those whose personal or corporate ox is being gored when they are forced to sustain financial losses because of a sensible (or occasionally senseless) application of regulations.

Of course, it is natural for many people to feel personally threatened by efforts to preserve a healthy environment. Consider a car salesperson who makes a bigger commission selling a large car than a small one, an executive of a petrochemical company that is liable for damage done by toxic chemicals released into the environment, a logger whose job is jeopardized by enforcement of the Endangered Species Act, a rancher whose way of life may be threatened by higher grazing fees on public lands, a farmer about to lose the farm because of environmentalists' attacks on subsidies for irrigation water, or a developer who wants to continue building subdivisions and is sick and tired of dealing with inconsistent building codes or U.S. Fish and Wildlife Service bureaucrats. In such situations, resentment of some of the rules, regulations, and recommendations designed to enhance human well-being and protect life-support systems is understandable.

Unfortunately, many of these dissatisfied individuals and companies have been recruited into the self-styled "wise-use" movement, which has attracted a surprisingly diverse coalition of people, including representatives of extractive and polluting industries who are motivated by corporate interests as well as private property rights activists and right-wing ideologues. Although some of these individuals simply believe that environmental regulations unfairly distribute the costs of environmental protection, some others are doubtless motivated more by a greedy desire for unrestrained economic expansion.

At a minimum, the wise-use movement firmly opposes most government efforts to maintain environmental quality in the belief that environmental regulation creates unnecessary and burdensome bureaucratic hurdles which stifle economic growth. Wise-use advocates see little or no need for constraints on the exploitation of resources for short-term economic benefits and argue that such exploitation can be accelerated with no adverse long-term consequences. Thus they espouse unrestricted drilling in the Arctic National Wildlife Refuge, logging in national forests, mining in protected areas or next door to national parks, and full compensation for any loss of actual or potential property value resulting from environmental restrictions.

In promoting the view that immediate economic interests are best served by continuing business as usual, the wise-use movement works to stir up discontent among everyday citizens who, rightly or wrongly, feel abused by environmental regulations. This tactic is described in detail in David Helvarg's book, *The War Against the Greens:*

> To date the Wise Use/Property Rights backlash has been a bracing if dangerous reminder to environmentalists that power concedes nothing without a demand and that no social movement, be it ethnic, civil, or environmental, can rest on its past laurels.... If the anti-enviros' links to the Farm Bureau, Heritage Foundation, NRA, logging companies, resource trade associations, multinational gold-mining companies, [and] ORV manufacturers... proves anything, it's that large industrial lobbies and transnational corporations have learned to play the grassroots game.

Wise-use proponents are not always candid about their motivations and intentions. Many of the organizations representing them masquerade as groups seemingly attentive to environmental quality. Adopting a strategy biologists call "aggressive mimicry," they often give themselves names resembling those of genuine environmental or scientific public-interest groups: National Wetland Coalition, Friends of Eagle Mountain, the Sahara Club, the Alliance for Environment and Resources, the Abundant Wildlife Society of North America, the Global Climate Coalition, the National Wilderness Institute, and the American Council on Science and Health. In keeping with aggressive mimicry, these organizations often actively work *against* the interests implied in their names—a practice sometimes called *greenscamming.*

One such group, calling itself Northwesterners for More Fish, seeks to limit federal protection of endangered fish species so the activities of utilities, aluminum companies, and timber outfits utilizing the region's rivers are not hindered. Armed with a $2.6 million budget, the group aims to discredit environmentalists who say industry is destroying the fish habitats of the Columbia and other rivers, threatening the Northwest's valuable salmon fishery, among others.

Representative George Miller, referring to the wise-use movement's support of welfare ranching, overlogging, and government giveaways of mining rights, stated: "What you have . . . is a lot of special interests who are trying to generate some ideological movement to try and disguise what it is individually they want in the name of their own profits, their own greed in terms of the use and abuse of federal lands."

Wise-use sentiments have been adopted by a number of deeply conservative legislators, many of whom have received campaign contributions from these organizations. One member of the House of Representatives recently succeeded in gaining passage of a bill that limited the annual budget for the Mojave National Preserve, the newest addition to the National Parks System, to one dollar—thus guaranteeing that the park would have no money for upkeep or for enforcement of park regulations.

These same conservative legislators are determined to slash funding for scientific research, especially on such subjects as endangered species, ozone depletion, and global warming, and have legislated for substantial cutbacks in funds for the National Science Foundation, the U.S. Geological Survey, the National Aeronautics and Space Administration, and the Environmental Protection Agency. Many of them and their supporters see science as self-indulgent, at odds with economic interests, and inextricably linked to regulatory excesses.

The scientific justifications and philosophical underpinnings for the positions of the wise-use movement are largely provided by the brownlash. Prominent promoters of the wise-use viewpoint on a number of issues include such conservative think tanks as the Cato Institute and the Heritage Foundation. Both organizations help generate and disseminate erroneous brownlash ideas and information. Adam Myerson, editor of the Heritage Foundation's journal *Policy Review,* pretty much summed up the brownlash perspective by saying: "Leading scientists have done major work disputing the current henny-pennyism about global warming, acid rain, and other purported environmental catastrophes." In reality, however, most "leading" scientists support what Myerson calls henny-pennyism; the scientists he refers to are a small group largely outside the mainstream of scientific thinking.

In recent years, a flood of books and articles has advanced the notion that all is well with the environment, giving credence to this anti-scientific "What, me worry?" outlook. Brownlash writers often pepper their works with code phrases such as *sound science* and *balance*—words that suggest objectivity while in fact having little connection to what is presented. *Sound science* usually means science that is interpreted to support the brownlash view. *Balance* generally means giving undue prominence to the opinions of one or a handful of contrarian scientists who are at odds with the consensus of the scientific community at large.

Of course, while pro-environmental groups and environmental scientists in general may sometimes be dead wrong (as can anybody confronting environmental complexity), they ordinarily are not acting on behalf of narrow economic interests. Yet one of the remarkable triumphs of the wise-use movement and its allies in the past decade has been their ability to define public-interest organizations, in the eyes of many legislators, as "special interests"—not different in kind from the American Tobacco Institute, the Western Fuels Association, or other organizations that represent business groups.

But we believe there is a very real difference in kind. Most environmental organizations are funded mainly by membership donations; corporate funding is at most a minor factor for public-interest advocacy groups. There are no monetary profits to be gained other than attracting a bigger membership. Environmental scientists have even less to gain; they usually are dependent upon university or research institute salaries and research funds from peer-reviewed government grants or sometimes (especially in new or controversial areas where government funds are largely unavailable) from private foundations.

One reason the brownlash messages hold so much appeal to many people, we think, is the fear of further change. Even though the American frontier closed a century ago, many Americans seem to believe they still live in what the great economist Kenneth Boulding once called a "cowboy economy." They still think they can figuratively throw their garbage over the backyard fence with impunity. They regard the environmentally protected public land as "wasted" and think it should be available for their self-beneficial appropriation. They believe that private property rights are absolute (despite a rich economic and legal literature showing they never have been). They do not understand, as Pace University law professor John Humbach wrote in 1993, that "the Constitution does not guarantee that land speculators will win their bets."

The anti-science brownlash provides a rationalization for the short-term economic interests of these groups: old-growth forests are decadent and should be harvested; extinction is natural, so there's no harm in overharvesting economically important animals; there is abundant undisturbed habitat, so human beings have a right to develop land anywhere and in any way they choose; global warming is a hoax or even will benefit agriculture, so there's no need to limit the burning of fossil fuels; and so on. Anti-science basically claims we can keep the good old days by doing business as usual. But the problem is we can't.

Thus the brownlash helps create public confusion about the character and magnitude of environmental problems, taking advantage of the lack of consensus among individuals and social groups on the urgency of enhancing environmental protection. A widely shared social consensus, such as the United States saw during World War II, will be essential if we are to maintain environmental quality while meeting the nation's other needs. By emphasizing dissent, the brownlash works against the formation of any such consensus; instead it has helped thwart the development of a spirit of cooperation mixed with concern for society as a whole. In our opinion, the brownlash fuels conflict by claiming the environmental problems are overblown or nonexistent and that unbridled economic development will propel the world to new levels of prosperity with little or no risk to the natural systems that support society. As a result, environmental groups and wise-use proponents are increasingly polarized.

Unfortunately, some of that polarization has led to ugly confrontations and activities that are not condoned by the brownlash or by most environmentalists, including us. As David Helvarg stated, "Along with the growth of Wise Use/Property Rights, the last six years have seen a startling increase in intimidation, vandalism, and violence directed against grassroots environmental activists." And while confrontations and threats have been generated by both sides—most notably (but by no means exclusively) over the northern spotted owl protection plan—the level of intimidation engaged in by wise-use proponents is disturbing, to say the least....

Fortunately, despite all the efforts of the brownlash to discourage it, environmental concern in the United States is widespread. Thus a public-opinion survey in 1995 indicated that slightly over half of all Americans felt that environmental problems in the United States were "very serious." Indeed, 85 percent were concerned "a fair amount" and 38 percent "a great deal" about the environment. Fifty-eight percent would choose protecting the environment over economic growth, and 65 percent said they would be willing to pay higher prices so that industry could protect the environment better. Responses in other rich nations have been similar, and people in developing nations have shown, if anything, even greater environmental concerns. These responses suggest that the notion that caring about the environment is a luxury of the rich is a myth. Furthermore, our impression is that young people care especially strongly about environmental quality—a good omen if true.

Nor is environmental concern exclusive to Democrats and "liberals." There is a strong Republican and conservative tradition of environmental protection dating back to Teddy Roosevelt and even earlier. Many of our most important environmental laws were passed with bipartisan support during the Nixon and Ford administrations. Recently, some conservative environmentalists have been speaking out against brownlash rhetoric. And public concern is rising about the efforts to cripple environmental laws and regulations posed by right-wing leaders in Congress, thinly disguised as "deregulation" and "necessary budget-cutting." In January 1996, a Republican pollster, Linda Divall, warned that "our party is out of sync with mainstream American opinion when it comes to the environment."

Indeed, some interests that might be expected to sympathize with the wise-use movement have moved beyond such reactionary views. Many leaders in corporations such as paper companies and chemical manufacturers, whose activities are directly harmful to the environment, are concerned about their firms' environmental impacts and are shifting to less damaging practices. Our friends in the ranching community in western Colorado indicate their concern to us every summer. They want to preserve a way of life and a high-quality environment—and are as worried about the progressive suburbanization of the area as are the scientists at the Rocky Mountain Biological Laboratory. Indeed, they have actively participated in discussions with environmentalists and officials of the Department of the Interior to set grazing fees at levels that wouldn't force them out of business but also wouldn't subsidize overgrazing and land abuse.

Loggers, ranchers, miners, petrochemical workers, fishers, and professors all live on the same planet, and all of us must cooperate to preserve a sound environment for our descendants. The environmental problems of the planet can be solved only in a spirit of cooperation, not one of conflict. Ways must be found to allocate fairly both the benefits and the costs of environmental quality.

POSTSCRIPT

Are Environmental Regulations Too Restrictive?

The effects of "wise use" policies can be seen in Douglas Gantenbein's "Old Growth for Sale," *Audubon* (May/June 1998), in which the author says that efforts to reduce logging of centuries-old trees on federal lands have failed. Such results would not surprise the Ehrlichs, who, in "Ehrlich's Fables," *Technology Review* (January 1997), write about "a sampling of the myths, or fables, that the promoters of 'sound science' and 'balance' are promulgating about issues relating to population and food, the atmosphere and climate, toxic substances, and economics and the environment."

The issue has found new life in the debate over whether or not the Arctic National Wildlife Refuge (ANWR) should be opened to oil drilling. Supporters of drilling put human needs first, with one writer saying that the ANWR should be fair game because it is a forlorn, unattractive, uninhabitable hellhole (see Jonah Goldberg, "Ugh, Wilderness!" *National Review* [August 6, 2001]).

In the Worldwatch Institute's *State of the World 1999* (W. W. Norton, 1999), Lester R. Brown says, "As we look forward to the twenty-first century, it is clear that satisfying the projected needs of an ever larger world population with the economy we now have is simply not possible.... Human societies cannot continue to prosper while the natural world is progressively degraded." Catastrophe can be avoided, but not if we insist on doing things the way we always have.

Since the 1992 Earth Summit (the UN Conference on Environment and Development) in Rio de Janeiro, the world's environmental problems have actually gotten worse. But according to Christopher Flavin, in "The Legacy of Rio," *State of the World 1997* (W. W. Norton, 1997), the answer does not lie in some centralized world government but "in an eclectic mix of international agreements, sensible government policies, efficient use of private resources, and bold initiatives by grassroots organizations and local governments."

In those nations that have ignored environmental issues and refused to regulate, the problems have grown much worse. See Mike Edwards, "Lethal Legacy: Pollution in the Former USSR," *National Geographic* (August 1994), which begins, "The story on these pages is not a pretty one. It stems from decades of neglect and abuse of a vast and beautiful land.... In their ruthless drive to exploit and industrialize their nation, Soviet leaders gave little thought to the health of the people or to the lands that they ruled."

ISSUE 4

Should Environmental Policy Attempt to Cure Environmental Racism?

YES: Robert D. Bullard, from "Dismantling Environmental Racism in the USA," *Local Environment* (vol. 4, no. 1, 1999)

NO: David Friedman, from "The 'Environmental Racism' Hoax," *The American Enterprise* (November/December 1998)

ISSUE SUMMARY

YES: Professor of sociology Robert D. Bullard argues that environmental racism is a genuine phenomenon and that the government must live up to its mandate to protect all people.

NO: Writer and social analyst David Friedman denies the existence of environmental racism. He argues that the environmental justice movement is a government-sanctioned political ploy that will hurt urban minorities by driving away industrial jobs.

Archeologists delight in our forebears' habit of dumping their trash behind the house or barn. Today, however, most people try to arrange for their junk to be disposed of as far away from home as possible. Landfills, junkyards, recycling centers, and other operations with large negative environmental impacts tend to be sited in low-income and minority areas. Is this mere coincidence? Or is it deliberate? Does the paucity of poor people and minorities in the environmental movement indicate that these people do not really care? (See Robert Emmett Jones, "Blacks Just Don't Care: Unmasking Popular Stereotypes About Concern for the Environment Among African-Americans," *International Journal of Public Administration* [vol. 25, nos. 2 & 3, 2002]).

The environmental movement has, in fact, been charged with having been created to serve the interests of white middle- and upper-income people. Native Americans, blacks, Hispanics, and poor whites were not well represented among early environmental activists. It has been suggested that the reason for this is that these people were more concerned with more basic needs, such as jobs, food, health, and safety. However, the situation has been changing. In 1982, for example, in Warren County, North Carolina, poor black and Native American communities held demonstrations in protest of a poorly planned PCB (polychlorinated biphenyl) disposal site. This incident kicked off the environmental justice movement, which has since grown to include numerous local, regional, national, and international groups. The movement's target is systematic discrimination in the setting of environmental goals and in the siting of polluting industries and waste disposal facilities—also known as environmental racism. The global reach of the problem is discussed by Jan Marie Fritz in "Searching for Environmental Justice: National Stories, Global Possibilities," *Social Justice* (Fall 1999).

In 1990 the Environmental Protection Agency (EPA) published "Environmental Equity: Reducing Risks for All Communities," a report that acknowledged the need to pay attention to many of the concerns raised by environmental justice activists. At the 1992 United Nations Earth Summit in Rio de Janeiro, a set of "Principles of Environmental Justice" was widely discussed. In 1993 the EPA opened an Office of Environmental Equity (now the Office of Environmental Justice) with plans for cleaning up sites in several poor communities. In February 1994 President Bill Clinton made environmental justice a national priority with an executive order. Since then, many complaints of environmental discrimination have been filed with the EPA under Title VI of the federal Civil Rights Act of 1964; and in March 1998 the EPA issued guidelines for investigating those complaints. However, in April 2001 the U.S. Supreme Court ruled that individuals cannot sue states by charging that federally funded policies unintentionally violate the Civil Rights Act of 1964. The decision is expected to limit environmental justice lawsuits (see Franz Neil, "Supreme Court Ruling May Hurt Environmental Justice Claims," *Chemical Week* [May 2, 2001]).

Critics of the environmental justice movement contend that inequities in the siting of sources of pollution are the natural consequence of market forces that make poor neighborhoods (whether occupied by whites or minorities) the economically logical choice for locating such facilities. Critics also charge that such facilities depress property values and drive more prosperous people away while attracting a poorer population. In the following selections, Robert D. Bullard describes the history of the environmental justice movement, argues that the inequities are not just economic, and calls for nondiscriminatory environmental enforcement. David Friedman, on the other hand, asserts that the environmental justice movement is a politically inspired movement that is unsupported by scientific facts. He calls environmental racism a hoax and argues that attacking it will harm the urban poor by denying them the industrial jobs they need.

Robert D. Bullard

 **YES**

Dismantling Environmental Racism in the USA

Introduction

Despite significant improvements in environmental protection over the past several decades, millions of Americans continue to live in unsafe and unhealthy physical environments. Many economically impoverished communities and their inhabitants are exposed to greater health hazards in their homes, in their jobs and in their neighbourhoods when compared to their more affluent counterparts. This paper examines the root causes and consequences of differential exposure of some US populations to elevated environmental health risks.

Defining Environmental Racism

In the real world, all communities are not created equal. All communities do not receive equal protection. Economics, political clout and race play an important part in sorting out residential amenities and disamenities. Environmental racism is as real as the racism found in housing, employment, education and voting. *Environmental racism refers to any environmental policy, practice or directive that differentially affects or disadvantages (whether intended or unintended) individuals, groups or communities based on race or colour.* Environmental racism is just one form of environmental injustice and is reinforced by government, legal, economic, political and military institutions. Environmental racism combines with public policies and industry practices to provide *benefits* for whites while shifting *costs* to people of colour.

From New York to Los Angeles, grassroots community resistance has emerged in response to practices, policies and conditions that residents have judged to be unjust, unfair and illegal. Some of these conditions include: (1) unequal enforcement of environmental, civil rights and public health laws; (2) differential exposure of some populations to harmful chemicals, pesticides and other toxins in the home, school, neighbourhood and workplace; (3) faulty assumptions in calculating, assessing and managing risks; (4) discriminatory

From Robert D. Bullard, "Dismantling Environmental Racism in the USA," *Local Environment*, vol. 4, no. 1 (1999), pp. 5–10, 12–18. Copyright © 1999 by Carfax Publishing Ltd. Reprinted by permission of Taylor & Francis Ltd. http://www.tandf.co.uk/journals. References omitted.

zoning and land-use practices; and (5) exclusionary practices that limit some individuals and groups from participation in decision making.

The Environmental Justice Paradigm

During its 28-year history, the US EPA [Environmental Protection Agency] has not always recognised that many government and industry practices (whether intended or unintended) have adverse impacts on poor people and people of colour. Growing grassroots community resistance has emerged in response to practices, policies and conditions that residents have judged to be unjust, unfair and illegal. The EPA is mandated to enforce the nation's environmental laws and regulations equally across the board. It is required to protect all Americans —not just individuals or groups who can afford lawyers, lobbyists and experts. Environmental protection is a right, not a privilege reserved for a few who can 'vote with their feet' and escape or fend off environmental stressors.

The current environmental protection apparatus is broken and needs to be fixed. The current apparatus manages, regulates and distributes risks. The dominant environmental protection paradigm institutionalises unequal enforcement, trades human health for profit, places the burden of proof on the 'victims' and not the polluting industry, legitimates human exposure to harmful chemicals, pesticides and hazardous substances, promotes 'risky' technologies, exploits the vulnerability of economically and politically disenfranchised communities, subsidises ecological destruction, creates an industry around risk assessment and risk management, delays clean-up actions and fails to develop pollution prevention as the overarching and dominant strategy.

Environmental justice is defined as the fair treatment and meaningful involvement of all people regardless of race, colour, national origin or income with respect to the development, implementation and enforcement of environmental laws, regulations and policies. Fair treatment means that no group of people, including racial, ethnic or socio-economic groups, should bear a disproportionate share of the negative environmental consequences resulting from industrial, municipal and commercial operations or the execution of federal, state, local and tribal programmes and policies.

A growing body of evidence reveals that people of colour and low-income persons have borne greater environmental and health risks than the society at large in their neighbourhoods, workplaces and playgrounds. On the other hand, the environmental justice paradigm embraces a holistic approach to formulating environmental health policies and regulations, developing risk reduction strategies for multiple, cumulative and synergistic risks, ensuring public health, enhancing public participation in environmental decision-making, promoting community empowerment, building infrastructure for achieving environmental justice and sustainable communities, ensuring inter-agency co-operation and co-ordination, developing innovative public/private partnerships and collaboratives, enhancing community-based pollution prevention strategies, ensuring community-based sustainable economic development and developing geographically oriented community-wide programming.

The question of environmental justice is not anchored in a debate about whether or not decision makers should tinker with risk assessment and risk management. The environmental justice framework rests on an ethical analysis of strategies to eliminate unfair, unjust and inequitable conditions and decisions. The framework attempts to uncover the underlying assumptions that may contribute to and produce differential exposure and unequal protection. It also brings to the surface the *ethical* and *political* questions of 'who gets what, when, why and how much'. Some general characteristics of this framework include the following.

- The environmental justice framework adopts a public health model of prevention (i.e. elimination of the threat before harm occurs) as the preferred strategy.
- The environmental justice framework shifts the burden of proof to polluters/dischargers who do harm, who discriminate or who do not give equal protection to people of colour, low-income persons and other 'protected' classes.
- The environmental justice framework allows disparate impact and statistical weight or an 'effect' test, as opposed to 'intent', to infer discrimination.
- The environmental justice framework redresses disproportionate impact through 'targeted' action and resources. In general, this strategy would target resources where environmental and health problems are greatest (as determined by some ranking scheme but not limited to risk assessment).

Endangered Communities

Numerous studies reveal that low-income persons and people of colour have borne greater health and environmental risk burdens than the society at large. Elevated public health risks have been found in some populations even when social class is held constant. For example, race has been found to be independent of class in the distribution of air pollution, contaminated fish consumption, municipal landfills and incinerators, abandoned toxic waste dumps, the clean-up of superfund sites and lead poisoning in children. . . .

Impetus for Policy Shift

The impetus behind the environmental justice movement did not come from within government or academia, or from within largely white middle-class nationally based environmental and conservation groups. The impetus for change came from people of colour, grassroots activists and their 'bottom-up' leadership approach. Grassroots groups organised themselves, educated themselves and empowered themselves to make fundamental change in the way environmental protection is performed in their communities.

The environmental justice movement has come a long way since its humble beginning in rural, predominantly African-American, Warren County,

North Carolina, where a polychlorinated biphenyl landfill ignited protests and where over 500 arrests were made. The Warren County protests provided the impetus for a US General Accounting Office (1983) study, *Siting of Hazardous Waste Landfills and their Correlation with Racial and Economic Status of Surrounding Communities.* That study revealed that three out of four of the off-site, commercial hazardous waste landfills in Region 4 (which comprises eight states in the South) happened to be located in predominantly African-American communities, although African-Americans made up only 20% of the region's population.

The protests also led the Commission for Racial Justice (1987) to produce *Toxic Wastes and Race in the United States,* the first national study to correlate waste facility sites and demographic characteristics. Race was found to be the most potent variable in predicting where these facilities were located—more powerful than poverty, land values and home ownership. In 1990, *Dumping in Dixie: Race, Class, and Environmental Quality* chronicled the convergence of two social movements—social justice and environmental movements—into the environmental justice movement. This book highlighted African-Americans' environmental activism in the South, the same region that gave birth to the modern civil rights movement. What started out as local and often isolated community-based struggles against toxics and facility siting blossomed into a multi-issue, multi-ethnic and multi-regional movement.

The First National People of Color Environmental Leadership Summit (1991) was probably the most important single event in the movement's history. The Summit broadened the environmental justice movement beyond its anti-toxics focus to include issues of public health, worker safety, land use, transportation, housing, resource allocation and community empowerment. The meeting, organised by and for people of colour, demonstrated that it is possible to build a multi-racial grassroots movement around environmental and economic justice....

Federal, state and local policies and practices have contributed to residential segmentation and unhealthy living conditions in poor, working-class and people of colour communities. Several recent cases in California bring this point to life. Disparate highway siting and mitigation plans were challenged by community residents, churches and the NAACP LDF [National Association for the Advancement of Colored People Legal Defense and Education Fund], in *Clear Air Alternative Coalition v. United States Department of Transportation* (ND Cal. C-93-0721-VRW), involving the reconstruction of the earthquake-damaged Cypress Freeway in West Oakland. The plaintiffs wanted the downed Cypress Freeway (which split their community in half) rebuilt further away. Although the plaintiffs were not able to get their plan implemented, they did change the course of the freeway in their out-of-court settlement.

The NAACP LDF has filed an administrative complaint, *Mothers of East Los Angeles, El Sereno Neighborhood Action Committee, El Sereno Organizing Committee et al. v. California Transportation Commission* et al. (before the US Department of Transportation and US Housing and Urban Development), challenging the construction of the 4.5 mile extension of the Long Beach Freeway in East Los Angeles through El Sereno, Pasadena and South Pasadena. The plaintiffs argue that the mitigation measures proposed by the state agencies to address noise, air and visual pollution discriminate against the mostly Latino El Sereno community. For example, all of the freeway in Pasadena and 80% of that in South Pasadena will be below ground level. On the other hand, most of the freeway in El Sereno will be above-grade. White areas were favoured over the mostly Latino El Sereno in the allocation of covered freeway, historic preservation measures and accommodation to local schools. . . .

Making Government More Responsive

Many of the nation's environmental policies distribute costs in a regressive pattern while providing disproportionate benefits for whites and individuals who fall at the upper end of the education and income scales. Lavelle & Coyle uncovered glaring inequities in the way the federal EPA enforces its laws:

> There is a racial divide in the way the US government cleans up toxic waste sites and punishes polluters. White communities see faster action, better results and stiffer penalties than communities where blacks, Hispanics and other minorities live. This unequal protection often occurs whether the community is wealthy or poor.

This study reinforced what many grassroots activists have known for decades: all communities are not treated the same. Communities that are located on the 'wrong side of the tracks' are at greater risk from exposure to lead, pesticides (in the home and the workplace), air pollution, toxic releases, water pollution, solid and hazardous waste, raw sewage and pollution from industries.

Government has been slow to ask the questions of who gets help and who does not, who can afford help and who can not, why some contaminated communities get studied while others get left off the research agenda, why industry poisons some communities and not others, why some contaminated communities get cleaned up while others are not, why some populations are protected and others are not protected, and why unjust, unfair and illegal policies and practices are allowed to go unpunished.

Struggles for equal environmental protection and environmental justice did not magically appear in the 1990s. Many communities of colour have been engaged in life and death struggles for more than a decade. In 1990, the Agency for Toxic Substances and Disease Registry (ATSDR) held a historic conference in Atlanta. The ATSDR National Minority Health Conference focused on contamination. In 1992, after meeting with community leaders, academicians and civil rights leaders, the US EPA (under the leadership of William Reilly) admitted there was a problem, and established the Office of Environmental Equity. The name was change to the Office of Environmental Justice under the Clinton Administration.

In 1992, the US EPA produced one of the first comprehensive documents to examine the whole question of risk and environmental hazards in their equity report, *Environmental Equity: reducing risk for all communities.* The report, and its Office of Environmental Equity, were initiated only after prodding from people of colour, environmental justice leaders, activists and a few academicians.

The EPA also established a 25-member National Environmental Justice Advisory Council (NEJAC) under the Federal Advisory Committee Act. The NEJAC divided its environmental justice work into six sub-committees: Health and Research, Waste and Facility Siting, Enforcement, Public Participation and Accountability, Native American and Indigenous Issues, and International Issues. The NEJAC is comprised of stakeholders representing grassroots community groups, environmental groups, NGOs [nongovernmental organizations], state, local and tribal governments, academia and industry.

In February 1994, seven federal agencies, including the ATSDR, the National Institute for Environmental Health Sciences, the EPA, the National Institute of Occupational Safety and Health, the National Institutes of Health, the Department of Energy and Centers for Disease Control and Prevention sponsored a National Health Symposium entitled 'Health and research needs to ensure environmental justice'. The conference planning committee was unique in that it included grassroots organisation leaders, affected community residents and federal agency representatives. The goal of the February conference was to bring diverse stakeholders and those most affected to the decision-making table. Some of the recommendations from that symposium included the following:

- Conduct meaningful health research in support of people of colour and low-income communities.
- Promote disease prevention and pollution prevention strategies.
- Promote inter-agency co-ordination to ensure environmental justice.
- Provide effective outreach, education and communications.
- Design legislative and legal remedies.

In response to growing public concern and mounting scientific evidence, President Clinton on 11 February 1994 (the second day of the National Health Symposium) issued Executive Order 12898, 'Federal actions to address environmental justice in minority populations and low-income populations'. This Order attempts to address environmental injustice within existing federal laws and regulations.

Executive Order 12898 reinforces the 30-year-old Civil Rights Act of 1964, Title VI, which prohibits discriminatory practices in programmes receiving federal funds. The Order also focuses the spotlight back on the National Environmental Policy Act (NEPA), a 25-year-old law that sets policy goals for the protection, maintenance and enhancement of the environment. The NEPA's goal is to ensure for all Americans a safe, healthful, productive and aesthetically and culturally pleasing environment. The NEPA requires federal agencies to prepare a detailed statement on the environmental effects of proposed federal actions that significantly affect the quality of human health. . . .

The Case of Citizens Against Nuclear Trash Versus Louisiana Energy Services

Executive Order 12898 was put to the test in rural north-west Louisiana. Since 1989, the Nuclear Regulatory Commission had under review a proposal from Louisiana Energy Services (LES) to build the nation's first privately owned uranium enrichment plant. A national search was undertaken by LES to find the 'best' site for a plant that would produce 17% of the nation's enriched uranium. LES supposedly used an objective scientific method in designing its site selection process.

The southern USA, Louisiana and Claiborne Parish ended up being the dubious 'winners' of the site selection process. Residents from Homer and the nearby communities of Forest Grove and Center Springs—two communities closest to the proposed site—disagreed with the site selection process and outcome. They organised themselves into a group called Citizens Against Nuclear Trash (CANT). CANT charged LES and the federal Nuclear Regulatory Commission (NRC) staff with practising environmental racism. CANT hired the Sierra Club Legal Defense Fund and sued LES.

The lawsuit dragged on for more than 8 years. On 1 May 1997, a three-judge panel of the NRC Atomic Safety and Licensing Board issued a final decision on the case. The judges concluded that 'racial bias played a role in the selection process'. The precedent-setting federal court ruling came some 2 years after President Clinton signed Executive Order 12898. The judges, in a 38-page written decision, also chastised the NRC staff for not addressing the provision called for under Executive Order 12898. The court decision was upheld on appeal on 4 April 1998.

A clear racial pattern emerged during the so-called national search and multi-stage screening and selection process. For example, African-Americans comprise about 13% of the US population, 20% of the Southern states' population, 31% of Louisiana's population, 35% of the population of Louisiana's northern parishes and 46% of the population of Claiborne Parish. This progressive trend, involving the narrowing of the site selection process to areas of increasingly high poverty and African-American representation, is also evident from an evaluation of the actual sites that were considered in the 'intermediate' and 'fine' screening stages of the site selection process. The aggregate average percentage of black population for a 1-mile radius around all of the 78 sites examined (in 16 parishes) was 28.35%. When LES completed its initial site cuts, and reduced the list to 37 sites within nine parishes (i.e. the same as counties in other states), the aggregate percentage of black population rose to 36.78%. When LES then further limited its focus to six sites in Claiborne Parish, the aggregate average percentage of black population rose again, to 64.74%. The final site selected, the 'LeSage' site, has a 97.1% black population within a 1-mile radius.

The plant was proposed on Parish Road 39 between two African-American communities, just 0.25 miles from Center Springs (founded in 1910) and 1.25 miles from Forest Grove (founded in the 1860s just after slavery). The proposed site was in a Louisiana parish that has a per capita earnings average of only

$5800 per year (just 45% of the national average, $12,800), and where over 58% of the African-American population is below the poverty line. The two African-American communities were rendered 'invisible' since they were not even mentioned in the NRC's draft environmental impact statement.

Only after intense public comments did the NRC staff attempt to address environmental justice and disproportionate impact implications, as required under the NEPA and called for under Environmental Justice Executive Order 12898. For example, the NEPA requires that the government consider the environmental impacts and weigh the costs and benefits of the proposed action. These include health and environmental effects, the risk of accidental but foreseeable adverse health and environmental effects and socio-economic impacts.

The NRC staff devoted less than a page to addressing the environmental justice concerns of the proposed uranium enrichment plant in its final environmental impact statement (FEIS). Overall, the FEIS and the environmental report are inadequate in the following respects: (1) they assess inaccurately the costs and benefits of the proposed plant; (2) they fail to consider the inequitable distribution of costs and benefits of the proposed plant between the white and African-American populations; (3) they fail to consider the fact that the siting of the plant in a community of colour follows a national pattern in which institutionally biased decision-making leads to the siting of hazardous facilities in communities of colour, which results in the inequitable distribution of costs and benefits to those communities.

Among the distributive costs not analysed in relationship to Forest Grove and Center Springs are the disproportionate burden of health and safety, effects on property values, fire and accidents, noise, traffic, radioactive dust in the air and water, and the dislocation from a road closure that connects the two communities. Overall, the CANT legal victory points to the utility of combining environmental and civil rights laws and the requirement of governmental agencies to consider Executive Order 12898 in their assessments.

In addition to the remarkable victory over LES, a company that had the backing of powerful US and European nuclear energy companies, CANT members and their allies won much more. They empowered themselves and embarked on a path of political empowerment and self-determination. During the long battle, CANT member Roy Madris was elected to the Claiborne Parish Jury (i.e. county commission), and CANT member Almeter Willis was elected to the Claiborne Parish School Board. The town of Homer, the nearest incorporated town to Forest Grove and Center Springs, elected its first African-American mayor, and the Homer town council now has two African-American members. In autumn 1998, LES sold the land on which the proposed uranium enrichment plant would have been located. The land is going back into timber production —as it was before LES bought it. . . .

Conclusion

The environmental protection apparatus in the USA does not provide equal protection for all communities. The current paradigm institutionalises unequal enforcement, trades human health for profit, places the burden of proof on the 'victims' and not on the polluting industry, legitimates human exposure to harmful chemicals, pesticides and hazardous wastes, promotes 'risky' technologies, exploits the vulnerability of economically and politically disenfranchised communities and nations, subsidises ecological destruction, creates an industry around risk assessment and delays clean-up actions, and fails to develop pollution prevention, waste minimisation and cleaner production strategies as the overarching and dominant goal.

The environmental justice movement emerged in response to environmental inequities, threats to public health, unequal protection, differential enforcement and disparate treatment received by the poor and people of colour. This movement has redefined environmental protection as a basic right. It has also emphasised pollution prevention, waste minimisation and cleaner production techniques as strategies to achieve environmental justice for all Americans without regard to race, colour, national origin or income.

Both race and class factors place low-income and people of colour communities at special risk. Unequal political power arrangements have also allowed poisons of the rich to be offered as short-term economic remedies for poverty of the poor. However, there is little or no correlation between the proximity of industrial plants in communities of colour and the employment of nearby residents. Having industrial facilities in one's community does not automatically translate into jobs for nearby residents. More often than not, communities of colour are stuck with the polluting industries and poverty, while other people commute in for the jobs.

Governments must live up to their mandate of *protecting* all peoples and the environment. The call for environmental and economic justice does not stop at US borders but extends to all communities and nations that are threatened by hazardous wastes, toxic products and environmentally unsound technology. The environmental justice movement has set out the clear goal of eliminating the unequal enforcement of environmental, civil rights and public health laws, the differential exposure of some populations to harmful chemicals, pesticides and other toxins in the home, school, neighbourhood and workplace, faulty assumptions in calculating, assessing and managing risks, discriminatory zoning and land-use practices, and exclusionary policies and practices that limit some individuals and groups from participation in decision-making.

The solution to environmental injustice lies in the realm of equal protection for all individuals, groups and communities. Many of these problems could be eliminated if existing environmental, health, housing and civil rights laws were vigorously enforced in a non-discriminatory way. No community, rich or poor, urban or suburban, black or white, should be allowed to become a 'sacrifice zone' or dumping ground.

NO

David Friedman

The "Environmental Racism" Hoax

When the U.S. Environmental Protection Agency (EPA) unveiled its heavily criticized environmental justice "guidance" earlier this year, it crowned years of maneuvering to redress an "outrage" that doesn't exist. The agency claims that state and local policies deliberately cluster hazardous economic activities in politically powerless "communities of color." The reality is that the EPA, by exploiting every possible legal ambiguity, skillfully limiting debate, and ignoring even its own science, has enshrined some of the worst excesses of racialist rhetoric and environmental advocacy into federal law.

"Environmental justice" entered the activist playbook after a failed 1982 effort to block a hazardous-waste landfill in a predominantly black North Carolina county. One of the protesters was the District of Columbia's congressional representative, who returned to Washington and prodded the General Accounting Office (GAO) to investigate whether noxious environmental risks were disproportionately sited in minority communities.

A year later, the GAO said that they were. Superfund and similar toxic dumps, it appeared, were disproportionately located in non-white neighborhoods. The well-heeled, overwhelmingly white environmentalist lobby christened this alleged phenomenon "environmental racism," and ethnic advocates like Ben Chavis and Robert Bullard built a grievance over the next decade.

Few of the relevant studies were peer-reviewed; all made critical errors. Properly analyzed, the data revealed that waste sites are just as likely to be located in white neighborhoods, or in areas where minorities moved only after permits were granted. Despite sensational charges of racial "genocide" in industrial districts and ghastly "cancer alleys," health data don't show minorities being poisoned by toxic sites. "Though activists have a hard time accepting it," notes Brookings fellow Christopher H. Foreman, Jr., a self-described black liberal Democrat, "racism simply doesn't appear to be a significant factor in our national environmental decision-making."

⋅◆⋅

This reality, and the fact that the most ethnically diverse urban regions were desperately trying to *attract* employers, not sue them, constrained the environmental racism movement for a while. In 1992, a Democrat-controlled Congress ignored environmental justice legislation introduced by then-Senator Al Gore. Toxic racism made headlines, but not policy.

All of that changed with the Clinton-Gore victory. Vice President Gore got his former staffer Carol Browner appointed head of the EPA and brought Chavis, Bullard, and other activists into the transition government. The administration touted environmental justice as one of the symbols of its new approach.

Even so, it faced enormous political and legal hurdles. Legislative options, never promising in the first place, evaporated with the 1994 Republican takeover in Congress. Supreme Court decisions did not favor the movement.

So the Clinton administration decided to bypass the legislative and judicial branches entirely. In 1994, it issued an executive order—ironically cast as part of Gore's "reinventing government" initiative to streamline bureaucracy—which directed that every federal agency "make achieving environmental justice part of its mission."

At the same time, executive branch lawyers generated a spate of legal memoranda that ingeniously used a poorly defined section of the Civil Rights Act of 1964 as authority for environmental justice programs. Badly split, confusing Supreme Court decisions seemed to construe the 1964 Act's "nondiscrimination" clause (prohibiting federal funds for states that discriminate racially) in such a way as to allow federal intervention wherever a state policy ended up having "disparate effects" on different ethnic groups.

Even better for the activists, the Civil Rights Act was said to authorize private civil rights lawsuits against state and local officials on the basis of disparate impacts. This was a valuable tool for environmental and race activists, who are experienced at using litigation to achieve their ends.

Its legal game plan in place, the EPA then convened an advocate-laden National Environmental Justice Advisory Council (NEJAC), and seeded activist groups (to the tune of $3 million in 1995 alone) to promote its policies. Its efforts paid off. From 1993, the agency backlogged over 50 complaints, and environmental justice rhetoric seeped into state and federal land-use decisions.

꿈🌀꿈

Congress, industry, and state and local officials were largely unaware of these developments because, as subsequent news reports and congressional hearings established, they were deliberately excluded from much of the agency's planning process. Contrary perspectives, including EPA-commissioned studies highly critical of the research cited by the agency to justify its environmental justice initiative in the first place; were ignored or suppressed.

The EPA began to address a wider audience in September 1997. It issued an "interim final guidance" (bureaucratese for regulation-like rules that agencies can claim are not "final" so as to avoid legal challenge) which mandated that environmental justice be incorporated into all projects that file federal environmental impact statements. The guidance directed that applicants pay particular attention to potential "disparate impacts" in areas where minorities live in "meaningfully greater" numbers than surrounding regions.

The new rules provoked surprisingly little comment. Many just "saw the guidance as creating yet another section to add to an impact statement," explains Jennifer Hernandez, a San Francisco environmental attorney. In response, companies wanting to build new plants had to start "negotiating with community advocates and federal agencies, offering new computers, job training, school or library improvements, and the like" to grease their projects through.

In December 1997, the Third Circuit Court of Appeals handed the EPA a breathtaking legal victory. It overturned a lower court decision against a group of activists who sued the state of Pennsylvania for granting industrial permits in a town called Chester, and in doing so the appeals court affirmed the EPA's extension of Civil Rights Act enforcement mechanisms to environmental issues.

(When Pennsylvania later appealed, and the Supreme Court agreed to hear the case, the activists suddenly argued the matter was moot, in order to avoid the Supreme Court's handing down an adverse precedent. This August, the Court agreed, but sent the case back to the Third Circuit with orders to dismiss the ruling. While activists may have dodged a decisive legal bullet, they also wiped from the books the only legal precedent squarely in their favor.)

Two months after the Third Circuit's decision, the EPA issued a second "interim guidance" detailing, for the first time, the formal procedures to be used in environmental justice complaints. To the horror of urban development, business, labor, state, local, and even academic observers, the guidance allows the federal agency to intervene at any time up to six months (subject to extension) after any land-use or environmental permit is issued, modified, or renewed anywhere in the United States. All that's required is a simple allegation that the permit in question was "an act of intentional discrimination or has the effect of discriminating on the basis of race, creed, or national origin."

The EPA will investigate such claims by considering "multiple, cumulative, and synergistic risks." In other words, an individual or company might not itself be in violation, but if, combined with previous (also legal) land-use decisions, the "cumulative impact" on a minority community is "disparate," this could suddenly constitute a federal civil rights offense. The guidance leaves important concepts like "community" and "disparate impact" undefined, leaving them to "case by case" determination. "Mitigations" to appease critics will likewise be negotiated with the EPA case by case.

This "guidance" subjects virtually any state or local land-use decision—made by duly elected or appointed officials scrupulously following validly enacted laws and regulations—to limitless ad hoc federal review, any time there is the barest allegation of racial grievance. Marrying the most capricious elements of wetlands, endangered species, and similar environmental regulations with the interest-group extortion that so profoundly mars urban ethnic politics, the guidance transforms the EPA into the nation's supreme land-use regulator.

◦◦◦

Reaction to the Clinton administration's gambit was swift. A coalition of groups usually receptive to federal interventions, including the U.S. Conference of Mayors, the National Association of Counties, and the National Association of Black County Officials, demanded that the EPA withdraw the guidance. The House amended an appropriations bill to cut off environmental justice enforcement until the guidance was revised. This August, EPA officials were grilled in congressional hearings led by Democratic stalwarts like Michigan's John Dingell.

Of greatest concern is the likelihood the guidance will dramatically increase already-crippling regulatory uncertainties in urban areas where ethnic populations predominate. Rather than risk endless delay and EPA-brokered activist shakedowns, businesses will tacitly "redline" minority communities and shift operations to white, politically conservative, less-developed locations.

Stunningly, this possibility doesn't bother the EPA and its environmentalist allies. "I've heard senior agency officials just dismiss the possibility that their policies might adversely affect urban development," says lawyer Hernandez. Dingell, a champion of Michigan's industrial revival, was stunned when Ann Goode, the EPA's civil rights director, said her agency never considered the guidance's adverse economic and social effects. "As director of the Office of Civil Rights," she lectured House lawmakers, "local economic development is not something I can help with."

Perhaps it should be. Since 1980, the economies of America's major urban regions, including Cleveland, Chicago, Milwaukee, Detroit, Pittsburgh, New Orleans, San Francisco, Newark, Los Angeles, New York City, Baltimore, and Philadelphia, grew at only one-third the rate of the overall American economy. As the economies of the nation's older cities slumped, 11 million new jobs were created in whiter areas.

Pushing away good industrial jobs hurts the pocketbook of urban minorities, and, ironically, harms their health in the process. In a 1991 *Health Physics* article, University of Pittsburgh physicist Bernard L. Cohen extensively analyzed mortality data and found that while hazardous waste and air pollution exposure takes from three to 40 days off a lifespan, poverty reduces a person's life expectancy by an average of 10 *years*. Separating minorities from industrial plants is thus not only bad economics, but bad health and welfare policy as well.

❧

Such realities matter little to environmental justice advocates, who are really more interested in radical politics than improving lives. "Most Americans would be horrified if they saw NEJAC [the EPA's environmental justice advisory council] in action," says Brookings's Foreman, who recalls a council meeting derailed by two Native Americans seeking freedom for an Indian activist incarcerated for killing two FBI officers. "Because the movement's main thrust is toward . . . 'empowerment' . . . , scientific findings that blunt or conflict with that goal are ignored or ridiculed."

Yet it's far from clear that the Clinton administration's environmental justice genie can be put back in the bottle. Though the Supreme Court's dismissal of the Chester case eliminated much of the EPA's legal argument for the new rules, it's likely that more lawsuits and bureaucratic rulemaking will keep the program alive. The success of the environmental justice movement over the last six years shows just how much a handful of ideological, motivated bureaucrats and their activist allies can achieve in contemporary America unfettered by fact, consequence, or accountability, if they've got a President on their side.

POSTSCRIPT

Should Environmental Policy Attempt to Cure Environmental Racism?

The problems that led to the environmental justice movement have been documented in many reports. For example, in "Who Gets Polluted? The Movement for Environmental Justice," *Dissent* (Spring 1994), Ruth Rosen presents a history of the environmental justice movement, stressing how the movement has woven together strands of the civil rights and environmental struggles. Rosen argues that racial discrimination plays a significant role in the unusually intense exposure to industrial pollutants experienced by disadvantaged minorities, and she expresses the hope that "greening the ghetto will be the first step in greening our entire society." In addition, Bullard's *Dumping in Dixie: Race, Class and Environmental Quality* (Westview Press, 1990, 1994, 2000) has become a standard text in the environmental justice field. Also see his *Unequal Protection: Environmental Justice and Communities of Color* (Sierra Club Books, 1994); Michael Heiman's "Waste Management and Risk Assessment: Environmental Discrimination Through Regulation," *Urban Geography* (vol. 17, no. 5, 1996); and Luke W. Cole and Sheila R. Foster's *From the Ground Up: Environmental Racism and the Rise of the Environmental Justice Movement* (New York University Press, 2000). David W. Allen, in "Social Class, Race, and Toxic Releases in American Counties, 1995," *Social Science Journal* (vol. 38, no. 1, 2001), finds that the data support the existence of environmental racism but that the effect is strongest in the southern portion of the United States (the Sun Belt).

Those who criticize the environmental justice movement tend to focus on other studies. In "Green Redlining: How Rules Against 'Environmental Racism' Hurt Poor Minorities Most of All," *Reason* (October 1998), Henry Payne labels the Environmental Protection Agency's efforts to impose environmental equity "redlining" and, like Friedman, argues that the practice reduces job opportunities and economic benefits for minorities.

There is great contrast in the sides to this debate. In such cases, the reader must not ignore the social values and political commitments of the debaters. The reader must also be careful to consider the data relied on by the debaters and to watch for unsupported claims and simplistic explanations for events whose causes are likely to be more complicated.

Where is government policy going? Jim Motavalli, in "Toxic Targets: Polluters That Dump on Communities of Color Are Finally Being Brought to Justice," *E: The Environmental Magazine* (July–August 1998), states that although minorities and the poor have been forced to bear a disproportionate share of the burden of industrial pollution, changes in environmental policy and law are finally offering remedies. And in an August 9, 2001, memorandum regarding the Environmental Protection Agency's stance on environmental justice, EPA administrator Christine Todd Whitman wrote, "The Environmental Protection Agency has a firm commitment to the issue of environmental justice and its integration into all programs, policies, and activities, consistent with existing environmental laws and their implementing regulations.... [E]nvironmental justice is the goal to be achieved for all communities and persons across this Nation. Environmental justice is achieved when everyone, regardless of race, culture, or income, enjoys the same degree of protection from environmental and health hazards and equal access to the decision-making process to have a healthy environment in which to live, learn, and work."

ISSUE 5

Is the Precautionary Principle a Sound Basis for International Policy?

YES: Paul L. Stein, from "Are Decision-Makers Too Cautious With the Precautionary Principle?" Paper Delivered at the Land and Environment Court of New South Wales Annual Conference (October 14 & 15, 1999)

NO: Henry I. Miller and Gregory Conko, from "The Perils of Precaution," *Policy Review* (June & July 2001)

ISSUE SUMMARY

YES: Paul L. Stein, a justice of the New South Wales Court of Appeals, argues that the precautionary principle is now a cornerstone of international environmental law and that the courts have a duty to implement the principle even beyond the requirements of legislation.

NO: Henry I. Miller, a research fellow at Stanford University's Hoover Institution, and policy analyst Gregory Conko argue that the precautionary principle leads "regulators to abandon the careful balancing of risks *and* benefits," blocks progress, limits the freedom of scientific researchers, and restricts consumer choice.

T he traditional approach to environmental problems has been reactive. That is, first the problem becomes apparent—wildlife or people sicken and die, or drinking water or air tastes foul. Then researchers seek the cause of the problem, and regulators seek to eliminate or reduce that cause. The burden is on society to demonstrate that harm is being done and that a particular cause is to blame.

An alternative approach is to presume that *all* human activities—construction projects, new chemicals, new technologies, etc.—have the potential to cause environmental harm. Therefore, those responsible for these activities should prove in advance that they will not do harm and should take suitable steps to prevent any harm from happening. This "precautionary principle" has played an increasingly important part in environmental law ever since it first appeared in Germany in the mid-1960s. On the international scene, it has been applied to climate change, hazardous waste management, ozone depletion, biodiversity, and fisheries management. In 1992 the Rio Declaration on Environment and Development, listing it as Principle 15, codified it thus:

> In order to protect the environment, the precautionary approach shall be widely applied by States according to their capabilities. When there are threats of serious or irreversible damage, lack of full scientific certainty shall not be used as a reason for postponing cost-effective measures to prevent environmental degradation.

Other versions of the principle also exist, but all agree that when there is reason to think—not absolute proof—that some human activity is or might be harming the environment, precautions should be taken. This has come to be broadly accepted as a basic tenet of ecologically or environmentally sustainable development.

The precautionary principle also contributes to thinking in the areas of risk assessment and risk management in general. Human activities—the manufacture of chemicals and other products; the use of pesticides, drugs, and fossil fuels; the construction of airports and shopping malls; and even agriculture—can damage health and the environment. Some people insist that action need not be taken against any particular activity until and unless there is solid, scientific proof that it is doing harm, and even then risks must be weighed against each other. See, for instance, Wendy Cleland-Hamnett, "The Role of Comparative Risk Analysis," *EPA Journal* (January–March 1993). Others insist that mere suspicion should be grounds enough for action and that there is a broad middle ground. For instance, Robert Costanza and Laura Cornwell, in "The 4P Approach to Dealing With Scientific Uncertainty," *Environment* (November 1992), argue that when uncertainty about potential harm is high, those who are potentially responsible should be required to post in advance a bond sufficient to cover the costs associated with the worst possible results.

Since solid, scientific proof can be very difficult to obtain, the question of just how much proof is needed to justify action is vital. Not surprisingly, if action threatens an industry, that industry's advocates will argue against taking precautions, generally saying that more proof is needed. A good example can be found in Stuart Pape, "Watch Out for the Precautionary Principle," *Prepared Foods* (October 1999): "In recent months, U.S. food manufacturers have experienced a rude introduction to the 'Precautionary Principle.' . . . European regulators have begun to adopt extreme definitions of the Principle in order to protect domestic industries and place severe restrictions on the use of both old and new materials without justifying their action upon sound science."

In the following selection, Paul L. Stein argues that the precautionary principle is now a cornerstone of international environmental law and that the question is no longer whether to implement the principle but how. He maintains that the courts must not shirk their responsibility to apply it, even when legislation is vague. In the second selection, Henry I. Miller and Gregory Conko argue that the precautionary principle blocks progress, limits the freedom of scientific researchers, and restricts consumer choice.

Paul L. Stein

 **YES**

Are Decision-Makers Too Cautious With the Precautionary Principle?

Precaution, (1603) a measure taken beforehand to ward off an evil.

— Shorter Oxford English Dictionary

Overview

Over the last decade the principles of ecologically sustainable development (ESD) have permeated inexorably into the interstices of environmental law. Many of the principles, particularly the precautionary principle, have become part and parcel of international, national and domestic laws and custom.

The core principles of ESD have come into regular use by decision-makers at a federal, state and local government level. This is partly because of governmental policies and practices and in part because of statute law, the highest form of expression of government policy. The legislation of all nine governments in Australia contain numerous references to ESD and its core principles. There are more Acts which include ESD in New South Wales [NSW] than anywhere else in Australia. Most important for our purposes are those now contained in the objects of the *Environmental Planning and Assessment Act* 1979 and the *Protection of the Environment Administration Act* 1991, as well as the new federal environmental legislation.

What may be noted, however, is that the inclusion of the principles in Australian legislation has been largely confined to objectives of statutes or agencies without any real guidance to decision-makers as to whether and how to apply the core principles or what weight to give them. Moreover, some of the principles contain vague statements, some might call them aspirations, as well as ambiguities, inconsistencies and uncertainties. Difficulties of interpretation and application are manifest. There is even discussion on whether the principles are merely guiding or whether they are also operational. In these circumstances, who can blame the courts for proceeding, like the precautionary principle, with a degree of caution. Nonetheless, my thesis is that there is the opportunity, if not the obligation, in the absence of clear legislative guidance, to apply the common law and assist in the development and fleshing out of the principles.

From Paul L. Stein, "Are Decision-Makers Too Cautious With the Precautionary Principle?" Paper Delivered at the Land and Environment Court of New South Wales Annual Conference (October 14 & 15, 1999). Subsequently published in *Environmental and Planning Law Journal*, vol. 17, no. 1 (February 2000). Copyright © 2000 by Lawbook Co., part of Thomson Legal & Regulatory Limited. Reprinted by permission. http://www.thomson.com.au. Notes omitted.

Our task is to turn soft law into hard law. This is an opportunity to be bold spirits rather than timorous souls and provide a lead for the common law world. It will make a contribution to the ongoing development of environmental law.

Introduction

The Origins of the Precautionary Principle

The origin of the precautionary principle lies in the German concept of *Vorsorgeprinzip*, literally translated as meaning the 'foresight principle' or 'precautionary principle'. The principle first appeared in the mid 1960's when environmental issues were becoming a major political theme in Germany. At around the same time the hypothesis of 'implementation shortfalls' emerged. The hypothesis identified that there existed a clear discrepancy between legal provisions and the goals of environmental policy, on the one hand, and its practical application on the other. The precautionary principle was originally used as a yardstick by which to judge political decisions. By the early 1970's the principle could be found in domestic West German legislation in respect of environmental policies aimed at combating the problems of global warming, acid rain and maritime pollution.

The precautionary principle has played an instrumental role in the policy reform of marine pollution. Despite regulation of both land based pollution and ocean dumping by regional bodies, the quality of the North Sea was seen to be continuing to decline. The German government, when calling the first North Sea meeting in 1984, had as a negotiating aim, the inclusion of the precautionary principle, *vorsorgeprinzip*.

The earliest international agreement which explicitly refers to the precautionary principle is the Ministerial Declaration of the Second International Conference on the Protection of the North Sea, issued in London in November 1987. It was accepted that:

> ... in order to protect the North Sea from possibly damaging effects of the most dangerous substances, a precautionary approach is necessary which may require action to control inputs of such substances even before a causal link has been established by absolutely clear scientific evidence.

The precautionary principle has since been widely used in international environmental law and has been applied to areas such as climate change, hazardous waste and ozone layer depletion, biodiversity, fisheries management and general environmental management. Many treaties, some of which are extracted below, illustrate the various circumstances in which the precautionary principle has been utilised.

The precautionary principle received strong endorsement in the Rio Declaration on Environment and Development (adopted in 1992 by the United Nations Conference on Environment and Development [UNCED] in Rio de Janeiro). The Rio Declaration contains 27 principles to guide the International Community in the promotion of sustainable development.

Principle 15 states:

> In order to protect the environment, the precautionary approach shall be widely applied by States according to their capabilities. Where there are threats of serious or irreversible damage, lack of full scientific certainty shall not be used as a reason for postponing cost-effective measures to prevent environmental degradation.

The revision to the Treaty of Rome as agreed at Maastricht states:

> The Community policy on the environment *shall* be based on the precautionary principle and on the principle that preventative action should be taken, that environmental damage should as a priority be rectified at source and that the polluter should pay. Environmental protection requirements must be integrated into the definition and implementation of other Community policies.[Emphasis added]

Article 3.3 of the 1992 U.N. Framework Convention on Climate Change states:

> The parties should take precautionary measures to anticipate, prevent or minimise the causes of climate change and mitigate its adverse effects. Where there are threats of serious or irreversible damage, lack of full scientific certainty should not be used as a reason for postponing such measures, taking into account that policies and measures to deal with climate change should be cost effective so as to ensure global benefits at the lowest possible cost.

Agenda 21, agreed to at the 1992 Rio conference, recommends in relation to radioactive waste that States should not:

> ... promote or allow the storage or disposal of high-level, intermediate level and low-level radioactive waste near the marine environment unless they determine that scientific evidence, consistent with the internationally agreed principles and guidelines, shows that such storage or disposal poses no unacceptable risk to people and the marine environment or does not interfere with other legitimate uses of the sea, making, in the process of consideration, appropriate use of the concept of the precautionary approach.

Agenda 21 on the Protection of the Oceans expressly requires:

> new approaches to marine and coastal area management and development at the national, subregional, regional and global levels, approaches that are integrated in content and are precautionary and anticipatory in ambit.

The June 1990 Amendments to the Montreal Protocol on Substances that Deplete the Ozone Layer states:

> [The Parties to this Protocol are] determined to protect the ozone layer by taking precautionary measures to control equitably total global emissions of substances that deplete it, with the ultimate objective of their elimination on the basis of developments in scientific knowledge, taking into account technical and economic considerations and bearing in mind the developmental needs of developing countries.

The 1992 OSPAR Convention (Convention for the Protection of the Marine Environment of the North East Atlantic) provides in Article 2 that Contracting Parties *shall* apply:

> ... the precautionary principle, by virtue of which preventative measures are to be taken when there are reasonable grounds for concern that substances or energy introduced, directly or indirectly, into the marine environment may bring about hazards to human health, harm living resources and marine ecosystems, damage amenities or interfere with other legitimate uses of the sea, even when there is no conclusive evidence of a causal relationship between the inputs and effects.

The Convention on Biological Diversity signed at the United Nations Conference on Environment and Development in 1992 notes in its preamble:

> ... that where there is a threat of significant reduction or loss of biological diversity, lack of full scientific certainty should not be used as a reason for postponing measures to avoid or minimise such a threat.

These are but a few of the international instruments which have incorporated the precautionary principle. Australia has ratified almost all of these environmental treaties and conventions which are relevant to our part of the world.

Defining the Precautionary Principle

The Intergovernmental Agreement on the Environment (the IGAE) endorses the precautionary principle in the following terms:

> Where there are *threats of serious or irreversible environmental damage, lack of full scientific certainty* should not be used as a reason for *postponing measures to prevent environmental degradation.* In the application of the precautionary principle, public and private decisions should be guided by:
>   (i) careful evaluation to avoid, wherever practicable, serious or irreversible damage to the environment; and
>   (ii) an assessment of the risk-weighted consequences of various options [Emphasis added]

Defining the application of the precautionary principle with any degree of precision has proved problematic because of the rapidly evolving nature of the concept. While the precautionary principle has proved to be useful in reformulating the way in which the law structures decision-making processes, 'ambiguity in the conceptualisation of the precautionary principle at the policy level has led to it being given a wide range of divergent meanings, providing a fundamental barrier to attempts at implementation'.

The precautionary principle has been described as a decision-making approach which ensures that a substance or activity posing a threat to the environment is prevented from adversely affecting the environment, even if there is no conclusive scientific proof linking that particular substance or activity to environmental damage. Briefly stated, the precautionary principle, both in its conceptual core and its practical implications, is preventative. The principle provides the philosophical authority to make decisions in the face of uncertainty. In this way, it is symbolic of the need for change in human behaviour towards the ecological sustainability of the environment....

The Precautionary Principle and Ecologically Sustainable Development

The precautionary principle needs to be considered in the broader context of the wider principles and philosophies forming the concept of ecologically sustainable development (ESD). It is accepted that ESD should be treated as a complete package where no one principle should dominate over any other. This requires that the precautionary principle be applied with consideration of other principles forming part of ESD.

The modern manifestation of ESD stems from the 1987 report of the World Commission of Environment and Development (The Brundtland Report) where development was defined as sustainable:

> ... if it meets the needs of the present without compromising the ability of future generations to meet their own needs.

The idea is premised on the integration of economic and environmental processes in decision-making. In 1992, the IGAE committed all nine Australian governments to the concept, as well as local government. ESD has since been incorporated into almost all Australian environmental legislation as an appropriate objective for environmental agencies and decision-makers. Often core principles are extracted for particular emphasis and utilisation, especially the precautionary principle....

In essence, ESD is development which aims to conserve and effectively manage the environment for the benefit of future generations. In 1990 the Commonwealth Government suggested the following definition for ESD:

> ... using, conserving and enhancing the community's resources so that ecological processes, on which life depends, are maintained, and the total quality of life, now and in the future, can be increased.

Two features are characteristic of an ESD approach. First, decision-makers need to consider the economic, social and environmental implications of actions for the local and international community and biosphere. Second, in reaching decisions, decision-makers must adopt a long-term rather than short-term view. In this sense, the precautionary principle ensures a better integration of environmental considerations in decision-making....

ESD represents a delicate balancing of the often competing interests of development and environmental protection. Application of the precautionary principle is considered appropriate in circumstances where a proposed activity carries with it a risk of potentially serious environmental damage which may threaten the interests of present and future generations. Properly evaluating risks is likely to be aided by the guiding principles and indicators of sustainability.

Legislation Incorporating ESD and the Precautionary Principle

... [A]n astounding number of federal, state and territory statutes have expressly referred to or incorporated ESD principles. However, an analysis of the legislation reveals that much of it adopts ESD in general terms without necessarily assigning a specific role to the principles. The following examples of centrally relevant environmental legislation are indicative of the lack of consistency in the approach to inclusion of ESD principles within Acts of Parliament. It will be readily appreciated that ESD is often included among the objects of an Act without further reference, whereas some legislation requires all decisions or specific decisions to take into consideration core principles or to have regard to principles of ESD. It will be seen that no statute gives any precise guidance as to the weight to be given to the principles, nor their particular role in the balancing of considerations in arriving at a decision....

International Cases

In the *Danish Bees* case the European Court of Justice indirectly applied the precautionary principle to justify a measure having equivalent effect to a quantitative restriction in EC [European Community] law. The case involved a decision made by the Danish Minister for Agriculture which prohibited the keeping of bees on the island of Laeso and certain neighbouring islands other than those of the sub-species, *Apis Mellifera Mellifera* (the Laeso Brown Bee).

The issue before the Court was whether the keeping on the islands of any species of bee other than the sub-species, *Apis Mellifera Mellifera* constituted a measure having equivalent effect to a quantitative restriction within the meaning of Article 30 of the European Community Treaty (the EC Treaty) and whether, if that were the case, such legislation was justified on the ground of the protection and health and life of animals. The Danish Government maintained that the establishment of pure breeding areas for the sub-species, in a particular area within a Members' State, did not affect trade between Member States. It was contended that this did not constitute discrimination in respect of bees originating in other Member States and was not intended to regulate trade between Member States. Further, the effects on trade flowing from the Minister's prohibition were too hypothetical and uncertain to be regarded as a measure likely to obstruct it.

Notwithstanding the lack of conclusive scientific evidence establishing both the nature of the sub-species and its risk of extinction, the Court concluded that the decision made by the Minister constituted a measure having an effect equivalent to a quantitative restriction within the meaning of Article 30 of the EC Treaty and that the prohibition was also justified under Article 36 of the Treaty:

> ... measures to preserve an indigenous animal population with distinct characteristics contribute to the maintenance of biodiversity by ensuring the survival of the population concerned. By so doing, they are aimed at

protecting the life of those animals and are capable of being justified under Article 36 of the Treaty.

The legislation was also justified under the Biodiversity Convention ratified by the EC. In so holding, the Court took a precautionary approach to the preservation of indigenous animal populations and the conservation of biodiversity....

In *AP Pollution Control Board v Nayudu,* the Supreme Court of India was considering a petition claiming that certain hazardous industries proposed to be established by the respondents without the necessary certificate from the State Pollution Control Board could not proceed. M. Jagannadha Rao, J discussed the difficulties faced by environmental courts globally in dealing with scientific data. He cited articles by Lord Woolf and Carnworth on the desirability of a specialist environmental court. In particular, his Honour discussed the status and application of the precautionary principle citing Barton and other articles.

His Honour said:

> The 'uncertainty' of scientific proof and its changing frontiers from time to time has led to great changes in environment concepts during the period between the Stockholm Conference of 1972 and the Rio Conference of 1992. In *Vellore Citizens' Welfare Forum v Union of India and others, 1995 (5) SCC 647,* a three Judge Bench of this Court referred to these changes, to the 'precautionary principle' and the new concept of 'burden of proof' in environmental matters. Kuldip Singh, J after referring to the principles evolved in various international Conferences and to the concept of 'Sustainable Development', stated that the Precautionary Principle, the Polluter-Pays Principle and the special concept of Onus of Proof have now emerged and govern the law in our country too, as is clear from Articles 47, 48-A and 51-A(g) of our Constitution and that, in fact, in the various environmental statutes, such as the *Water Act,* 1974 and other statutes, including *The Environment (Protection) Act* 1986, these concepts are already implied. The learned Judge declared that these principles have now become part of our law.

The relevant observations in the *Vellore Case* in this behalf read as follows:

> In view of the above-mentioned constitutional and statutory provisions we have no hesitation in holding that the *Precautionary Principle* and the Polluter-Pays Principle are part of the environmental law of this country.

The Supreme Court discussed the development of the precautionary principle in replacing the Assimilative Capacity Principle adopted at an earlier point of time.

Rao J stated:

> The principle of precaution involves the anticipation of environmental harm and taking measures to avoid it or to choose the least environmentally harmful activity. It is *based* on Scientific uncertainty. Environmental protection should not only aim at protecting health, property and economic interest but also protect the environment for its own sake. Precautionary duties must not only be triggered by the suspicion of concrete danger but also by (justified) concern or risk potential. The precautionary principle was

recommended by the UNEP [United Nations Environment Program] Governing Council (1989). The Bomako Convention also lowered the threshold at which scientific evidence might require action by not referring to 'serious' or 'irreversible' as adjectives qualifying harm. However, summing up the legal status of the precautionary principle, one commentator characterised the principle as still 'evolving' for though it is accepted as part of the international customary law, 'the consequences of its application in any potential situation will be influenced by the circumstances of each case'.

The Court also discussed the issue of burden of proof in cases involving the application of the precautionary principle:

> ... Therefore, it is necessary that the party attempting to preserve the *status quo* by maintaining a less-polluted state should not carry the burden of proof and the party who wants to alter it, must bear this burden. (See James M. Olson, Shifting the Burden of Proof, 20 Envtl. Law p.891 at 898 (1990). (Quoted in Vol 22 (1998) Harv. Env. Law Review p. 509 at 519, 550).

The precautionary principle suggested that where there is an identifiable risk of serious or irreversible harm, including, for example, extinction of species, widespread toxic pollution in major threats to essential ecological processes, it may be appropriate to place the burden of proof on the person or entity proposing the activity that is potentially harmful to the environment.

The case of *Ashburton Acclimatisation Society v Federated Farmers of New Zealand Inc* was determined well before ESD principles became included in legislation. It is referred to by Burton and picked up by the Supreme Court of India in *Nayudu*. It involved an appeal, referring back to the Planning Tribunal for consideration, its report for a national water conservation order affecting the Raikaia River. The contest was between conservationists, who wished the flow and characteristics of the river to be conserved, and farmers who wished to use the water from the river for irrigation. It was submitted that if implemented the report would unduly prejudice the rights and expectation of the Farmers Federation.

At the heart of the appeal was the ground that the Tribunal had misconstrued of the Act by placing undue emphasis upon protection of outstanding features of the river and by failing to pay sufficient regard to the competing need of out of stream users, in particular the needs of primary industry and the community. The Tribunal had regarded the sustainability of the amenity afforded by the waters in their natural state as being the overriding consideration under the *Water and Soil Conservation Act* 1967 (NZ).

The Court of Appeal held that the *Water and Soil Conservation Act*, as amended, placed emphasis on conservation of natural waters. Once it was determined that the amenity afforded by the waters in their natural state should be recognised and sustained, *primacy* was to be accorded to that object and it should not be defeated by striving to achieve a balance for other users of water. The needs of primary industry were to be given weight in considering an application for a conservation order, but this was to be done bearing in mind that the primary object of the Act was the conservation of waters in their natural state.

The case is a good illustration of a court adopting a precautionary approach given the scope, purpose and subject matter of the legislation.

The New Zealand High Court case of *Greenpeace New Zealand Inc v Minister for Fisheries* involved a total allowable commercial catch (TACC) for orange roughy set by the Minister of Fisheries. *Greenpeace* applied for judicial review of the decision on the basis that the orange roughy fishery was depleted and that overfishing had endangered its survival. The New Zealand Fishing Industry Association and others argued that:

> ... the research into the fishery has not yet been sufficient to establish that the concerns of the applicant or the Ministry scientists are justified and sees an excessive reduction as being not only unjustified, but as imposing serious and unnecessary losses on the industry.

Greenpeace argued that, in considering the TACC, the Minister was required to apply the precautionary approach. Counsel drew attention to a statement of the Minister referring to decisions of the kind under consideration, when he had said:

> It must be a fundamental starting point that management decisions are based on the best data and science available and, in the absence of adequate data, upon the appropriate application of precautionary approaches to management.

After referring to the decision in *Leatch*, Gallen J recognised that the precautionary approach would also apply in New Zealand. His Honour noted that in the case under consideration, there was no statutory obligation for the precautionary approach to be adopted under the *Fisheries Act* 1983, but the statute reflected international obligations accepted by New Zealand and that 'there is in that context at least a movement towards the view that in questions of such moment, a degree of caution is appropriate'. His Honour went on to say that:

> The fact that a dispute exists as to the basic material upon which the decision must rest, does not mean that necessarily the most conservative approach must be adopted. The obligation is to consider the material and decide upon the weight which can be given it with such care as the situation requires. . . .
> At the same time I note, as counsel did, that in the end this is a weighing and not a decisive factor.

It was held that the precautionary approach must be applied by the Minister in formulating a TACC:

> In assessing the information upon which a decision must be based, the precautionary principle ought to be applied so that where uncertainty or ignorance exists, decision-makers should be cautious.

As noted by Mascher, the Court's finding signals an important landmark in New Zealand environmental law, with implications for fisheries law worldwide, as well as environmental law in general. . . .

Of particular importance to the development of ESD and the precautionary principle is the *Case Concerning the Gabcikovo-Nagymaros Project (Hungary v Slovakia)* in the International Court of Justice, otherwise known as the *Danube*

Dam case. The Separate Opinion of Judge Weeramantry, Vice President of the ICJ [International Court of Justice], is of signal importance, if not inspirational. While his Honour espoused the principle in commendable detail, the main Opinion has come under attack by some commentators as not taking the many opportunities presenting themselves (at different points of time) to apply the principle, describing the Opinion as a missed opportunity. The Vice President, however, referred to the duty on States to carry out 'continuing environmental impact assessment' because of the potential for significant impact on the environment and that this was 'a specific application of the larger general principle of caution'. . . .

Some Practical Examples of the Application of the Precautionary Principle

The application of the precautionary principle is becoming a daily occurrence for decision-makers, especially local government, given the requirements of the *Local Government Act* and an increasing number of local environmental plans incorporating ESD. Central Agencies are also having to consider the relevance of the principle in their decisions and recommendations. Both Commonwealth and NSW Commissioners of Inquiry have considered and applied the precautionary principle in their reports. The NSW Minister for Planning utilised the precautionary principle in refusing the proposed Lake Cowell gold mine in the central west of the state—'the application of the precautionary principle means that the unknown risks to this significant environment can only be avoided by refusing this mining proposal'. . . .

The Industry Commission Report of the Inquiry into Ecologically Sustainable Land Management examined ESD and the precautionary principle. Its centrepiece recommendation was the establishment of a statutory duty of care to the environment. The proposed duty would require everyone who influences the management of the risks to the environment to take all reasonable and practical steps to prevent harm to the environment that could have been reasonably foreseen.

Conclusion

[David] Freestone sees the emergence of the precautionary principle as one of the most remarkable developments of the last decade and arguably one of the most significant in the emergence of international environmental law itself. The great preponderance of opinion nowadays is that the principle has become part of international customary law.

How the rhetoric of the principle can be operationalised is one of the challenges for the first decade of the 21st Century. However, what is slowly occurring is that the bones of the principle are starting to be fleshed out. It must be remembered that the precautionary principle is not absolute or extreme. It does not prohibit an activity until the science is clear. It does however change the underlying presumption from freedom of exploitation to one of conservation.

One thing is clear—the precautionary principle will not go away. It is here to stay, with or without legislative prescription. Decision-makers and courts (hearing appeals or challenges) will not be able to dodge it or merely pay lip-service to it. Undeniably the courts will be required to review its application and attempt to apply it. In doing so, we will be called upon to evaluate the principle and its place in environmental decision-making. We must not shirk this responsibility.

NO

Henry I. Miller and Gregory Conko

The Perils of Precaution

Environmental and public health activists have clashed with scholars and risk-analysis professionals for decades over the appropriate regulation of various risks, including those from consumer products and manufacturing processes. Underlying the controversies about various specific issues—such as clorinated water, pesticides, gene-spliced foods, and hormones in beef—has been a fundamental, almost philosophical question: How should regulators, acting as society's surrogate, approach risk in the absence of certainty about the likelihood or magnitude of potential harm?

Proponents of a more risk-averse approach have advocated a "precautionary principle" to reduce risks and make our lives safer. There is no widely accepted! definition of the principle, but in its most common formulation, governments should implement regulatory measures to prevent or restrict actions that raise even conjectural threats of harm to human health or the environment, even though there may be incomplete scientific evidence as to the potential significance of these dangers. Use of the precautionary principle is sometimes represented as "erring on the side of safety," or "better safe than sorry"—the idea being that the failure to regulate risky activities sufficiently could result in severe harm to human health or the environment, and that "overregulation" causes little or no harm. Brandishing the precautionary principle, environmental groups have prevailed upon governments in recent decades to assail the chemical industry and, more recently, the food industry.

Potential risks should, of course, be taken into consideration before proceeding with any new activity or product, whether it is the siting of a power plant or the introduction of a new drug into the pharmacy. But the precautionary principle focuses solely on the *possibility* that technologies could pose unique, extreme, or unmanageable risks, even after considerable testing has already been conducted. What is missing from precautionary calculus is an acknowledgment that even when technologies introduce new risks, most confer net benefits—that is, their use reduces many other, often far more serious, hazards. Examples include blood transfussions, MRI scans, and automobile air bags, all of which offer immense benefits and only minimal risk.

Several subjective factors can cloud thinking about risks and influence how nonexperts view them. Studies of risk perception have shown that people

From Henry I. Miller and Gregory Conko, "The Perils of Precaution," *Policy Review* (June & July 2001), pp. 25–39. Copyright © 2001 by The Hoover Institution, Stanford University. Reprinted by permission.

tend to overestimate risks that are unfamiliar, hard to understand, invisible, involuntary, and/or potentially catastrophic—and vice versa. Thus, they *over*estimate invisible "threats" such as electromagnetic radiation and trace amounts of pesticides in foods, which inspire uncertainty and fear sometimes verging on superstition. Conversely, they tend to *under*estimate risks the nature of which they consider to be clear and comprehensible, such as using a chain saw or riding a motorcycle.

These distorted perceptions complicate the regulation of risk, for if democracy must eventually take public opinion into account, good government must also discount heuristic errors or prejudices. Edmund Burke emphasized government's pivotal role in making such judgments: "Your Representative owes you, not only his industry, but his judgment; and he betrays, instead of serving you, if he sacrifices it to your opinion." Government leaders should *lead;* or putting it another way, government officials should make decisions that are rational and in the public interest even if they are unpopular at the time. This is especially true if, as is the case for most federal and state regulators, they are granted what amounts to lifetime job tenure in order to shield them from political manipulation or retaliation. Yet in too many cases, the precautionary principle has led regulators to abandon the careful balancing of risks *and* benefits—that is, to make decisions, in the name of precaution, that cost real lives due to forgone benefits.

The Danger of Precaution

The danger in the precautionary principle is that it distracts consumers and policymakers from known, significant threats to human health and diverts limited public health resources from those genuine and far greater risks. Consider, for example, the environmental movement's campaign to rid society of chlorinated compounds.

By the late 1980s, environmental activists were attempting to convince water authorities around the world of the possibility that carcinogenic byproducts from chlorination of drinking water posed a potential cancer risk. Peruvian officials, caught in a budget crisis, used this supposed threat to public health as a justification to stop chlorinating much of the country's drinking water. That decision contributed to the acceleration and spread of Latin America's 1991–96 cholera epidemic, which afflicted more than 1.3 million people and killed at least 11,000.

Activists have since extended their antichlorine campaign to so-called "endocrine disrupters," or modulators, asserting that certain primarily man-made chemicals mimic or interfere with human hormones (especially estrogens) in the body and thereby cause a range of abnormalities and diseases related to the endocrine system.

The American Council on Science and Health has explored the endocrine disrupter hypothesis and found that while *high* doses of certain environmental contaminants produce toxic effects in laboratory test animals—in some cases involving the endocrine system—humans' actual exposure to these suspected endocrine modulators is many orders of magnitude lower. It is well documented that while a chemical administered at high doses may cause cancer in certain laboratory animals, it does not necessarily cause cancer in humans—both because of different susceptibilities and because humans are subjected to far lower exposures to synthetic environmental chemicals.

No consistent, convincing association has been demonstrated between real-world exposures to synthetic chemicals in the environment and increased cancer in hormonally sensitive human tissues. Moreover, humans are routinely exposed through their diet to many estrogenic substances (substances having an effect similar to that of the human hormone estrogen) found in many plants. Dietary exposures to these plant estrogens, or phytoestrogens, are far greater than exposures to supposed synthetic endocrine modulators, and no adverse health effects have been associated with the overwhelming majority of these dietary exposures.

Furthermore, there is currently a trend toward *lower* concentrations of many contaminants in air, water, and soil—including several that are suspected of being endocrine disrupters. Some of the key research findings that stimulated the endocrine disrupter hypothesis originally have been retracted or are not reproducible. The available human epidemiological data do not show any consistent, convincing evidence of negative health effects related to industrial chemicals that are suspected of disrupting the endocrine system. In spite of that, activists and many government regulators continue to invoke the need for precautionary (over-) regulation of various products, and even outright bans.

Antichlorine campaigners more recently have turned their attacks to phthalates, liquid organic compounds added to certain plastics to make them softer. These soft plastics are used for important medical devices, particularly fluid containers, blood bags, tubing, and gloves; children's toys such as teething rings and rattles; and household and industrial items such as wire coating and flooring. Waving the banner of the precautionary principle, activists claim that phthalates *might* have numerous adverse health effects—even in the face of significant scientific evidence to the contrary. Governments have taken these unsupported claims seriously, and several formal and informal bans have been implemented around the world. As a result, consumers have been denied product choices, and doctors and their patients deprived of life-saving tools.

In addition to the loss of beneficial products, there are more indirect and subtle perils of government overregulation established in the name of the precautionary principle. Money spent on implementing and complying with regulation (justified or not) exerts an "income effect" that reflects the correlation between wealth and health, an issue popularized by the late political scientists Aaron Wildavsky. It is no coincidence, he argued, that richer societies have lower mortality rates than poorer ones. To deprive communities of wealth, therefore, is to enhance their risks.

Wildavsky's argument is correct: Wealthier individuals are able to purchase better health care, enjoy more nutritious diets, and lead generally less stressful lives. Conversely, the deprivation of income itself has adverse health effects—for example an increased incidence of stress-related problems including ulcers, hypertension, heart attacks, depression, and suicides.

It is difficult to quantify precisely the relationship between mortality and the deprivation of income, but academic studies suggest, as a conservative estimate, that every $7.25 million of regulatory costs will induce one additional fatality through this "income effect." The excess costs in the tens of billions of dollars required annually by precautionary regulation for various classes of consumer products would, therefore, be expected to cause thousands of deaths per year. These are the real costs of "erring on the side of safety." The expression "regulatory overkill" is not merely a figure of speech.

Rationalizing Precaution

During the past few years, skeptics have begun more actively to question the theory and practice of the precautionary principle. In response to those challenges, the European Commission (EC), a prominent advocate of the precautionary principle, last year published a formal communication to clarify and to promote the legitimacy of the concept. The EC resolved that, under its auspices, precautionary restrictions would be "proportional to the chosen level of protection," "non-discriminatory in their application," and "consistent with other similar measures." The commission also avowed that EC decision makers would carefully weigh "potential benefits and costs." EC Health Commissioner David Byrne, repeating these points [recently] in an article on food and agriculture regulation in *European Affairs,* asked rhetorically, "How could a Commissioner for Health and Consumer Protection reject or ignore well-founded, independent scientific advice in relation to food safety?"

Byrne should answer his own question: The ongoing dispute between his European Commission and the United States and Canada over restrictions on hormone-treated beef cattle is exactly such a case of rejecting or ignoring well-founded research. The EC argued that the precautionary principle permits restriction of imports of U.S. and Canadian beef from cattle treated with certain growth hormones.

In their rulings, a WTO [World Trade Organization] dispute resolution panel and its appellate board both acknowledged that the general "look before you leap" sense of the precautionary principle could be found within WTO agreements, but that its presence did not relieve the European Commission of its obligation to base policy on the outcome of a scientific risk assessment. And the risk assessment clearly favored the U.S.-Canadian position. A scientific committee assembled by the WTO dispute resolution panel found that even the scientific studies cited by the EC in its own defense did not indicate a safety risk when the hormones in question were used in accordance with accepted animal husbandry practices. Thus, the WTO ruled in favor of the Untied States and Canada because the European Commission had failed to demonstrate a real or imminent harm. Nevertheless, the EC continues to enforce restrictions on hormone-treated beef, a blatantly unscientific and protectionist policy that belies the commission's insistence that the precautionary principle will not be abused.

Precaution Meets Biotech

Perhaps the most egregious application by the European Commission of the precautionary principle is in its regulation of the products of the new biotechnology, or gene-splicing. By the early 1990s, many of the countries in Western Europe, as well as the EC itself, had erected strict rules regarding the testing and commercialization of gene-spliced crop plants. In 1999, the European Commission explicitly invoked the precautionary principle in establishing a moratorium on the approval of all new gene-spliced crop varieties, pending approval of an even more strict EU-wide regulation.

Notwithstanding the EC's promises that the precautionary principle would not be abused, all of the stipulations enumerated by the commission have been flagrantly ignored or tortured in its regulatory approach to gene-spliced (or in their argot, "genetically modified" or "GM") foods. Rules for gene-spliced plants and microorganisms are inconsistent, discriminatory, and bear no proportionality to risk. In fact, there is arguably *inverse* proportionality to risk, in that the more crudely crafted organisms of the old days of mutagenesis and gene transfers are subject to less stringent regulation than those organisms more precisely crafted by biotech. This amounts to a violation of a cardinal principle of regulation: that the degree of regulatory scrutiny should be commensurate with risk.

Dozens of scientific bodies—including the U.S. National Academy of Sciences (NAS), the American Medical Association, the UK's Royal Society, and the World Health Organization—have analyzed the oversight that is appropriate for gene-spliced organisms and arrived at remarkably congruent conclusions: The newer molecular techniques for genetic improvement are an extension, or refinement, of earlier, far less precise ones; adding genes to plants or microorganisms does not make them less safe either to the environment or to eat; the risks associated with gene-spliced organisms are the same in kind as those associated with conventionally modified organisms and unmodified ones; and regulation should be based upon the risk-related characteristics of individual products, regardless of the techniques used in their development.

An authoritative 1989 analysis of the modern gene-splicing techniques published by the NAS's research arm, the National Research Council, concluded that "the same physical and biological laws govern the response of organisms modified by modern molecular and cellular methods and those produced by classical methods," but it went on to observe that gene-splicing is more precise, circumscribed, and predictable than other techniques.

> [Gene-splicing] methodology makes it possible to introduce pieces of DNA, consisting of either single or multiple genes, that can be defined in function and even in nucleotide sequence. With classical techniques of gene transfer, a variable number of genes can be transferred, the number depending on the mechanism of transfer; but predicting the precise number or the traits that have been transferred is difficult, and we cannot always predict the [characteristics] that will result. With organisms modified by molecular methods, we are in a better, if not perfect, position to predict the [characteristics].

In other words, gene-splicing technology is a refinement of older, less precise techniques, and its use generates less uncertainty. But for gene-spliced plants, both the fact and degree of regulation are determined by the production methods—that is, if gene-splicing techniques have been used, the plant is immediately subject to extraordinary pre-market testing requirements for human health and environmental safety, regardless of the level of risk posed. Throughout most of the world, gene-spliced crop plants such as insect-resistant corn and cotton are subject to a lengthy and hugely expensive process of mandatory testing before they can be brought to market, while plants with similar properties but developed with older, less precise genetic techniques are exempt from such requirements. . . .

Another striking example of the disproportionate regulatory burden borne only by gene-spliced plants involves a process called induced-mutation breeding, which has been in common use since the 1950s. This technique involves exposing crop plants to ionizing radiation or toxic chemicals to induce random genetic mutations. These treatments most often kill the plants (or seeds) or cause detrimental genetic changes, but on rare occasions, the result is a desirable mutation—for example, one producing a new trait in the plant that is agronomically useful, such as altered height, more seeds, or larger fruit. In these cases, breeders have no real knowledge of the exact nature of the genetic mutation(s) that produced the useful trait, or of what other mutations might have occurred in the plant. Yet the approximately 1,400 mutation-bred plant varieties from a range of different species that have been marketed over the past half century have been subject to no formal regulation before reaching the market—even though several, including two varieties of squash and one of potato, have contained dangerous levels of endogenous toxins and had to be banned afterward.

What does this regulatory inconsistency mean in practice? If a student doing a school biology project takes a packet of "conventional" tomato or pea seed to be irradiated at the local hospital x-ray suite and plants them in his backyard in order to investigate interesting mutants, he need not seek approval from any local, national, or international authority. However, if the seeds have been modified by the addition of one or a few genes via gene-splicing techniques —and even if the genetic change is merely to remove a gene—this would-be Mendel faces a mountain of bureaucratic paperwork and expense (to say nothing of the very real possibility of vandalism, since the site of the experiment must be publicized and some opponents of biotech are believers in "direct action"). The same would apply, of course, to professional agricultural scientists in industry and academia. In the United States, Department of Agriculture requirements for paperwork and field trial design make field trials with gene-spliced organisms 10 to 20 times more expensive than the same experiments with virtually identical organisms that have been modified with conventional genetic techniques.

Why are new genetic constructions crafted with these older techniques exempt from regulation, from the dirt to the dinner plate? Why don't regulatory regimes require that new genetic variants made with older techniques be evaluated for increased weediness or invasiveness, or for new allergens that could show up in food? The answer is based on millennia of experience with ge-netically improved crop plants from the era before gene-splicing: Even the use of relatively crude and unpredictable genetic techniques for the improvement of crops and microorganisms poses minimal—but, as noted above, not zero—risk to human health and the environment.

If the proponents of the precautionary principle were applying it ratio-nally and fairly, surely greater precautions would be appropriate not to gene-splicing but to the cruder, less precise, less predictable "conventional" forms of genetic modifications. Furthermore, in spite of the assurance of the European Commission and other advocates of the precautionary principle, regulators of gene-spliced products seldom take into consideration the potential risk-*reducing* benefits of the technology. For example, some of the most successful of the gene-spliced crops, especially cotton and corn, have been constructed by splic-ing in a bacterial gene that produces a protein toxic to predatory insects, but not to people or other mammals. Not only do these gene-spliced corn varieties repel pests, but grain obtained from them is less likely to contain *Furarium,* a toxic fungus often carried into the plants by the insects. That, in turn, significantly reduces the levels of the fungal toxin fumonisin, which is known to cause fatal diseases in horses and swine that eat infected corn, and esophageal cancer in humans. When harvested, these gene-spliced varieties of grain also end up with lower concentrations of insect parts than conventional varieties. Thus, gene-spliced corn is not only cheaper to produce but yields a higher quality product and is a potential boon to public health. Moreover, by reducing the need for spraying chemical pesticides on crops, it is environmentally friendly.

Other products, such as gene-spliced herbicide-resistant crops, have per-mitted farmers to reduce their herbicide use and to adopt more environment-friendly no-till farming practices. Crops now in development with improved yields would allow more food to be grown on less acreage, saving more land area for wildlife or other uses. And recently developed plant varieties with en-hanced levels of vitamins, minerals, and dietary proteins could dramatically improve the health of hundreds of millions of malnourished people in develop-ing countries. These are the kinds of tangible environmental and health benefits that invariably are given little or no weight in precautionary risk calculations.

In spite of incontrovertible benefits and greater predictability and safety of gene-spliced plants and foods, regulatory agencies have regulated them in a discriminatory, unnecessarily burdensome way. They have imposed require-ments that could not possibly be met for conventionally bred crop plants. And, as the European Commission's moratorium on new product approvals demon-strates, even when that extraordinary burden of proof is met via monumental amounts of testing and evaluation, regulators frequently declare themselves unsatisfied.

Biased Decision Making

While the European Union is a prominent practitioner of the precautionary principle on issues ranging from toxic substances and the new biotechnology to climate change and gun control, U.S. regulatory agencies also commonly practice excessively precautionary regulation. The precise term of art "precautionary principle" is not used in U.S. public policy, but the regulation of such products as pharmaceuticals, food additives, gene-spliced plants and microorganisms, synthetic pesticides, and other chemicals is without question "precautionary" in nature. U.S. regulators actually appear to be more precautionary than the Europeans towards several kinds of risks, including the licensing of new medicines, lead in gasoline, nuclear power, and others. They have also been highly precautionary towards gene-splicing, although not to the extremes of their European counterparts. The main difference between precautionary regulation in the United States and the use of the precautionary principle in Europe is largely a matter of degree—with reference to products, technologies, and activities—and of semantics.

In both the United States and Europe, public health and environmental regulations usually require a risk assessment to determine the extent of potential hazards and of exposure to them, followed by judgments about how to regulate. The precautionary principle can distort this process by introducing a systematic bias into decision making. Regulators face an asymmetrical incentive structure in which they are compelled to address the potential harms from new products, but are free to discount the hidden risk-reducing properties of unused or underused ones. The result is a lopsided process that is inherently biased against change and therefore against innovation.

To see why, one must understand that there are two basic kinds of mistaken decisions that a regulator can make: First, a harmful product can be approved for marketing—called a Type I error in the parlance of risk analysis. Second, a useful product can be rejected or delayed, can fail to achieve approval at all, or can be inappropriately withdrawn from the market—a Type II error. In other words, a regulator commits a Type I error by permitting something harmful to happen and a Type II error by preventing something beneficial from becoming available. Both situations have negative consequences for the public, but the outcomes for the regulator are very different.

Examples of this Type I-Type II error dichotomy in both the U.S. and Europe abound, but it is perhaps illustrated most clearly in the FDA's [Food and Drug Administration] approval process for new drugs. A classic example is the FDA's approval in 1976 of the swine flu vaccine—generally perceived as a Type I error because while the vaccine was effective at preventing influenza, it had a major side effect that was unknown at the time of approval: A small number of patients suffered temporary paralysis from Guillain-Barré Syndrome. This kind of mistake is highly visible and has immediate consequences: The media pounce and the public and Congress are roused, and Congress takes up the matter. Both the developers of the product and the regulators who allowed it to be marketed are excoriated and punished in such modern-day pillories as congressional hearings, television newsmagazines, and newspaper editorials. Because a regulatory official's career might be damaged irreparably by his good-faith but mistaken approval of a high-profile product, decisions are often made defensively—in other words, above all to avoid Type I errors.

Former FDA Commissioner Alexander Schmidt aptly summarized the regulator's dilemma:

> In all our FDA history, we are unable to find a single instance where a Congressional committee investigated the failure of FDA to approve a new drug. But, the times when hearings have been held to criticize our approval of a new drug have been so frequent that we have not been able to count them. The message to FDA staff could not be clearer. Whenever a controversy over a new drug is resolved by approval of the drug, the agency and the individuals involved likely will be investigated. Whenever such a drug is disapproved, no inquiry will be made. The Congressional pressure for *negative* action is, therefore, intense. And it seems to be ever increasing....

Although they can dramatically compromise public health, Type II errors caused by a regulator's bad judgment, timidity, or anxiety seldom gain public attention. It may be only the employees of the company that makes the product and a few stock market analysts and investors who are knowledgeable about unnecessary delays. And if the regulator's mistake precipitates a corporate decision to abandon the product, cause and effect are seldom connected in the public mind. Naturally, the companies themselves are loath to complain publicly about a mistaken FDA judgment, because the agency has so much discretionary control over their ability to test and market products. As a consequence, there may be no direct evidence of, or publicity about, the lost societal benefits, to say nothing of the culpability of regulatory officials.

Exceptions exist, of course. A few activists, such as the AIDs advocacy groups that closely monitor the FDA, scrutinize agency review of certain products and aggressively publicize Type II errors. In addition, congressional oversight *should* provide a check on regulators' performance, but as noted above by former FDA Commissioner Schmidt, only rarely does oversight focus on their Type II errors. Type I errors make for more dramatic hearings, after all, including injured patients and their family members. And even when such mistakes are exposed, regulators frequently defend Type II errors as erring on the side of caution—in effect, invoking the precautionary principle.... Too often this euphemism is accepted uncritically by legislators, the media, and the public, and our system of pharmaceutical oversight becomes progressively less responsive to the public interest.

The FDA is not unique in this regard, of course. All regulatory agencies are subject to the same sorts of social and political pressures that a cause them to be castigated when dangerous products accidentally make it to market (even if, as is often the case, those products produce net benefits) but to escape blame when they keep beneficial products out of the hands of consumers. Adding the precautionary principle's bias against new products into the public policy mix further encourages regulators to commit Type II errors in their frenzy to avoid Type I errors. This is hardly conducive to enhancing overall public safety.

Extreme Precaution

For some antitechnology activists who push the precautionary principle, the deeper issue is not really safety at all. Many are more antibusiness and antitechnology than they are pro-safety. And in their mission to oppose business interests and disparage technologies they don't like or that they have decided we just don't need, they are willing to seize any opportunity that presents itself.

These activists consistently (and intentionally) confuse *plausibility* with *provability.* Consider, for example, *Our Stolen Future,* the bible of the proponents of the endocrine disrupter hypothesis discussed above. The book's premise—that estrogen-like synthetic chemicals damage health in a number of ways—is not supported by scientific data. Much of the research offered as evidence for its arguments has been discredited. The authors equivocate wildly: "Those exposed prenatally to endocrine-disrupting chemicals *may* have abnormal hormone levels as adults, and they *could* also pass on persistent chemicals they themselves have inherited—both factors that *could* influence the development of their own children [emphasis added]." The authors also assume, in the absence of any actual evidence, that exposures to small amounts of many chemicals create a synergistic effect—that is, that total exposure constitutes a kind of witches' brew that is far more toxic than the sum of the parts. For these anti-innovation ideologues, the mere fact that such questions have been asked requires that inventors or producers expend time and resources answering them. Meanwhile, the critics move on to yet another frightening plausibility and still more questions. No matter how outlandish the claim, the burden of proof is put on the innovator.

Whether the issue is environmental chemicals, nuclear power, or gene-spliced plants, many activists are motivated by their own parochial vision of what constitutes a "good society" and how to achieve it. One prominent biotechnology critic at the Union of Concerned Scientists rationalizes her organization's opposition to gene-splicing as follows: "Industrialized countries have few genuine needs for innovative food stuffs, regardless of the method by which they are produced"; therefore, society should not squander resources on developing them. She concludes that although "the malnourished homeless" are, indeed, a problem, the solution lies "in resolving income disparities, and educating ourselves to make better choices from among the abundant foods that are available."

Greenpeace, one of the principal advocates of the precautionary principle, offered in its 1999 IRS filings the organization's view of the role in society of safer, more nutritious, higher-yielding, environment-friendly, gene-spliced plants: There isn't any. By its own admission, Greenpeace's goal is not the prudent, safe use of gene-spliced foods or even their mandatory labeling, but rather these products' "complete elimination [from] the food supply and the environment." Many of the groups, such as Greenpeace, do not stop at demanding illogical and stultifying regulation or outright bans on product testing and commercialization; they advocate and carry out vandalism of the very field trials intended to answer questions about environmental safety.

Such tortured logic and arrogance illustrate that the metastasis of the precautionary principle generally, as well as the pseudocontroversies over the testing and use of gene-spliced organisms in particular, stem from a social vision that is not just strongly antitechnology, but one that poses serious challenges to academic, commercial, and individual freedom.

The precautionary principle shifts decision making power away from individuals and into the hands of government bureaucrats and environmental activists. Indeed, that is one of its attractions for many NGOs [nongovernmental organizations]. Carolyn Raffensperger, executive director of the Science and Environmental Health Network, a consortium of radical groups, asserts that discretion to apply the precautionary principle "is in the hands of the people." According to her, this devolution of power is illustrated by violent demonstrations against economic globalization such as those in Seattle at the 1999 meeting of the World Trade Organization. "This is [about] how they want to live their lives," Raffensperger said.

To be more precise, it is about how small numbers of vocal activists want the rest of us to live *our* lives. In other words, the issue here is freedom and its infringement by ideologues who disapprove, on principle, of a certain technology, or product, or economic system. . . .

Precaution *v.* Freedom

History offers compelling reasons to be cautious about societal risks, to be sure. These include the risk of incorrectly assuming the absence of danger (false negatives), overlooking low probability but high impact events in risk assessments, the danger of long latency periods before problems become apparent, and the lack of remediation methods in the event of an adverse event. Conversely, there are compelling reasons to be wary of excessive precaution, including the risk of too eagerly detecting a nonexistent danger (false positives), the financial cost of testing for or remediating low-risk problems, the opportunity costs of forgoing net-beneficial activities, and the availability of a contingency regime in case of an adverse event. The challenge for regulators is to balance these competing risk scenarios in a way that reduces overall harm to public health. This kind of risk balancing is often conspicuously absent from precautionary regulation.

It is also important that regulators take into consideration the degree of restraint generally imposed by society on individuals' and companies' freedom to perform legitimate activities (e.g., scientific research). In Western democratic societies, we enjoy long traditions of relatively unfettered scientific research and development, except in the very few cases where bona fide safety issues are raised. Traditionally, we shrink from permitting small, authoritarian minorities to dictate our social agenda, including what kinds of research are permissible and which technologies and products should be available in the marketplace.

Application of the precautionary principle has already elicited unscientific, discriminatory policies that inflate the costs of research, inhibit the development of new products, divert and waste resources, and restrict consumer choice. The excessive and wrong-headed regulation of the new biotechnology is one particularly egregious example. Further encroachment of precautionary regulation into other areas of domestic and international health and safety standards will create a kind of "open sesame" that government officials could invoke whenever they wish arbitrarily to introduce new barriers to trade, or simply to yield disingenuously to the demands of antitechnology activists. Those of us who both value the freedom to perform legitimate research and believe in the wisdom of market processes must not permit extremists acting in the name of "precaution" to dictate the terms of the debate.

POSTSCRIPT

Is the Precautionary Principle a Sound Basis for International Policy?

In their definition of the precautionary principle, Miller and Conko emphasize supposition: Precautions must be taken whenever there might be a problem. Stein emphasizes uncertainty: Lack of full scientific certainty shall not be used as a reason for postponing cost-effective precautions. The same tension is visible in Kenneth R. Foster, Paolo Vecchia, and Michael H. Repacholi, "Science and the Precautionary Principle," *Science* (May 12, 2000).

Other writers who oppose the precautionary principle have approached it in much the same way as Miller and Conko. For instance, Ronald Bailey, in "Precautionary Tale," *Reason* (April 1999), defines the precautionary principle as "precaution in the face of any actions that may affect people or the environment, no matter what science is able—or unable—to say about that action." "No matter what science says" is not quite the same thing as "lack of full scientific certainty." Indeed, Bailey turns the precautionary principle into a straw man and thereby endangers whatever points he makes that are worth considering. One of those points is that widespread use of the precautionary principle would hamstring the development of the Third World. Bonner R. Cohen, in "The Safety Nazis," *American Spectator* (July/August 2001), echoes this point, calling the precautionary principle a massive threat to human health in the less developed countries.

Yet the 1992 Rio Declaration emphasized that the precautionary principle should be "applied by States according to their capabilities" and that it should be applied in a cost-effective way. These provisions would seem to preclude the draconian interpretations that most alarm the critics. Turned around, however, these same provisions are what led to the World Trade Organization's efforts to keep developed nations from applying their own stringent environmental regulations to Third World trading partners, which alarms those who favor the precautionary principle. Hilary French takes a somewhat broader view in "Coping With Ecological Globalization," in Lester R. Brown et al., *State of the World 2000* (W. W. Norton, 2000). Conflicts of interest can take other forms as well. David Kriebel et al., in "The Precautionary Principle in Environmental Science," *Environmental Health Perspectives* (September 2001), state that "environmental scientists should be aware of the policy uses of their work and of their social responsibility to do science that protects human health and the environment." Businesses are also conflicted, writes Arnold Brown in "Suitable Precautions," *Across the Board* (January/February 2002), because the precautionary principle tends to slow decision making, but he maintains that "we will all have to learn and practice anticipation."

ISSUE 6

Do Environmentalists Overstate Their Case?

YES: Ronald Bailey, from "Debunking Green Myths," *Reason* (February 2002)

NO: David Pimentel, from "Skeptical of the Skeptical Environmentalist," *Skeptic* (vol. 9, no. 2, 2002)

ISSUE SUMMARY

YES: Environmental journalist Ronald Bailey argues that the natural environment is not in trouble, despite the arguments of many environmentalists that it is. He holds that the greatest danger facing the environment is not human activity but "ideological environmentalism, with its hostility to economic growth and technological progress."

NO: David Pimentel, a professor of insect ecology and agricultural sciences, argues that those who contend that the environment is not threatened are using data selectively and that the supply of basic resources to support human life is declining rapidly.

For over two centuries, seemingly everyone who has claimed to see environmental disaster in the offing has been challenged. In 1798, for example, English parson Thomas Malthus thought population must inevitably outstrip the ability of the environment to produce food. When the crisis he foretold did not come about, he was ridiculed; indeed, his failure has been held up ever since as a main reason why we need not be concerned about the consequences of population growth, urbanization, industrialization, and other human activities.

Yet the environmentalists have continued to find things to be concerned about. Rachel Carson (1907–1964) is famous for realizing the dangers of pesticides and other chemicals that we release to the environment, which she reported in her best-seller *Silent Spring* (Houghton Mifflin, 1962). Ecologists Paul Ehrlich and Garrett Hardin have reiterated Malthus's concern about population. A Massachusetts Institute of Technology team lead by Donella Meadows and Dennis Meadows used computer models to analyze population, development, and pollution trends and forecast a crisis of resource depletion and economic collapse before 2050 (see *The Limits to Growth* [Universe Books, 1972]). In 1992 the study was repeated with improved computer models, and even more pessimistic conclusions were reached (see *Beyond the Limits: Confronting Global Collapse, Envisioning a Sustainable Future* [Chelsea Green, 1992]). In 1980 the U.S. government published *The Global 2000 Report to the President*, which projected increased environmental degradation, loss of resources, and a widening gap between the rich and the poor.

No one likes such conclusions. Nor does anyone like the implications for what must be done: limit industrial development and population growth. Conservatives object to proposals to regulate industrial development, for only unchecked industry can generate the wealth needed to solve problems, and the free market can be trusted to produce solutions for all problems that truly need solutions. They also object to proposals to limit family size. Liberals object that restricting development will harm the poor much more than it will the rich. Some also object that the true problem is modern capitalism, which emphasizes short-term economic payoffs over longer-term benefits.

The growing sense that we do indeed face environmental crises lies behind the long series of international conferences arranged by the United Nations, from 1968's Biosphere Conference through 1972's Conference on the Human Environment and 1992's Earth Summit (or the Conference on Environment and Development) to 2002's World Summit on Sustainable Development. The concept of sustainable development became prominent after 1992 and has now taken center stage. Yet the debate is hardly over. Analysts such as Niles Eldredge (*Life in the Balance: Humanity and the Biodiversity Crisis* [Princeton University Press, 1998]) have argued that development in the traditional sense cannot be sustained in a world whose resources are finite; sustainable development must mean development without growth in industrial activity or population.

One prominent contrary voice was that of economist Julian L. Simon (1932–1998), who argued that environmental problems could only be short-term problems; increased population and the free market would ensure an ever-improving standard of living and an ever-healthier environment (see *The Ultimate Resource* [Princeton University Press, 1981]). In 1998 statistician and political scientist Bjorn Lomborg joined the fray with *The Skeptical Environmentalist: Measuring the Real State of the World* (Cambridge University Press, 2001). In it, he accuses environmentalists of distorting the truth in a litany of disaster. The truth, he says, is that "mankind's lot has actually improved in terms of practically every measurable indicator."

Is Lomborg right? The following selections represent two of the many reviews that have discussed this question. In the first selection, Ronald Bailey argues that Lomborg is indeed correct. Despite the claims of environmentalists, he contends, the natural environment is not in trouble from human activity but is, in fact, more threatened by ideological environmentalism. In the second selection, David Pimentel argues that Lomborg misrepresents the truth by selecting only data that support his case. He maintains that human activities do threaten the environment and that mankind's lot is at the mercy of a rapidly declining supply of basic resources.

Ronald Bailey

 **YES**

Debunking Green Myths

$\mathbf{M}$odern environmentalism, born of the radical movements of the 1960s, has often made recourse to science to press its claims that the world is going to hell in a handbasket. But this environmentalist has never really been a matter of objectively describing the world and calling for the particular social policies that the description implies.

Environmentalism is an ideology, very much like Marxism, which pretended to base its social critique on a "scientific" theory of economic relations. Like Marxists, environmentalists have had to force the facts to fit their theory. Environmentalism is an ideology in crisis: The massive, accumulating contradictions between its pretensions and the actual state of the world can no longer be easily explained away.

The publication of *The Skeptical Environmentalist,* a magnificent and important book by a former member of Greenpeace, deals a major blow to that ideology by superbly documenting a response to environmental doomsaying. The author, Bjorn Lomborg, is an associate professor of statistics at the University of Aarhus in Denmark. On a trip to the United States a few years ago, Lomborg picked up a copy of *Wired* that included an article about the late "doomslayer" Julian Simon.

Simon, a professor of business administration at the University of Maryland, claimed that by most measures, the lot of humanity is improving and the world's natural environment was not critically imperiled. Lomborg, thinking it would be an amusing and instructive exercise to debunk a "right-wing" anti-environmentalist American, assigned his students the project of finding the "real" data that would contradict Simon's outrageous claims.

Lomborg and his students discovered that Simon was essentially right, and that the most famous environmental alarmists (Stanford biologist Paul Ehrlich, Worldwatch Institute founder Lester Brown, former Vice President Al Gore, *Silent Spring* author Rachel Carson) and the leading environmentalist lobbying groups (Greenpeace, the World Wildlife Fund, Friends of the Earth) were wrong. It turns out that the natural environment is in good shape, and the prospects of humanity are actually quite good.

From Ronald Bailey, "Debunking Green Myths," *Reason*, vol. 33, no. 9 (February 2002). Copyright © 2002 by The Reason Foundation. Reprinted by permission of The Reason Foundation, 3415 S. Sepulveda Boulevard, Suite 400, Los Angeles, CA 90034. http://www.reason.com.

❧

Lomborg beings with "the Litany" of environmentalist doom, writing: "We are all familiar with the Litany.... Our resources are running out. The population is ever growing, leaving less and less to eat. The air and water are becoming ever more polluted. The planet's species are becoming extinct in vast numbers.... The world's ecosystem is breaking down.... We all know the Litany and have heard it so often that yet another repetition is, well, almost reassuring." Lomborg notes that there is just one problem with the Litany: "It does not seem to be backed up by the available evidence."

Lomborg then proceeds to demolish the Litany. He shows how, time and again, ideological environmentalists misuse, distort, and ignore the vast reams of data that contradict their dour visions. In the course of *The Skeptical Environmentalist*, Lomborg demonstrates that the environmentalist lobby is just that, a collection of interest groups that must hype doom in order to survive monetarily and politically.

Lomborg notes, "As the industry and farming organizations have an obvious interest in portraying the environment as just-fine and no-need-to-do-anything, the environmental organizations also have a clear interest in telling us that the environment is in a bad state, and that we need to act now. And the worse they can make this state appear, the easier it is for them to convince us we need to spend more money on the environment rather than on hospitals, kindergartens, etc. Of course, if we were equally skeptical of both sorts of organization there would be less of a problem. But since we tend to treat environmental organizations with much less skepticism, this might cause a grave bias in our understanding of the state of the world." Lomborg's book amply shows that our understanding of the state of the world is indeed biased.

❧

So what is the real state of humanity and the planet?

Human life expectancy in the developing world has more than doubled in the past century, from 31 years to 65. Since 1960, the average amount of food per person in the developing countries has increased by 38 percent, and although world population has doubled, the percentage of malnourished poor people has fallen globally from 35 percent to 18 percent, and will likely fall further over the next decade, to 12 percent. In real terms, food costs a third of what it did in the 1960s. Lomborg points out that increasing food production trends show no sign of slackening in the future.

What about air pollution? Completely uncontroversial data show that concentrations of sulfur dioxide are down 80 percent in the U.S. since 1962, carbon monoxide levels are down 75 percent since 1970, nitrogen oxides are down 38 percent since 1975, and ground level ozone is down 30 percent since 1977. These trends are mirrored in all developed countries.

Lomborg shows that claims of rapid deforestation are vastly exaggerated. One United Nations Food and Agriculture survey found that globally, forest cover has been reduced by a minuscule 0.44 percent since 1961. The World Wildlife Fund claims that two-thirds of the world's forests have been lost since the dawn of agriculture; the reality is that the world still has 80 percent of its forests. What about the Brazilian rainforests? Eighty-six percent remain uncut, and the rate of clearing is falling. Lomborg also debunks the widely circulated claim that the world will soon lose up to half of its species. In fact, the best evidence indicates that 0.7 percent of species might be lost in the next 50 years if nothing is done. And of course, it is unlikely that nothing will be done.

Finally, Lomborg shows that global warming caused by burning fossil fuels is unlikely to be a catastrophe. Why? First, because actual measured temperatures aren't increasing nearly as fast as the computer climate models say they should be—in fact, any increase is likely to be at the low end of the predictions, and no one thinks that would be a disaster. Second, even in the unlikely event that temperatures were to increase substantially, it will be far less costly and more environmentally sound to adapt to the changes rather than institute draconian cuts in fossil fuel use. The best calculations show that adapting to global warming would cost $5 trillion over the next century. By comparison, substantially cutting back on fossil fuel emissions in the manner suggested by the Kyoto Protocol would cost between $107 and $274 trillion over the same period. (Keep in mind that the current yearly U.S. gross domestic product is $10 trillion.) Such costs would mean that people living in developing countries would lose over 75 percent of their expected increases in income over the next century. That would be not only a human tragedy, but an environmental one as well, since poor people generally have little time for environmental concerns.

Where does Lomborg fall short? He clearly understands that increasing prosperity is the key to improving human and environmental health, but he often takes for granted the institutions of property and markets that make progress and prosperity possible. His analysis, as good as it is, fails to identify the chief cause of most environmental problems. In most cases, imperiled resources such as fisheries and airsheds are in open-access commons where the incentive is for people to take as much as possible of the resource before someone else beats them to it. Since they don't own the resource, they have no incentive to protect and conserve it.

Clearly, regulation has worked to improve the state of many open-access commons in developed countries such as the U.S. Our air and streams are much cleaner than they were 30 years ago, in large part due to things like installing catalytic converters on automobiles and building more municipal sewage treatment plants. Yet there is good evidence that assigning private property rights to these resources would have resulted in a faster and cheaper cleanup. Lomborg's analysis would have been even stronger had he more directly taken on ideological environmentalism's bias against markets. But perhaps that is asking for too much in an already superb book.

"Things are *better* now," writes Lomborg, "but they are still not *good* enough." He's right. Only continued economic growth will enable the 800 million people who are still malnourished to get the food they need; only continued economic growth will let the 1.2 billion who don't have access to clean water and sanitation obtain those amenities. It turns out that ideological environmentalism, with its hostility to economic growth and technological progress, is the biggest threat to the natural environment and to the hopes of the poorest people in the world for achieving better lives.

"The very message of the book," Lomborg concludes, is that "children born today—in both the industrialized world and the developing countries—will live longer and be healthier, they will get more food, a better education, a higher standard of living, more leisure time and far more possibilities—without the global environment being destroyed. And that is a beautiful world."

David Pimentel  **NO**

Skeptical of the Skeptical Environmentalist

Bjorn Lomborg discusses a wide range of topics in his book and implies, through his title, that he will inform readers exactly what the real state of world is. In this effort, he criticizes countless world economists, agriculturists, water specialists, and environmentalists, and furthermore, accuses them of misquoting and/or organizing published data to mislead the public concerning the status of world population, food supplies, malnutrition, disease, and pollution. Lomborg bases his optimistic opinion on his selective use of data. Some of Lomborg's assertions will be examined in this review, and where differing information is presented, extensive documentation will be provided.

Lomborg reports that "we now have more food per person than we used to." In contrast, the Food and Agricultural Organization (FAO) of the United Nations reports that food per capita has been declining since 1984, based on available cereal grains (Figure 1). Cereal grains make up about 80% of the world's food. Although grain yields per hectare (abbreviated ha) in both developed and developing countries are still increasing, these increases are slowing while the world population continues to escalate. Specifically from 1950 to 1980, U.S. grains yields increased at about 3% per year, but after 1980 the rate of increase for corn and other grains has declined to only about 1% (Figure 2).

Obviously fertile cropland is an essential resource for the production of foods but Lomborg has chosen not to address this subject directly. Currently, the U.S. has available nearly 0.5 ha of prime cropland per capita, but it will not have this much land if the population continues to grow at its current rapid rate. Worldwide the average cropland available for food production is only 0.25 ha per person. Each person added to the U.S. population requires nearly 0.4 ha (1 acre) of land for urbanization and transportation. One example of the impact of population growth and development is occurring in California where an average of 156,000 ha of agricultural land is being lost each year. At this rate it will not be long before California ceases to be the number one state in U.S. agricultural production.

In addition to the quantity of agricultural land, soil quality and fertility is vital for food production. The productivity of the soil is reduced when it is eroded by rainfall and wind. Soil erosion is not a problem, according to

From David Pimentel, "Skeptical of the Skeptical Environmentalist," *Skeptic*, vol. 9, no. 2 (2002). Copyright © 2002 by *Skeptic*. Reprinted by permission. Notes omitted.

Figure 1

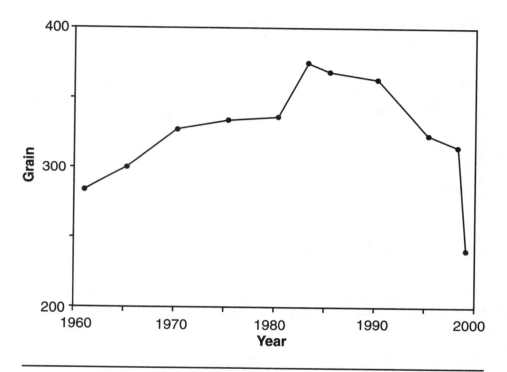

Cereal Grain Production Per Capita in the World From 1961 to 1999

FAO, 1961–1999

Lomborg, especially in the U.S. where soil erosion has declined during the past decade. Yes, as Lomborg states, instead of losing an average of 17 metric tons per hectare per year on cropland, the U.S. cropland is now losing an average of 13 t/ha/yr. However, this average loss is 13 times the sustainability rate of soil replacement. Exceptions occur, as during the 1995–96 winter in Kansas, when it was relatively dry and windy, and some agricultural lands lost as much as 65 t/ha of productive soil. This loss is 65 times the natural soil replacement in agriculture.

Worldwide soil erosion is more damaging than in the United States. For instance, the India soil is being lost at 30 to 40 times its sustainability. Rate of soil loss in Africa is increasing not only because of livestock overgrazing but also because of the burning of crop residues due to the shortages of wood fuel. During the summer of 2000, NASA published a satellite image of a cloud of soil from Africa being blown across the Atlantic Ocean, further attesting to the massive soil erosion problem in Africa. Worldwide evidence concerning

Figure 2

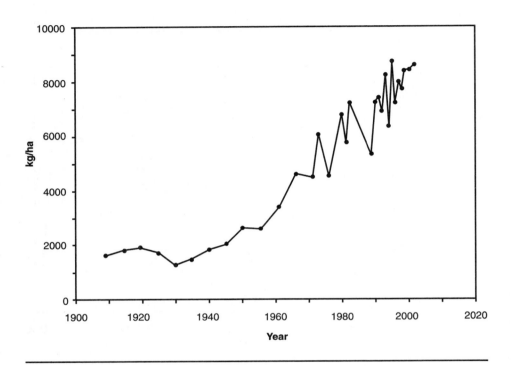

Corn Grain Yields From 1910 to 1999

USDA, 1910–2000

soil loss is substantiated and it is difficult to ignore its effect on sustainable agricultural production.

Contrary to Lomborg's belief, crop yields cannot continue to increase in response to the increased applications of more fertilizers and pesticides. In fact, field tests have demonstrated that applying excessive amounts of nitrogen fertilizer stresses the crop plants, resulting in declining yields. The optimum amount of nitrogen for corn, one of the crops that require heavy use of nitrogen, is approximately 120 kg/ha.

Although U.S. farmers frequently apply significantly more nitrogen fertilizer than 120 kg/ha, the extra is a waste and pollutant. The corn crop can only utilize about one-third of the nitrogen applied, while the remainder leaches either into the ground or surface waters. This pollution of aquatic ecosystems in agricultural areas results in the high levels of nitrogen and pesticides occurring in many U.S. water bodies. For example, nitrogen fertilizer has found its way into 97% of the well-water supplies in some regions, like North Carolina. The concentrations of nitrate are above the U.S. Environmental Protection Agency drinking-water standard of 10 milligrams per liter (nitrogen) and are a toxic threat to young children and young livestock. In the last 30 years, the nitrate content has tripled in the Gulf of Mexico, where it is reducing the Gulf fishery.

In an undocumented statement Lomborg reports that pesticides cause very little cancer. Further, he provides no explanation as to why human and other nontarget species are not exposed to pesticides when crops are treated. There is abundant medical and scientific evidence that confirms that pesticides cause significant numbers of cancers in the U.S. and throughout the world. Lomborg also neglects to report that some herbicides stimulate the production of toxic chemicals in some plants, and that these toxicants can cause cancer.

In keeping with Lomborg's view that agriculture and the food supply are improving, he states that "fewer people are starving." Lomborg criticizes the validity of the two World Health Organization [WHO] reports that confirm more than 3 billion people are malnourished. This is the largest number and proportion of malnourished people ever in history! Apparently Lomborg rejects the WHO data because they do not support his basic thesis. Instead, Lomborg argues that only people who suffer from calorie shortages are malnourished, and ignores the fact that humans die from deficiencies of protein, iron, iodine, and vitamin A, B, C, and D.

Further confirming a decline in food supply, the FAO reports that there has been a three-fold decline in the consumption of fish in the human diet during the past seven years. This decline in fish per capita is caused by overfishing, pollution, and the impact of a rapidly growing world population that must share the diminishing fish supply.

In discussing the status of water supply and sanitation services, Lomborg is correct in stating that these services were improved in the developed world during the 19th century, but he ignores the available scientific data when he suggests that these trends have been "replicated in the developing world" during the 20th century. Countless reports confirm that developing countries discharge most of their untreated urban sewage directly into surface waters. For example, of India's 3,119 towns and cities, only eight have full waste water treatment facilities. Furthermore, 114 Indian cities dump untreated sewage and partially cremated bodies directly into the sacred Ganges River. Downstream the untreated water is used for drinking, bathing, and washing. In view of the poor sanitation, it is no wonder that water borne infectious diseases account for 80% of all infections worldwide and 90% of all infections in developing countries.

Contrary to Lomborg's view, most infectious diseases are increasing worldwide. The increase is due not only to population growth but also because of increasing environmental pollution. Food-borne infections are increasing rapidly worldwide and in the United States. For example, during 2000 in the U.S. there were 76 million human food-borne infections with 5,000 associated deaths. Many of these infections are associated with the increasing contamination of food and water by livestock wastes in the United States.

In addition, a large number of malnourished people are highly susceptible to infectious diseases, like tuberculosis (TB), malaria, schistosomiasis, and AIDS. For example, the number of people infected with tuberculosis in the U.S.

and the world is escalating, in part because medicine has not kept up with the new forms of TB. Currently, according to the World Health Organization, more than 2 billion people in the world are infected with TB, with nearly 2 million people dying each year from it.

Consistent with Lomborg's thesis that world natural resources are abundant, he reports that the U.S. Energy Information Agency for the period 2000 to 2020 projects an almost steady oil price over the next two decades at about $22 per barrel. This optimistic projection was crossed late in 2000 when oil rose to $30 or more per barrel in the United States and the world. The best estimates today project that world oil reserves will last approximately 50 years, based on current production rates.

Figure 3

Number of Hectares in Forests Worldwide (x 1 million ha) From 1961 to 1994

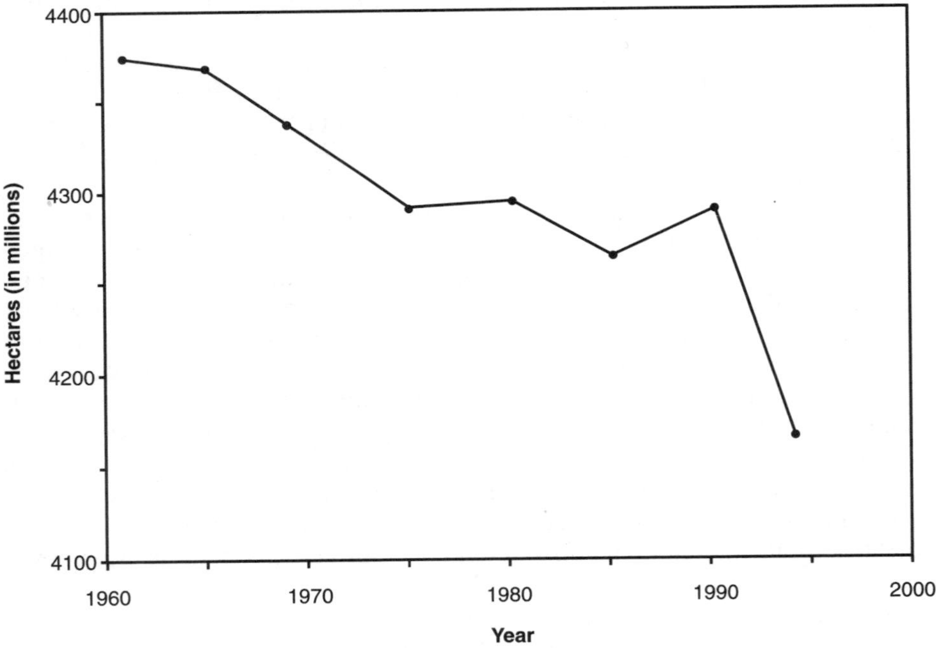

FAOSTAT Database, consulted September 3, 2001

Lomborg takes the World Wildlife Fund (WWF) to task for their estimates on the loss of world forests during the past decade and their emphasis on resulting ecological impacts and loss of biodiversity. Whether the loss of forests is slow, as Lomborg suggests, or rapid as WWF reports, there is no question that forests are disappearing worldwide (Figure 3). Forests not only provide valuable products but they harbor a vast diversity of species of plants, animals and microbes. Progress in medicine, agriculture, genetic engineering, and environmental quality depend on maintaining the species diversity in the world.

This reviewer takes issue with Lomborg's underlying thesis that the size and growth of the human population is not a major problem. The difference between Lomborg's figure that 76 million humans were added to the world population in 2000, or the 80 million reported by the Population Reference Bureau, is not the issue, thought the magnitude of both projections is of serious concern. Lomborg neglects to explain that the major problem with world population growth is the young age structure that now exists. Even if the world adopted a policy of only two children per couple tomorrow, the world population would continue to increase for more than 70 years before stabilizing at more than 12 billion people. As an agricultural scientist and ecologist, I wish I could share Lomborg's optimistic views, but my investigations and those of countless scientists lead me to a more conservative outlook. The supply of basic resources, like fertile cropland, water, energy, and an unpolluted atmosphere that support human life is declining rapidly, as nearly a quarter million people are daily added to the Earth. We all desire a high standard of living for each person on Earth, but with every person added, the supply of resources must be divided and shared. Current losses and degradation of natural resources suggest concern and a need for planning for future generations of humans. Based on our current understanding of the real state of the world and environment, there is need for conservation and protection of vital world resources.

POSTSCRIPT

Do Environmentalists Overstate Their Case?

One of the basic issues at work in the debate between "Malthusians" (who believe that the environment can only support a limited number of people and amount of industrial activity) and "cornucopians" (who believe that there are no such limits) is whether or not past trends can be extrapolated reliably into the future. Cornucopians see little problem in such extrapolation: If the per capita food supply has continued to increase for the last half century, it will continue to do so ad infinitum. Ecologists point out that natural environments have a finite supply of resources (water, soil nutrients, and so on). In nature, population growth follows an S-shaped curve with a steep rise phase followed by a leveling-off at what is known as the "carrying capacity." In "Anticipating Environmental 'Surprise,'" in Lester R. Brown et al., *State of the World 2000* (W. W. Norton, 2000), Chris Brown stresses that straight-line extrapolation cannot be trusted in the real world, because real-world trends are not linear; straight-line trends level off, and they may even peak and fall.

Bjorn Lomborg's *The Skeptical Environmentalist*, with its essentially optimistic message that conditions will continue to get better and better and that we do not need to worry about environmental issues, received very positive responses from the press. However, the scientific community reacted very differently. The January 2002 issue of *Scientific American*, for example, features a series of articles by leading scientists under the heading "Misleading Math About the Earth." Stephen Schneider deals with Lomborg's comments on global warming, John P. Holdren deals with energy, John Bongaarts addresses population, and Thomas Lovejoy discusses biodiversity. All four document the numerous ways in which they feel that Lomborg distorts the truth. Richard B. Norgaard, in "Optimists, Pessimists, and Science," *Bioscience* (March 2002), calls the *Scientific American* article series "devastating" and develops an explanation for Lomborg's approach in the very different ways that economics and science approach the world. Russ Baker, in "The Lomborg File: When the Press Is Lured by a Contrarian's Tale," *Columbia Journalism Review* (March/April 2002), bemoans the uncritical reaction of the nonscientific media, quoting Winston Churchill's famous remark "A lie gets halfway around the world before the truth has a chance to get its pants on."

In "The Skeptical Environmentalist Replies," *Scientific American* (May 2002), Lomborg rebuts his detractors. However, John Rennie, editor in chief of *Scientific American*, charges that Lomborg fails to address the scientists' points and continues to be selective about data.

On the other hand, a review in the May/June 2002 issue of *Foreign Affairs* calls the responses of the scientists "highly critical, even petulant" and says

114

that Lomborg's own "occasionally tendentious examination of sources pales in comparison with the errors of Lomborg's targets." In "Green With Ideology," *Reason* (May 2002), Ronald Bailey argues that the scientists' responses amply demonstrate the truth of his point that environmentalism is an ideology. The debate between Lomborg and his opponents failed to affect planning for the Johannesburg World Summit on Sustainable Development, which is premised on the belief that humanity needs to adjust its behavior if it is to continue to thrive on earth.

Environmental Defense

Environmental Defense (formerly the Environmental Defense Fund) "is dedicated to protecting the environmental rights of all people, including future generations. Guided by science, Environmental Defense evaluates environmental problems and works to create and advocate solutions that win lasting political, economic and social support because they are nonpartisan, cost-efficient and fair."

http://www.environmentaldefense.org/home.cfm

National Wilderness Preservation System

Operated by representatives of the Arthur Carhart National Wilderness Training Center, the Aldo Leopold Wilderness Research Institute, and the Wilderness Institute at the University of Montana's School of Forestry, the National Wilderness Preservation System provides information, news, and Internet links related to wilderness. It includes a database of information on all of the 644 wilderness areas that make up the system.

http://www.wilderness.net/nwps/default.cfm

Intergovernmental Panel on Climate Change

The Intergovernmental Panel on Climate Change (IPCC) was formed by the World Meteorological Organization (WMO) and the United Nations Environment Programme (UNEP) to assess the scientific, technical, and socioeconomic information relevant to understanding the risks of human-induced climate change.

http://www.ipcc.ch

Agriculture Network Information Center

The Agriculture Network Information Center (AgNIC) is a guide to quality agricultural (including biotechnologal) information on the Internet, as selected by the National Agricultural Library, Land-Grant Universities, and other institutions.

http://www.agnic.org

Environmental Estrogens and Other Hormones

This Web site, maintained by the Center for Bioenvironmental Research at Tulane and Xavier Universities in New Orleans, provides accurate, timely information on environmental hormones and their impacts.

http://www.som.tulane.edu/cbr/ecme/eehome/default.html

PART 2

Environmental Impacts

*A*great many of today's environmental issues have to do with industrial development, which expanded greatly during the twentieth century.
Just since World War II, many thousands of synthetic chemicals—such as
pesticides, plastics, and antibiotics—have flooded the environment. We
have become dependent on the production and use of energy, particularly
fossil fuels. Air and water pollution have become global problems. And we
have discovered that our actions may change the world for generations
to come.

Some blame the increasing environmental problems on human
greed. It is perhaps more realistic to say that people generally want a
better standard of living. For most of the world's people, this does not
mean air conditioning or an SUV but rather clean water, a reliable food
supply, health, a job, education, electric light, indoor plumbing—all the
things that the industrialized West tends to take for granted. Can such
basics be extended to everyone on earth? Can standards of living for all
continue to improve beyond that basic level? Such questions are at the
heart of many environmental concerns.

- Should the Arctic National Wildlife Refuge Be Opened to Oil
  Drilling?

- Should DDT Be Banned Worldwide?

- Is Genetic Engineering an Environmentally Sound Way to
  Increase Food Production?

- Do Environmental Hormone Mimics Pose a Potentially Serious
  Health Threat?

- Is the Environmental Protection Agency's Decision to Tighten Air
  Quality Standards for Ozone and Particulates Justified?

- Do Human Activities Threaten to Change the Global Climate?

ISSUE 7

Should the Arctic National Wildlife Refuge Be Opened to Oil Drilling?

YES: Dwight R. Lee, from "To Drill or Not to Drill: Let the Environmentalists Decide," *The Independent Review* (Fall 2001)

NO: Amory B. Lovins and L. Hunter Lovins, from "Fool's Gold in Alaska," *Foreign Affairs* (July/August 2001)

ISSUE SUMMARY

YES: Professor of economics Dwight R. Lee argues that the economic and other benefits of Arctic National Wildlife Refuge (ANWR) oil are so great that even environmentalists should agree to permit drilling—and they probably would if they stood to benefit directly.

NO: Physicist Amory B. Lovins and lawyer L. Hunter Lovins assert that recovering ANWR oil is too costly and too vulnerable to disruption. They hold that alternatives such as developing greater fuel efficiency are wiser choices for meeting future energy needs.

T he birth of environmental consciousness in the United States was marked by two strong, opposing views. Late in the nineteenth century, John Muir (1838–1914) called for the preservation of natural wilderness, untouched by human activities. At about the same time, Gifford Pinchot (1865–1946) became a strong voice for conservation (not to be confused with preservation; Gifford's conservation allowed the use of nature but in such a way that it was not destroyed; his aim was "the greatest good of the greatest number in the long run"). Both views agree that nature has value; however, they disagree on the form of that value. The preservationist says that nature has value in its own right and has a right to be left alone, neither developed with houses and roads nor exploited with farms, dams, mines, and oil wells. The conservationist says that nature's value lies chiefly in the benefits it provides to human beings.

The first national parks date back to the 1870s. Parks and the national forests are managed for "multiple use" on the premise that wildlife protection, recreation, timber cutting, and even oil drilling and mining can coexist. The first "primitive areas," where all development is barred, were created by the U.S.

Forest Service in the 1920s. However, pressure from commercial interests (the timber and mining industries, among others) led to the reclassification of many such areas and their opening to exploitation. In 1964 the Federal Wilderness Act provided a mechanism for designating "wilderness" areas, defined as areas "where the earth and its community of life are untrammeled by man, where man himself is a visitor who does not remain." Since then it has become clear that pesticides and other man-made chemicals are found everywhere on earth, drifting on winds and ocean currents and traveling in migrant birds even to areas without obvious human presence. Humans might not be present in these places, but their effects are. And commercial interests are just as interested in the wealth that may be extracted from these areas as they ever were. There is continual pressure to expand commercial use of national forests and parks and to open wilderness areas to exploitation.

The Arctic National Wildlife Refuge (ANWR) provides a good illustration. It is not a "wilderness" area, for it was designated a wildlife preserve in 1960 and enlarged and renamed in 1980 with the proviso that its coastal plain be evaluated for its potential value in terms of oil and gas production. In 1987 the Department of the Interior recommended that the coastal plain be opened for oil and gas exploration. In 1995 Congress approved doing so, but President Bill Clinton vetoed the legislation. In 2001, after California experienced electrical blackouts, President George W. Bush declared that opening the ANWR to oil exploitation was essential to national energy security. The National Energy Security Act of 2001 was promptly put before Congress and debated vigorously. Industry representatives favored the bill; environmental groups opposed it. For a good review, see Norman Chance, "The Arctic National Wildlife Refuge: A Special Report" at http://arcticcircle.uconn.edu/ANWR/anwrindex.html.

Strict preservationists still remain, but the debate over protecting wilderness areas generally centers on economic arguments. In the following selections, Dwight R. Lee argues that the economic and other benefits of Arctic National Wildlife Refuge oil are so great that drilling should be permitted. Amory B. Lovins and L. Hunter Lovins argue that ANWR oil should not be developed, because it is too limited in quantity to relieve energy needs for long, it is too costly to exploit compared to alternative oil sources, and its delivery system is too vulnerable to disruption.

Dwight R. Lee

 **YES**

To Drill or Not to Drill

High prices of gasoline and heating oil have made drilling for oil in Alaska's Arctic National Wildlife Refuge (ANWR) an important issue. ANWR is the largest of Alaska's sixteen national wildlife refuges, containing 19.6 million acres. It also contains significant deposits of petroleum. The question is, Should oil companies be allowed to drill for that petroleum?

The case for drilling is straightforward. Alaskan oil would help to reduce U.S. dependence on foreign sources subject to disruptions caused by the volatile politics of the Middle East. Also, most of the infrastructure necessary for transporting the oil from nearby Prudhoe Bay to major U.S. markets is already in place. Furthermore, because of the experience gained at Prudhoe Bay, much has already been learned about how to mitigate the risks of recovering oil in the Arctic environment.

No one denies the environmental risks of drilling for oil in ANWR. No matter how careful the oil companies are, accidents that damage the environment at least temporarily might happen. Environmental groups consider such risks unacceptable; they argue that the value of the wilderness and natural beauty that would be spoiled by drilling in ANWR far exceeds the value of the oil that would be recovered. For example, the National Audubon Society characterizes opening ANWR to oil drilling as a threat "that will destroy the integrity" of the refuge (see statement at www.audubon.org/campaign/refuge).

So, which is more valuable, drilling for oil in ANWR or protecting it as an untouched wilderness and wildlife refuge? Are the benefits of the additional oil really less than the costs of bearing the environmental risks of recovering that oil? Obviously, answering this question with great confidence is difficult because the answer depends on subjective values. Just how do we compare the convenience value of using more petroleum with the almost spiritual value of maintaining the "integrity" of a remote and pristine wilderness area? Although such comparisons are difficult, we should recognize that they can be made. Indeed, we make them all the time.

We constantly make decisions that sacrifice environmental values for what many consider more mundane values, such as comfort, convenience, and material well-being. There is nothing wrong with making such sacrifices because up to some point the additional benefits we realize from sacrificing a little more

From Dwight R. Lee, "To Drill or Not to Drill: Let the Environmentalists Decide," *The Independent Review*, vol. 6, no. 2 (Fall 2001). Copyright © 2001 by The Independent Institute, Oakland, CA. Reprinted by permission of *The Independent Review*, a journal of political economy. Notes and references omitted.

environmental "integrity" are worth more than the necessary sacrifice. Ideally, we would somehow acquire the information necessary to determine where that point is and then motivate people with different perspectives and preferences to respond appropriately to that information.

Achieving this ideal is not as utopian as it might seem; in fact, such an achievement has been reached in situations very similar to the one at issue in ANWR. In this article, I discuss cases in which the appropriate sacrifice of wilderness protection for petroleum production has been responsibly determined and harmoniously implemented. Based on this discussion, I conclude that we should let the Audubon Society decide whether to allow drilling in ANWR. That conclusion may seem to recommend a foregone decision on the issue because the society has already said that drilling for oil in ANWR is unacceptable. But actions speak louder than words, and under certain conditions I am willing to accept the actions of environmental groups such as the Audubon Society as the best evidence of how they truly prefer to answer the question, To drill or not to drill in ANWR?

Private Property Changes One's Perspective

What a difference private property makes when it comes to managing multiuse resources. When people make decisions about the use of property they own, they take into account many more alternatives than they do when advocating decisions about the use of property owned by others. This straightforward principle explains why environmental groups' statements about oil drilling in ANWR (and in other publicly owned areas) and their actions in wildlife areas they own are two very different things.

For example, the Audubon Society owns the Rainey Wildlife Sanctuary, a 26,000-acre preserve in Louisiana that provides a home for fish, shrimp, crab, deer, ducks, and wading birds, and is a resting and feeding stopover for more than 100,000 migrating snow geese each year. By all accounts, it is a beautiful wilderness area and provides exactly the type of wildlife habitat that the Audubon Society seeks to preserve. But, as elsewhere in our world of scarcity, the use of the Rainey Sanctuary as a wildlife preserve competes with other valuable uses.

Besides being ideally suited for wildlife, the sanctuary contains commercially valuable reserves of natural gas and oil, which attracted the attention of energy companies when they were discovered in the 1940s. Clearly, the interests served by fossil fuels do not have high priority for the Audubon Society. No doubt, the society regards additional petroleum use as a social problem rather than a social benefit. Of course, most people have different priorities: they place a much higher value on keeping down the cost of energy than they do on bird-watching and on protecting what many regard as little more than mosquito-breeding swamps. One might suppose that members of the Audubon Society have no reason to consider such "anti-environmental" values when deciding how to use their own land. Because the society owns the Rainey Sanctuary, it can ignore interests antithetical to its own and refuse to allow drilling.

Yet, precisely because the society owns the land, it has been willing to accommodate the interests of those whose priorities are different and has allowed thirty-seven wells to pump gas and oil from the Rainey Sanctuary. In return, it has received royalties of more than $25 million.

One should not conclude that the Audubon Society has acted hypocritically by putting crass monetary considerations above its stated concerns for protecting wilderness and wildlife. In a wider context, one sees that because of its ownership of the Rainey Sanctuary, the Audubon Society is part of an extensive network of market communication and cooperation that allows it to do a better job of promoting its objectives by helping others promote theirs. Consumers communicate the value they receive from additional gas and oil to petroleum companies through the prices they willingly pay for those products, and this communication is transmitted to owners of oil-producing land through the prices the companies are willing to pay to drill on that land. Money really does "talk" when it takes the form of market prices. The money offered for drilling rights in the Rainey Sanctuary can be viewed as the most effective way for millions of people to tell the Audubon Society how much they value the gas and oil its property can provide.

By responding to the price communication from consumers and by allowing the drilling, the Audubon Society has not sacrificed its environmental values in some debased lust for lucre. Instead, allowing the drilling has served to reaffirm and promote those values in a way that helps others, many of whom have different values, achieve their own purposes. Because of private ownership, the valuations of others for the oil and gas in the Rainey Sanctuary create an opportunity for the Audubon Society to purchase additional sanctuaries to be preserved as habitats for the wildlife it values. So the society has a strong incentive to consider the benefits as well as the costs of drilling on its property. Certainly, environmental risks exist, and the society considers them, but if also responsibly weighs the costs of those risks against the benefits as measured by the income derived from drilling. Obviously, the Audubon Society appraises the benefits from drilling as greater than the costs, and it acts in accordance with that appraisal.

Cooperation Between Bird-Watchers and Hot-Rodders

The advantage of private ownership is not just that it allows people with different interests to interact in mutually beneficial ways. It also creates harmony between those whose interests would otherwise be antagonistic. For example, most members of the Audubon Society surely see the large sport utility vehicles and high-powered cars encouraged by abundant petroleum supplies as environmentally harmful. That perception, along with the environmental risks associated with oil recovery, helps explain why the Audubon Society vehemently opposes drilling for oil in the ANWR as well as in the continental shelves in the Atlantic, the Pacific, and the Gulf of Mexico. Although oil companies promise to take extraordinary precautions to prevent oil spills when drilling in these areas, the Audubon Society's position is no off-shore drilling, none. One might expect to find Audubon Society members completely unsympathetic with hot-rodding enthusiasts, NASCAR racing fans, and drivers of Chevy Suburbans. Yet, as we have seen, by allowing drilling for gas and oil in the Rainey Sanctuary, the society is accommodating the interests of those with gas-guzzling lifestyles, risking the "integrity" of its prized wildlife sanctuary to make more gasoline available to those whose energy consumption it verbally condemns as excessive.

The incentives provided by private property and market prices not only motivate the Audubon Society to cooperate with NASCAR racing fans, but also motivate those racing enthusiasts to cooperate with the Audubon Society. Imagine the reaction you would get if you went to a stock-car race and tried to convince the spectators to skip the race and go bird-watching instead. Be prepared for some beer bottles tossed your way. Yet by purchasing tickets to their favorite sport, racing fans contribute to the purchase of gasoline that allows the Audubon Society to obtain additional wildlife habitat and to promote bird-watching. Many members of the Audubon Society may feel contempt for racing fans, and most racing fans may laugh at bird-watchers, but because of private property and market prices, they nevertheless act to promote one another's interests.

The Audubon Society is not the only environmental group that, because of the incentives of private ownership, promotes its environmental objectives by serving the interests of those with different objectives. The Nature Conservancy accepts land and monetary contributions for the purpose of maintaining natural areas for wildlife habitat and ecological preservation. It currently owns thousands of acres and has a well-deserved reputation for preventing development in environmentally sensitive areas. Because it owns the land, it has also a strong incentive to use that land wisely to achieve its objectives, which sometimes means recognizing the value of developing the land.

For example, soon after the Wisconsin chapter received title to 40 acres of beach-front land on St. Croix in the Virgin Islands, it was offered a much larger parcel of land in northern Wisconsin in exchange for its beach land. The Wisconsin chapter made this trade (with some covenants on development of the beach land) because owning the Wisconsin land allowed it to protect an entire watershed containing endangered plants that it considered of greater environmental value than what was sacrificed by allowing the beach to be developed.

Thanks to a gift from the Mobil Oil Company, the Nature Conservancy of Texas owns the Galveston Bay Prairie Preserve in Texas City, a 2,263-acre refuge that is home to the Attwater's prairie chicken, a highly endangered species (once numbering almost a million, its population had fallen to fewer than ten by the early 1990s). The conservancy has entered into an agreement with Galveston Bay Resources of Houston and Aspects Resources, LLC, of Denver to drill for oil and natural gas in the preserve. Clearly some risks attend oil drilling in the habitat of a fragile endangered species, and the conservancy has considered them, but it considers the gains sufficient to justify bearing the risks. According to Ray Johnson, East County program manager for the Nature Conservancy of Texas. "We believe this could provide a tremendous opportunity to raise funds to acquire additional habitat for the Attwater's prairie chicken, one of the most threatened birds in North America." Obviously the primary concern is to protect the endangered species, but the demand for gas and oil is helping achieve that objective. Johnson is quick to point out, "We have taken every precaution to minimize the impact of the drilling on the prairie chickens and to ensure their continued health and safety."

Back to ANWR

Without private ownership, the incentive to take a balanced and accommodating view toward competing land-use values disappears. So, it is hardly surprising that the Audubon Society and other major environmental groups categorically oppose drilling in ANWR. Because ANWR is publicly owned, the environmental groups have no incentive to take into account the benefits of drilling. The Audubon Society does not capture any of the benefits if drilling is allowed, as it does at the Rainey Sanctuary; in ANWR, it sacrifices nothing if drilling is prevented. In opposing drilling in ANWR, despite the fact that the precautions to be taken there would be greater than those required of companies operating in the Rainey Sanctuary, the Audubon Society is completely unaccountable for the sacrificed value of the recoverable petroleum.

Obviously, my recommendation to "let the environmentalists decide" whether to allow oil to be recovered from ANWR makes no sense if they are not accountable for any of the costs (sacrificed benefits) of preventing drilling. I am confident, however, that environmentalists would immediately see the advantages of drilling in ANWR if they were responsible for both the costs and the benefits of that drilling. As a thought experiment about how incentives work, imagine that a consortium of environmental organizations is given veto power over drilling, but is also given a portion (say, 10 percent) of what energy companies are willing to pay for the right to recover oil in ANWR. These organizations could capture tens of millions of dollars by giving their permission to drill. Suddenly the opportunity to realize important environmental objectives by favorably considering the benefits others gain from more energy consumption would come into sharp focus. The environmentalists might easily conclude that although ANWR is an "environmental treasure," other environmental treasures in other parts of the country (or the world) are even more valuable; moreover, with just a portion of the petroleum value of the ANWR, efforts might be made to reduce the risks to other natural habitats, more than compensating for the risks to the Arctic wilderness associated with recovering that value.

Some people who are deeply concerned with protecting the environment see the concentration on "saving" ANWR from any development as misguided even without a vested claim on the oil wealth it contains. For example, according to Craig Medred, the outdoor writer for the *Anchorage Daily News* and a self-described "development-phobic wilderness lover,"

> That people would fight to keep the scar of clearcut logging from the spectacular and productive rain-forests of Southeast Alaska is easily understandable to a shopper in Seattle or a farmer in Nebraska. That people would argue against sinking a few holes through the surface of a frozen wasteland,

however, can prove more than a little baffling even to development-phobic, wilderness lovers like me. Truth be known, I'd trade the preservation rights to any 100 acres on the [ANWR] slope for similar rights to any acre of central California wetlands.... It would seem of far more environmental concern that Alaska's ducks and geese have a place to winter in overcrowded, overdeveloped California than that California's ducks and geese have a place to breed each summer in uncrowded and undeveloped Alaska.

— (1996, C1)

Even a small share of the petroleum wealth in ANWR would dramatically reverse the trade-off Medred is willing to make because it would allow environmental groups to afford easily a hundred acres of central California wetlands in exchange for what they would receive for each acre of ANWR released to drilling.

We need not agree with Medred's characterization of the ANWR as "a frozen wasteland" to suspect that environmentalists are overstating the environmental amenities that drilling would put at risk. With the incentives provided by private property, environmental groups would quickly reevaluate the costs of drilling in wilderness refuges and soften their rhetoric about how drilling would "destroy the integrity" of these places. Such hyperbolic rhetoric is to be expected when drilling is being considered on public land because environmentalists can go to the bank with it. It is easier to get contributions by depicting decisions about oil drilling on public land as righteous crusades against evil corporations out to destroy our priceless environment for short-run profit than it is to work toward minimizing drilling costs to accommodate better the interests of others. Environmentalists are concerned about protecting wildlife and wilderness areas in which they have ownership interest, but the debate over any threat from drilling and development in those areas is far more productive and less acrimonious than in the case of ANWR and other publicly owned wilderness areas.

The evidence is overwhelming that the risks of oil drilling to the arctic environment are far less than commonly claimed. The experience gained in Prudhoe Bay has both demonstrated and increased the oil companies' ability to recover oil while leaving a "light footprint" on arctic tundra and wildlife. Oil-recovery operations are now sited on gravel pads providing foundations that protect the underlying permafrost. Instead of using pits to contain the residual mud and other waste from drilling, techniques are now available for pumping the waste back into the well in ways that help maintain well pressure and reduce the risks of spills on the tundra. Improvements in arctic road construction have eliminated the need for the gravel access roads used in the development of the Prudhoe Bay oil fields. Roads are now made from ocean water pumped onto the tundra, where it freezes to form a road surface. Such roads melt without a trace during the short summers. The oversize rubber tires used on the roads further minimize any impact on the land.

Improvements in technology now permit horizontal drilling to recover oil that is far from directly below the wellhead. This technique reduces further the already small amount of land directly affected by drilling operations. Of the more than 19 million acres contained in ANWR, almost 18 million acres have been set aside by Congress—somewhat more than 8 million as wilderness and 9.5 million as wildlife refuge. Oil companies estimate that only 2,000 acres would be needed to develop the coastal plain.

This carefully conducted and closely confined activity hardly sounds like a sufficient threat to justify the rhetoric of a righteous crusade to prevent the destruction of ANWR, so the environmentalists warn of a detrimental effect on arctic wildlife that cannot be gauged by the limited acreage directly affected. Given the experience at Prudhoe Bay, however, such warnings are difficult to take seriously. The oil companies have gone to great lengths and spent tens of millions of dollars to reduce any harm to the fish, fowl, and mammals that live and breed on Alaska's North Slope. The protections they have provided for wildlife at Prudhoe Bay have been every bit as serious and effective as those the Audubon Society and the Nature Conservancy find acceptable in the Rainey Sanctuary and the Galveston Bay Prairie Preserve. As the numbers of various wildlife species show, many have thrived better since the drilling than they did before.

Before drilling began at Prudhoe Bay, a good deal of concern was expressed about its effect on caribou herds. As with many wildlife species, the population of the caribou on Alaska's North Slope fluctuates (often substantially) from year to year for completely natural reasons, so it is difficult to determine with confidence the effect of development on the caribou population. It is noteworthy, however, that the caribou population in the area around Prudhoe Bay has increased greatly since that oil field was developed, from approximately 3,000 to a high of some 23,400. . . . Some argue that the increase has occurred because the caribou's natural predators have avoided the area—some of these predators are shot, whereas the caribou are not. But even if this argument explains some or even all of the increase in the population, the increase still casts doubt on claims that the drilling threatens the caribou. Nor has it been shown that the viability of any other species has been genuinely threatened by oil drilling at Prudhoe Bay.

Caribou Versus Humans

Although consistency in government policy may be too much to hope for, it is interesting to contrast the federal government's refusal to open ANWR with some of its other oil-related policies. While opposing drilling in ANWR, ostensibly because we should not put caribou and other Alaskan wildlife at risk for the sake of getting more petroleum, we are exposing humans to far greater risks because of federal policies motivated by concern over petroleum supplies.

For example, the United States maintains a military presence in the Middle East in large part because of the petroleum reserves there. It is doubtful that the U.S. government would have mounted a large military action and sacrificed American lives to prevent Iraq from taking over the tiny sheikdom of Kuwait except to allay the threat to a major oil supplier. Nor would the United States have lost the nineteen military personnel in the barracks blown up in Saudi Arabia in 1996 or the seventeen killed onboard the USS Cole in a Yemeni harbor in 2000. I am not arguing against maintaining a military presence in the Middle East, but if it is worthwhile to sacrifice Americans' lives to protect oil supplies in the Middle East, is it not worthwhile to take a small (perhaps nonexistent) risk of sacrificing the lives of a few caribou to recover oil in Alaska?

Domestic energy policy also entails the sacrifice of human lives for oil. To save gasoline, the federal government imposes Corporate Average Fuel Economy (CAFE) standards on automobile producers. These standards now require all new cars to average 27.5 miles per gallon and new light trucks to average 20.5 miles per gallon. The one thing that is not controversial about the CAFE standards is that they cost lives by inducing manufacturers to reduce the weight of vehicles. Even Ralph Nader has acknowledged that "larger cars are safer—there is more bulk to protect the occupant." An interesting question is, How many lives might be saved by using more (ANWR) oil and driving heavier cars rather than using less oil and driving lighter, more dangerous cars?

It has been estimated that increasing the average weight of passenger cars by 100 pounds would reduce U.S. highway fatalities by 200 a year. By determining how much additional gas would be consumed each year if all passenger cars were 100 pounds heavier, and then estimating how much gas might be recovered from ANWR oil, we can arrive at a rough estimate of how many human lives potentially might be saved by that oil. To make this estimate, I first used data for the technical specifications of fifty-four randomly selected 2001 model passenger cars to obtain a simple regression of car weight on miles per gallon. This regression equation indicates that every additional 100 pounds decreases mileage by 0.85 miles per gallon. So 200 lives a year could be saved by relaxing the CAFE standards to allow a 0.85 miles per gallon reduction in the average mileage of passenger cars. How much gasoline would be required to compensate for this decrease of average mileage? Some 135 million passenger cars are currently in use, being driven roughly 10,000 miles per year on average (1994–95 data from U.S. Bureau of the Census 1997, 843). Assuming these vehicles travel 24 miles per gallon on average, the annual consumption of gasoline by passenger cars is 56.25 billion gallons (= 135 million × 10,000/24). If instead of an average of 24 miles per gallon the average were reduced to 23.15 miles per gallon, the annual consumption of gasoline by passenger cars would be 58.32 billion gallons (= 135 million x 10,000/23.15). So, 200 lives could be saved annually by an extra 2.07 billion gallons of gas. It is estimated that ANWR contains from 3 to 16 billion barrels of recoverable petroleum. Let us take the midpoint in this estimated range, or 9.5 billion barrels. Given that on average each barrel of petroleum is refined into 19.5 gallons of gasoline, the ANWR oil could be turned into 185.25 billion additional gallons of gas, or enough to save 200 lives a year for almost ninety years (185.25/2.07 = 89.5). Hence, in total almost 18,000 lives could be saved by opening up ANWR to drilling and using the fuel made available to compensate for increasing the weight of passenger cars.

I claim no great precision for this estimate. There may be less petroleum in ANWR than the midpoint estimate indicates, and the study I have relied on may have overestimated the number of lives saved by heavier passenger cars. Still, any reasonable estimate will lead to the conclusion that preventing the recovery of ANWR oil and its use in heavier passenger cars entails the loss of thousands of lives on the highways. Are we willing to bear such a cost in order to avoid the risks, if any, to ANWR and its caribou?

Conclusion

I am not recommending that ANWR actually be given to some consortium of environmental groups. In thinking about whether to drill for oil in ANWR, however, it is instructive to consider seriously what such a group would do if it owned ANWR and therefore bore the costs as well as enjoyed the benefits of preventing drilling. Those costs are measured by what people are willing to pay for the additional comfort, convenience, and safety that could be derived from the use of ANWR oil. Unfortunately, without the price communication that is possible only by means of private property and voluntary exchange, we cannot be sure what those costs are or how private owners would evaluate either the costs or the benefits of preventing drilling in ANWR. However, the willingness of environmental groups such as the Audubon Society and the Nature Conservancy to allow drilling for oil an environmentally sensitive land they own suggests strongly that their adamant verbal opposition to drilling in ANWR is a poor reflection of what they would do if they owned even a small fraction of the ANWR territory containing oil.

NO

**Amory B. Lovins and
L. Hunter Lovins**

Fool's Gold in Alaska

The Bottom of the Barrel?

Oil prices have fluctuated randomly for well over a century. Heedless of this fact, oil's promoters are always offering opportunities that could make money —but on the flawed assumption that high prices will prevail. Leading the field of these optimists are Alaskan politicians. Eager to keep funding their state's de facto negative income tax—oil provides 80 percent of the state's unrestricted general revenue—they have used every major rise in oil prices since 1973 to advocate drilling beneath federal lands on the coastal plain of the Arctic National Wildlife Refuge. Just as predictably, environmentalists counter that the refuge is the crown jewel of the American wilderness and home to the threatened indigenous Gwich'in people. As some see it, drilling could raise human rights issues under international law. Canada, which shares threatened wildlife, also opposes drilling.

Both sides of this debate have largely overlooked the central question: Does drilling for oil in the refuge's coastal plain make sense for economic and security reasons? After all, three imperatives should shape a national energy policy: economic vitality, secure supplies, and environmental quality. To merit serious consideration, a proposal must meet at least one of these goals.

Drilling proponents claim that prospecting for refuge oil will enhance the first two while not unduly harming the third. In fact, not only does refuge oil fail to meet any of the three goals, it could even compromise the first two. First, the refuge is unlikely to hold economically recoverable oil. And even if it did, exploitation would only briefly reduce U.S. dependence on imported oil by just a few percentage points, starting in about a decade. Nor would the refuge yield significant natural gas. Despite some recent statements by the Bush administration, the North Slope's important natural-gas deposits are almost entirely outside the refuge. The gas-rich areas are already open to industry, and environmentalists would likely support a gas pipeline there, but its high cost— an estimated $10 billion—would make it seem uneconomical.

Furthermore, those who suppose that any domestic oil is more secure than imported oil should remember that oil reserves almost anywhere else on earth are more accessible and more reliably deliverable than those above the Arctic

Circle. Importing oil in tankers from the highly diversified world market is arguably better for energy security than delivering refuge oil to other U.S. states through one vulnerable conduit, the Trans-Alaska Pipeline System [TAPS]. Although proponents argue that exploiting refuge oil would make better use of TAPS (which is all paid for but only half-full), that pipeline is easy to disrupt and difficult to repair. More than half of it is elevated and indefensible; in fact, it has already been bombed twice. If one of its vital pumping stations were attacked in the winter, its nine million barrels of hot oil could congeal into the world's largest Chapstick. Nor has the 24-year-old TAPS aged gracefully: premature and accelerated corrosion, erosion, and stress are raising maintenance costs. [In 2000], the pipeline suffered two troubling accidents plus another that almost blew up the Valdez oil terminal. If TAPS were to start transporting refuge oil, it would start only around the end of its originally expected lifetime. That one fragile link, soon to be geriatric, would then bring as much oil to U.S. refineries as now flows through the Strait of Hormuz—a chokepoint that is harder to disrupt, is easier to fix, and has alternative routes.

Available and proven technological alternatives that use energy more productively can meet all three goals of energy policy with far greater effectiveness, speed, profit, and security than can drilling in the refuge. The untapped, inexpensive "reserves" of oil-efficiency technology exceed by more than 50 times the average projection of what refuge drilling might yield. The existence of such alternatives makes drilling even more economically risky.

In sum, even if drilling in the Arctic Wildlife Refuge posed no environmental or human rights concerns, it still could not be justified on economic or security grounds. These reasons remain as compelling as they were 14 years ago, when drilling there was last rejected, and they are likely to strengthen further with technological advances. Comparing all realistic ways to meet the goals of national energy policy suggests a simple conclusion: refuge oil is unnecessary, insecure, a poor business risk, and a distraction from a sound national debate over realistic energy priorities. If that debate is informed by the past quarter-century's experience of what works, a strong energy policy will seek the lowest-cost mix of demand- and supply-side investments that compete fairly at honest prices. It will not pick winners, bail out losers, substitute central planning for market forces, or forecast demand and then plan capacity to meet it. Instead, it will treat demand as a choice, not fate. If consumers can choose optimal levels of efficiency, demand can remain stable (as oil demand did during 1975–91) or even decline—and it will be possible to provide secure, safe, and clean energy services at the lowest cost. In this market-driven world, the time for costly refuge oil has passed.

Doing More With Less

Unstable oil prices have historically triggered the new energy strategies. In the years following the oil-price jump in 1973, Presidents Richard Nixon and Gerald Ford sought to reduce U.S. dependence on oil imports by stimulating domestic energy supplies. With the country beset by inflation, however, they also controlled oil and gas prices, so the new supplies often appeared cheaper than they really were. President Jimmy Carter repeated this supply mistake by promoting a costly flop in synthetic fuels, but he also trusted the market enough to deregulate oil and gas prices. (Paradoxically, he discouraged exploration for natural gas by prohibiting its use in most new power plants.) The fall of the shah of Iran again hiked oil prices in 1979 and contributed to Carter's political demise. Yet that second shock also stimulated a nationwide, seven-year drive for greater energy efficiency. Cheaper ways of delivering "energy services" (e.g., hot showers and cold beer) by using energy more productively left the energy-supply industries with costly surpluses as their prices collapsed in 1985–86. This crash benefited consumers but punished the same energy producers that the Reagan administration had sought to help. Underlying this energy glut was not just a response to higher prices but a basic policy shift: Carter had emphasized the efficient use of energy, especially in cars, and Americans then discovered how quickly demand-side policies can swing the global oil market.

Greater efficiency bore dramatic results. Carter's policies made new American-built cars more efficient by seven miles per gallon (mpg) over six years. During Carter's term and the five years following it, oil imports from the unstable Persian Gulf region fell by 87 percent. From 1977 to 1985, U.S. GDP [gross domestic product] rose 27 percent while total U.S. oil imports fell by 42 percent, or 3.74 million barrels a day. That savings took away from the Organization of Petroleum Exporting Countries [OPEC] an eighth of its market. The entire world oil market shrank by a tenth; OPEC's share of it was slashed from 52 percent to 30 percent, while OPEC's output fell by 48 percent. The United States accounted for one-fourth of that reduction. More-efficient cars —each driving one percent fewer miles on 20 percent fewer gallons—were the most important cause; 96 percent of those savings came from smarter design, whereas 4 percent came from smaller size. Other countries also improved car efficiency, but they used higher fuel taxes instead of higher efficiency standards to do so.

In those eight years, U.S. oil productivity soared by 52 percent, demonstrating an effective new source of energy security and a potent weapon against high oil prices and supply manipulations. The United States showed that a major nation could respond to supply disruptions by focusing on the demand side and boosting its energy productivity at will. It could thereby exercise more market power than suppliers, beat down prices, and enhance the relative importance of less vulnerable, more diversified sources of energy.

Drilling proponents today ignore this lesson. Instead, they cite the imperative of displacing Middle East oil to justify drilling in every U.S. site where oil might occur. But even if this imperative existed, refuge oil would be a poor solution. After a decade of drilling and preparation, it could provide only modest, brief relief—totaling less than one percent of projected U.S. oil needs —and would cost much more than the efficiency-boosting alternatives. Repaying refuge-oil investments would require oil prices so high that, in the ensuing decade, they would elicit far greater efficiency. Those efficiency gains, in turn, would depress oil prices, displace the targeted imports, and make refuge oil unnecessary. That was what happened in the mid-1980s; repeating the same experiment will yield the same result. . . .

Oil Roulette

The refuge is one of the planet's most inhospitable and remote locations. For oil companies to invest profitably there, it must hold a lot of oil. Furthermore, world oil prices must stay high enough for a long enough time to recover costs and earn profits. But even official proponents of drilling have found its economics dubious.

In 1998, the U.S. Geological Survey (USGS) found that better (and fourfold cheaper) production technologies could probably draw 3.2 billion barrels from the refuge. This oil would be worth recovering only if its long-term price were at least $22 per barrel in West Coast ports (the destinations that the USGS picked for its price calculations). But until it spiked up from $13 per barrel in 1998 to $30 per barrel in late 2000, Alaskan oil did not exceed that level for 8 years. That spike was a blip, not a trend. In April 2001, Alaska's Department of Revenue forecast a steady price drop from $22 per barrel in 2001–2 to less than $13 per barrel in 2009–10—the earliest that any refuge oil might flow. Alaska's latest price forecast for 2020 is $18 per barrel. The U.S. Department of Energy predicts that world oil prices will not reach $23 per barrel until 2020; nearly all industry forecasts are lower.

But it is no longer necessary to speculate which forecast is correct; they all tend to converge on the prices discovered in the futures market. Alaska's forecasters agree that this convergence is unaffected by price spikes such as the one in 2000. Their projection for 2004–10 accordingly stays under $16 per barrel. (One of the world's largest oil companies does not even consider any prospect requiring a delivered price of more than $14 per barrel.) According to the USCG, that price is also the threshold below which there is probably no economically recoverable oil beneath the refuge. Even that threshold may be too high; volatile oil prices make drilling especially risky, requiring higher returns and prices in any high-cost area where exploration and development will be slow and difficult. And if the federal government were to demand lease fees, such as the multi-billion-dollar revenues that the Alaskan delegation inserts into budget bills, or if TAPS needed more maintenance, the price threshold would rise.

Some drilling advocates argue that technological advances in finding and extracting oil can still make refuge oil profitable. Those advances are indeed real and astoundingly rapid. From 1987 to 1999, they increased the discovery of new U.S. oil resources by an estimated three-fifths. One-ninth of all U.S. oil reserves discovered since 1859 were found just in the past decade, even as oil prices fell. Better technologies could make extracting refuge oil cheaper—but those same advances would also cut costs everywhere else, and just about anywhere else is easier and more attractive. Better technology makes global oil more plentiful and therefore cheaper, so it renders high-cost areas less competitive. During the 1990s, this process combined with increasing competition from energy alternatives to halve long-term forecasts of oil prices, which are still falling. The

Department of Energy now forecasts that imported oil will cost three-fifths less by 2020 than what the Department of the Interior had forecast in 1987, when it predicted prices hitting $61 per barrel. If oil companies really believed in sustained high prices, they would be drilling everywhere—and they are not. On the contrary, when oil prices rose from $10 per barrel to $25 per barrel in 1998–99 and lifted the oil and gas revenues of major U.S. energy companies by more than 50 percent, those firms cut exploration and development outlays by 66 percent in the United States (onshore) and 38 percent worldwide. These companies believe that advancing technology will keep the world long awash in oil that is too cheap for refuge drilling to beat.

Who, then, is pushing for drilling—aside from the powerful Alaskan congressional delegation? Oil-service companies and Alaskan operations offices of major oil companies naturally want to extend and expand their activities and apply their special skills, but they would be risking others' money, not their own. Likewise, the TAPS consortium wants more revenue and a political commitment that might justify a later government bailout if the pipeline turned out to need costly repairs, but it too would not be the one making the huge investment. Conspicuously absent is a ringing endorsement from leaders of major oil firms. They understand the high risk and the prospect of poor rewards, and those that are more astute also fear global consumer boycotts. To the extent that any are interested, it is to seek a bargaining chip for other areas now off-limits or to avoid the social embarrassment of being left off the dance card if the government throws an oil party—not because there is a sound business case.

Finally, the rationale that refuge drilling is urgently needed to relieve U.S. dependence on OPEC oil is full of holes. Net U.S. oil imports have indeed risen past their 1977 peak, but OPEC's share of imports has fallen by one-third. Only a quarter of the oil consumed in the United States now comes from OPEC members. Imports are diversified and come mainly from western hemisphere countries that offer major opportunities for expanding both oil and gas supplies. The more that imports are a concern, however, the stronger the case for substituting not just any option but the cheapest one—slashing America's energy bills by a further $300 billion a year by raising energy productivity.

It's Easy (and Lucrative) Being Green

Oil is becoming more abundant but relatively less important. For each dollar of GDP, the United States used 49 percent less oil in 2000 than it did in 1975. Compared with 1975, the amount that energy efficiency now saves each year is more than five times the country's annual domestic oil production, twelve times its imports from the Persian Gulf, and twice its total oil imports. And the efficiency resource is far from tapped out; instead, it is constantly expanding. It is already far larger and cheaper than anyone had dared imagine.

Increased energy productivity now delivers two-fifths of all U.S. energy services and is also the fastest growing "source." (Abroad, renewable energy supply is growing even faster; it is expected to generate 22 percent of the European Union's electricity by 2010.) Efficient energy use often yields annual after-tax returns of 100 to 200 percent on investment. Its frequent fringe benefits are even more valuable: 6 to 16 percent higher labor productivity in energy-efficient buildings, 40 percent higher retail sales in stores with good natural lighting, and improved output and quality in efficient factories. Efficiency also has major policy advantages. It is here and now, not a decade away. It improves the environment and protects the earth's climate. It is fully secure, already delivered to customers, and immune to foreign potentates and volatile markets. It is rapidly and equitably deployable in the market. It supports jobs all across the United States rather than in a few firms in one state. Yet the energy options now winning in the marketplace seem oddly invisible, unimportant, and disfavored in current national strategy.

Those who have forgotten the power of energy efficiency should remember the painful business lessons learned from the energy policies of the early 1970s and the 1980s. Energy gluts rapidly recur whenever customers pay attention to efficiency—because the nationwide reserve of cheap, qualitatively superior savings from efficient energy use is enormous and largely accessible. That overhang of untapped and unpredictably accessed efficiency presents an opportunity for entrepreneurs and policymakers, but it also poses a risk to costly supply investments. That risk is now swelling ominously.

In the early 1980s, vigorous efforts to boost both supply and efficiency succeeded. Supply rose modestly while efficiency soared. From 1979 to 1986, GDP grew 20 percent while total energy use fell by 5 percent. Improved efficiency provided more than five times as much new energy service as the vaunted expansion of the coal and nuclear industries; domestic oil output rose only 1.5 percent while domestic natural gas output fell 18 percent. When the resulting glut slashed energy prices in 1985–86, attention strayed and efficiency slowed. But just in the past five years, the United States has quietly entered a second golden age of rapidly improving energy efficiency. Now, with another efficiency boom underway, the whole cycle is poised to repeat itself—threatening another energy-policy train wreck with serious economic consequences.

From 1996 to 2000, a complex mix of factors—such as competitive pressures, valuable side benefits, climate concerns, and e-commerce's structural shifts—unexpectedly pushed the pace of U.S. energy savings to nearly an all-time high, averaging 3.1 percent per year despite the record-low and falling energy prices of 1996–99. Meanwhile, investment in energy supply, which is slower to mature, lagged behind demand growth in some regions as the economy boomed. Then in 2000, Middle East political jitters, OPEC machinations, and other factors made world oil prices spike just as cold weather and turbulence in the utility industry coincidentally boosted natural gas prices. Gasoline prices are rising this year—even though crude-oil prices are softening—due to shortages not of crude oil but of refineries and additives. . . .

The higher fuel and electricity prices and occasional local shortages that have vexed many Americans this past year have rekindled a broader national interest in efficient use. The current economic slowdown will further dampen demand but should also heighten business interest in cutting costs. Efficiency also lets numerous actors harness the energy market's dynamism and speed— and it tends to bear results quickly. All these factors could set the stage for another price crash as burgeoning energy savings coincide, then collide, with the new administration's push to stimulate energy supplies. Producers who answer that call will risk shouldering the cost of added supply without the revenue to pay for it, for oil prices high enough to make refuge oil profitable would collapse before or as supply boomed.

Policymakers can avoid such overreaction and instability if they understand the full range of competing options, especially the ability of demand to react faster than supply and the need for balancing investment between them. As outlined above, in the first half of the 1980s, the U.S. economy grew while total energy use fell and oil imports from the Persian Gulf were nearly eliminated. This achievement showed the power of a demand-side national energy policy. Today, new factors—even more powerful technologies and better designs, streamlined delivery methods, and better understanding of how public policy can correct dozens of market failures in buying efficiency—can make the demand-side response even more effective. This can give the Untied States a more affordable and secure portfolio of diverse energy sources, not just a few centralized ones.

A Barrel Saved, a Barrel Earned

If oil were found and profitably extracted from the refuge, its expected peak output would equal for a few years about one percent of the world oil market. Senator Frank Murkowski (R-Alaska) has claimed that merely announcing refuge leasing would bring down world oil prices. Yet even a giant Alaskan discovery several times larger than the refuge would not stabilize world oil markets. Oil prices reached their all-time high, for example, just as such a huge field, in Alaska's Prudhoe Bay, neared its maximum output. Only energy efficiency can stabilize oil prices—as well as sink them. And only a tiny fraction of the vast untapped efficiency gains is needed to do so.

What could the refuge actually produce under optimal conditions? Starting about ten years from now, if oil prices did stay around $22 per barrel, if Congress approved the project, and if the refuge yielded the USGS's mean estimate of about 3.2 billion barrels of profitable oil, the 30-year output would average a modest 292,000 barrels of crude oil a day. (This estimate also assumes that such oil would feed U.S. refineries rather than go to Asian markets, as some Alaskan oil did in 1996–2000.) Once refined, that amount would yield 156,000 barrels of gasoline per day—enough to run 2 percent of American cars and light trucks. That much gasoline could be saved if light vehicles became 0.4 mpg more efficient. Compare that feat to the one achieved in 1979–85, when new light vehicles on average gained 0.4 mpg every 5 months.

Equipping cars with replacement tires as efficient as the original ones would save consumers several "refuges" full of crude oil. Installing superinsulating windows could save even more oil and natural gas while making buildings more comfortable and cheaper to construct. A combination of all the main efficiency options available in 1989 could save today the equivalent of 54 "refuges"—but at a sixth of the cost. New technologies for saving energy are being found faster than the old ones are being used up—just like new technologies for finding and extracting oil, only faster. As gains in energy efficiency continue to outpace oil depletion, oil will probably become uncompetitive even at low prices before it becomes unavailable even at high prices. This is especially likely because the latest efficiency revolution squarely targets oil's main users and its dominant growth market—cars and light trucks—where gasoline savings magnify crude-oil savings by 85 percent.

New American cars are hardly models of fuel efficiency. Their average rating of 24 mpg ties for a 20-year low. The auto industry can do much better—and is now making an effort. Briskly selling hybrid-electric cars such as the Toyota Pruis (a Corolla-class 5-seater) offer 49 mpg, and the Honda Insight (a CRX-class 2-seater) gets 67 mpg. A fleet that efficient, compared to the 24 mpg average, would save 26 or 33 refuges, respectively. General Motors, DaimlerChrysler, and Ford are now testing family sedans that offer 72–80 mpg. For Europeans who prefer subcompact city cars, Volkswagen is selling a 4-seater at 78 mpg and has announced a smaller 2003 model at 235 mpg. Still more efficient cars powered by clean and silent fuel cells are slated for production by at least eight major automakers starting in 2003–5. An uncompromised fuel-cell vehicle—the Hypercar^{SM}—has been designed and costed for production and would achieve 99 mpg; it is as roomy and safe as a midsized sport-utility vehicle but uses 82 percent less fuel and no oil. Such high-efficiency vehicles, which probably can be manufactured at competitive cost, could save globally as much oil as OPEC now sells; when parked, the cars' dual function as plug-in power stations could displace the world's coal and nuclear plants many times over.

As long as the world runs largely on oil, economics dictates a logical priority for displacing it. Efficient use of oil wins hands down on cost, risk, and speed. Costlier options thus incur an opportunity cost. Buying costly refuge oil instead of cheap oil productivity is not simply a bad business decision; it worsens the oil-import problem. Each dollar spent on the costly option of refuge oil could have bought more of the cheap option of efficient use instead. Choosing the expensive option causes more oil to be used and imported than if consumers had bought the efficiency option first. The United States made exactly this mistake when it spent $200 billion on unneeded (but officially encouraged) nuclear and coal plants in the 1970s and 1980s. The United States now imports oil, produces nuclear waste, and risks global climate instability partly because it bought those assets instead of buying far cheaper energy efficiency.

Drilling for refuge oil is a risk the nation should consider taking only if no other choice is possible. But other choices abound. If three or four percent of all U.S. cars were as efficient as today's popular hybrid models, they would save the equivalent of all the refuge's oil. In all, many tens of times more oil is available —sooner, more surely, and more cheaply—from proven energy efficiency. The cheaper, faster energy alternatives now succeeding in the marketplace are safe, clean, climate-friendly, and overwhelmingly supported by the public. Equally important, they remain profitable at any oil price. They offer economic, security, and environmental benefits rather than costs. If any oil is beneath the refuge, its greatest value just might be in holding up the ground beneath the people and animals that live there.

POSTSCRIPT

Should the Arctic National Wildlife Refuge Be Opened to Oil Drilling?

Those who see in nature only values that can be expressed in human terms are well represented by Jonah Goldberg, who, in "Ugh, Wilderness! The Horror of 'ANWR,' the American Elite's Favorite Hellhole," *National Review* (August 6, 2001), describes the ANWR as so bleak and desolate that development can only improve it. On the other hand, Adam Kolton, testifying before the House Committee on Resources on July 11, 2001, in opposition to the National Energy Security Act of 2001 (NESA), presented the coastal plain as "the site of one of our continent's most awe-inspiring wildlife spectacles" and, thus, deserving of protection from exploitation. Kennan Ward, in *The Last Wilderness: Arctic National Wildlife Refuge* (Wildlight Press, 2001), describes a realm where human impact is still minimal and wilderness endures. John G. Mitchell, in "Oil Field or Sanctuary?" *National Geographic* (August 2001), is more balanced in his appraisal but still sides with the Lovinses, concluding that better alternatives to developing the ANWR exist.

The House of Representatives approved the NESA in August 2001. The bill then stalled in the Senate, with pro-drilling senators attempting to woo votes with such measures as promising to use oil revenues to pay pension benefits for steelworkers. Their efforts failed in April 2002, when the bill was defeated and a competing energy bill took the lead. This alternative bill, introduced in December 2001 and sponsored by Senate Majority Leader Tom Daschle (D-South Dakota) and Senator Jeff Bingaman (D-New Mexico), does not allow for oil exploration in the ANWR.

In "ANWR Oil: An Alternative to War Over Oil," *American Enterprise* (June 2002), Walter J. Hickle, former U.S. secretary of the interior and twice the governor of Alaska, writes, "[T]he issue is not going to go away. Given our continuing precarious dependence on overseas oil suppliers ranging from Saddam Hussein to the Saudis to Venezuela's Castro-clone Hugo Chavez, sensible Americans will continue to press Congress in the months and years ahead to unlock America's great Arctic energy storehouse."

Similar debate has centered on mineral exploitation in the American Southwest. President Clinton created the Grand Staircase–Escalante National Monument by executive order to protect an important part of Utah's remaining wilderness, but opposition remains. See T. H. Watkins, *The Redrock Chronicles: Saving Wild Utah* (Johns Hopkins University Press, 2000). For a survey of the wilderness system created by the 1964 Wilderness Act, see John G. Mitchell and Peter Essick, "Wilderness: America's Land Apart," *National Geographic* (November 1998).

ISSUE 8

Should DDT Be Banned Worldwide?

YES: Anne Platt McGinn, from "Malaria, Mosquitoes, and DDT," *World Watch* (May/June 2002)

NO: Roger Bate, from "A Case of the DDTs," *National Review* (May 14, 2001)

ISSUE SUMMARY

YES: Anne Platt McGinn, a senior researcher at the Worldwatch Institute, argues that although DDT is still used to fight malaria, there are other, more effective and less environmentally harmful methods. She maintains that DDT should be banned or reserved for emergency use.

NO: Roger Bate, director of Africa Fighting Malaria, asserts that DDT is the cheapest and most effective way to combat malaria and that it should remain available for use.

D DT is a crucial element in the story of environmentalism. The chemical was first synthesized in 1874. Swiss entomologist Paul Mueller was the first to notice that DDT has insecticidal properties, which, it was quickly realized, implied that the chemical could save human lives. It had long been known that more soldiers died during wars because of disease than because of enemy fire. During World War I, for example, some 5 million lives were lost to typhus, a disease carried by body lice. DDT was first deployed during World War II to halt a typhus epidemic in Naples, Italy. It was a dramatic success, and DDT was soon used routinely as a dust for soldiers and civilians. During and after the war, DDT was also deployed successfully against the mosquitoes that carry malaria and other diseases. In the United States cases of malaria fell from 120,000 in 1934 to 72 in 1960, and cases of yellow fever dropped from 100,000 in 1878 to none. In 1948 Mueller received the Nobel Prize for medicine and physiology because DDT had saved so many civilian lives.

DDT was by no means the first pesticide. But its predecessors—arsenic, strychnine, cyanide, copper sulfate, and nicotine—were all markedly toxic to humans. DDT was not only more effective as an insecticide, it was also less hazardous to users. It is therefore not surprising that DDT was seen as a beneficial substance. It was soon applied routinely to agricultural crops and used to control mosquito populations in American suburbs. However, insects quickly became resistant to the insecticide. (In any population of insects, some will be more resistant than others; when the insecticide kills the more vulnerable members of the population, the resistant ones are left to breed and multiply. This is an example of natural selection.) In *Silent Spring* (Houghton Mifflin, 1962), marine scientist Rachel Carson demonstrated that DDT was concentrated in the food chain and affected the reproduction of predators such as hawks and eagles. In 1972 the U.S. Environmental Protection Agency banned almost all uses of DDT (it could still be used to protect public health). Other developed countries soon banned it as well, but developing nations, especially those in the tropics, saw it as an essential tool for fighting diseases such as malaria.

It soon became apparent that DDT is by no means the only pesticide or organic toxin with environmental effects. As a result, on May 24, 2001, the United States joined 90 other nations in signing the Stockholm Convention on Persistent Organic Pollutants (POPs). This treaty aims to eliminate from use the entire class of chemicals to which DDT belongs, beginning with the pesticides DDT, aldrin, dieldrin, endrin, chlordane, heptachlor, mirex, and toxaphene, and the industrial chemicals polychlorinated biphenyls (PCBs), hexachlorobenzene (HCB), dioxins, and furans. Fifty more countries signed during the next year, but according to the Pesticide Action Network North America (http://panna.igc.org), only eight (not including the United States) have formally ratified the treaty.

In the following selection, Anne Platt McGinn, granting that malaria remains a serious problem in the developing nations of the tropics, especially Africa, contends that although DDT is still used to fight malaria in these nations, it is far less effective than it used to be. She argues that the environmental effects are also serious concerns and that DDT should be banned or reserved for emergency use. In the second selection, Roger Bate argues that DDT remains the cheapest and most effective way to combat malaria and that it should remain available for use.

Anne Platt McGinn

 **YES**

Malaria, Mosquitoes, and DDT

This year, like every other year within the past couple of decades, uncountable trillions of mosquitoes will inject malaria parasites into human blood streams billions of times. Some 300 to 500 million full-blown cases of malaria will result, and between 1 and 3 million people will die, most of them pregnant women and children. That's the official figure, anyway, but it's likely to be a substantial underestimate, since most malaria deaths are not formally registered, and many are likely to have escaped the estimators. Very roughly, the malaria death toll rivals that of AIDS, which now kills about 3 million people annually.

But unlike AIDS, malaria is a low-priority killer. Despite the deaths, and the fact that roughly 2.5 billion people (40 percent of the world's population) are at risk of contracting the disease, malaria is a relatively low public health priority on the international scene. Malaria rarely makes the news. And international funding for malaria research currently comes to a mere $150 million annually. Just by way of comparison, that's only about 5 percent of the $2.8 billion that the U.S. government alone is considering for AIDS research in fiscal year 2003.

The low priority assigned to malaria would be at least easier to understand, though no less mistaken, if the threat were static. Unfortunately it is not. It is true that the geographic range of the disease has contracted substantially since the mid-20th century, but over the past couple of decades, malaria has been gathering strength. Virtually all areas where the disease is endemic have seen drug-resistant strains of the parasites emerge—a development that is almost certainly boosting death rates. In countries as various as Armenia, Afghanistan, and Sierra Leone, the lack or deterioration of basic infrastructure has created a wealth of new breeding sites for the mosquitoes that spread the disease. The rapidly expanding slums of many tropical cities also lack such infrastructure; poor sanitation and crowding have primed these places as well for outbreaks —even though malaria has up to now been regarded as predominantly a rural disease.

What has current policy to offer in the face of these threats? The medical arsenal is limited; there are only about a dozen antimalarial drugs commonly in use, and there is significant malaria resistance to most of them. In the absence

From Anne Platt McGinn, "Malaria, Mosquitoes, and DDT," *World Watch*, vol. 15, no. 3 (May/June 2002). Copyright © 2002 by The Worldwatch Institute. Reprinted by permission. http://www.worldwatch.org.

of a reliable way to kill the parasites, policy has tended to focus on killing the mosquitoes that bear them. And that has led to an abundant use of synthetic pesticides, including one of the oldest and most dangerous: dichlorodiphenyl trichloroethane, or DDT.

DDT is no longer used or manufactured in most of the world, but because it does not break down readily, it is still one of the most commonly detected pesticides in the milk of nursing mothers. DDT is also one of the "dirty dozen" chemicals included in the 2001 Stockholm Convention on Persistent Organic Pollutants [POPs]. The signatories to the "POPs Treaty" essentially agreed to ban all uses of DDT except as a last resort against disease-bearing mosquitoes. Unfortunately, however, DDT is still a routine option in 19 countries, most of them in Africa. (Only 11 of these countries have thus far signed the treaty.) Among the signatory countries, 31—slightly fewer than one-third—have given notice that they are reserving the right to use DDT against malaria. On the face of it, such use may seem unavoidable, but there are good reasons for thinking that progress against the disease is compatible with *reductions* in DDT use.

Malaria is caused by four protozoan parasite species in the genus *Plasmodium.* These parasites are spread exclusively by certain mosquitoes in the genus *Anopheles.* An infection begins when a parasite-laden female mosquito settles onto someone's skin and pierces a capillary to take her blood meal. The parasite, in a form called the *sporozoite,* moves with the mosquito's saliva into the human bloodstream. About 10 percent of the mosquito's lode of sporozoites is likely to be injected during a meal, leaving plenty for the next bite. Unless the victim has some immunity to malaria—normally as a result of previous exposure—most sporozoites are likely to evade the body's immune system and make their way to the liver, a process that takes less than an hour. There they invade the liver cells and multiply asexually for about two weeks. By this time, the original several dozen sporozoites have become millions of *merozoites* —the form the parasite takes when it emerges from the liver and moves back into the blood to invade the body's red blood cells. Within the red blood cells, the merozoites go through another cycle of asexual reproduction, after which the cells burst and release millions of additional merozoites, which invade yet more red blood cells. The high fever and chills associated with malaria are the result of this stage, which tends to occur in pulses. If enough red blood cells are destroyed in one of these pulses, the result is convulsions, difficulty in breathing, coma, and death.

As the parasite multiplies inside the red blood cells, it produces not just more merozoites, but also *gametocytes,* which are capable of sexual reproduction. This occurs when the parasite moves back into the mosquitoes; even as they inject sporozoites, biting mosquitoes may ingest gametocytes if they are feeding on a person who is already infected. The gametocytes reproduce in the insect's gut and the resulting eggs move into the gut cells. Eventually, more sporozoites emerge from the gut and penetrate the mosquito's salivary glands, where they await a chance to enter another human bloodstream, to begin the cycle again.

Of the roughly 380 mosquito species in the genus *Anopheles,* about 60 are able to transmit malaria to people. These malaria vectors are widespread throughout the tropics and warm temperate zones, and they are very efficient at spreading the disease. Malaria is highly contagious, as is apparent from a measurement that epidemiologists call the "basic reproduction number," or BRN. The BRN indicates, on average, how many new cases a single infected person is likely to cause. For example, among the nonvectored diseases (those in which the pathogen travels directly from person to person without an intermediary like a mosquito), measles is one of the most contagious. The BRN for measles is 12 to 14, meaning that someone with measles is likely to infect 12 to 14 other people. (Luckily, there's an inherent limit in this process: as a pathogen spreads through any particular area, it will encounter fewer and fewer susceptible people who aren't already sick, and the outbreak will eventually subside.) HIV/AIDS is on the other end of the scale: it's deadly, but it burns through a population slowly. Its BRN is just above 1, the minimum necessary for the pathogen's survival. With malaria, the BRN varies considerably, depending on such factors as which mosquito species are present in an area and what the temperatures are. (Warmer is worse, since the parasites mature more quickly.) But malaria can have a BRN in excess of 100: over an adult life that may last about a week, a single, malaria-laden mosquito could conceivably infect more than 100 people.

Seven Years, Seven Months

"Malaria" comes from the Italian "mal'aria." For centuries, European physicians had attributed the disease to "bad air." Apart from a tradition of associating bad air with swamps—a useful prejudice, given the amount of mosquito habitat in swamps—early medicine was largely ineffective against the disease. It wasn't until 1897 that the British physician Ronald Ross proved that mosquitoes carry malaria.

The practical implications of Ross's discovery did not go unnoticed. For example, the U.S. administration of Theodore Roosevelt recognized malaria and yellow fever (another mosquito-vectored disease) as perhaps the most serious obstacles to the construction of the Panama Canal. This was hardly a surprising conclusion, since the earlier and unsuccessful French attempt to build the canal—an effort that predated Ross's discovery—is thought to have lost between 10,000 and 20,000 workers to disease. So the American workers draped their water supplies and living quarters with mosquito netting, attempted to fill in or drain swamps, installed sewers, poured oil into standing water, and conducted mosquito-swatting campaigns. And it worked: the incidence of malaria declined. In 1906, 80 percent of the workers had the disease; by 1913, a year before the Canal was completed, only 7 percent did. Malaria could be suppressed, it seemed, with a great deal of mosquito netting, and by eliminating as much mosquito habitat as possible. But the labor involved in that effort could be enormous.

That is why DDT proved so appealing. In 1939, the Swiss chemist Paul Müller discovered that this chemical was a potent pesticide. DDT was first used during World War II, as a delousing agent. Later on, areas in southern Europe, North Africa, and Asia were fogged with DDT, to clear malaria-laden mosquitoes from the paths of invading Allied troops. DDT was cheap and it seemed to be harmless to anything other than insects. It was also long-lasting: most other insecticides lost their potency in a few days, but in the early years of its use, the effects of a single dose of DDT could last for up to six months. In 1948, Müller won a Nobel Prize for his work and DDT was hailed as a chemical miracle.

A decade later, DDT had inspired another kind of war—a general assault on malaria. The "Global Malaria Eradication Program," launched in 1955, became one of the first major undertakings of the newly created World Health Organization [WHO]. Some 65 nations enlisted in the cause. Funding for DDT factories was donated to poor countries and production of the insecticide climbed.

The malaria eradication strategy was not to kill every single mosquito, but to suppress their populations and shorten the lifespans of any survivors, so that the parasite would not have time to develop within them. If the mosquitoes could be kept down long enough, the parasites would eventually disappear from the human population. In any particular area, the process was expected to take three years—time enough for all infected people either to recover or die. After that, a resurgence of mosquitoes would be merely an annoyance, rather than a threat. And initially, the strategy seemed to be working. It proved especially effective on islands—relatively small areas insulated from reinfestation. Taiwan, Jamaica, and Sardinia were soon declared malaria-free and have remained so to this day. By 1961, arguably the year at which the program had peak momentum, malaria had been eliminated or dramatically reduced in 37 countries.

One year later, Rachel Carson published *Silent Spring*, her landmark study of the ecological damage caused by the widespread use of DDT and other pesticides. Like other organochlorine pesticides, DDT bioaccumulates. It's fat soluble, so when an animal ingests it—by browsing contaminated vegetation, for example—the chemical tends to concentrate in its fat, instead of being excreted. When another animal eats that animal, it is likely to absorb the prey's burden of DDT. This process leads to an increasing concentration of DDT in the higher links of the food chain. And since DDT has a high chronic toxicity—that is, long-term exposure is likely to cause various physiological abnormalities—this bioaccumulation has profound implications for both ecological and human health.

With the miseries of malaria in full view, the managers of the eradication campaign didn't worry much about the toxicity of DDT, but they were greatly concerned about another aspect of the pesticide's effects: resistance. Continual exposure to an insecticide tends to "breed" insect populations that are at least partially immune to the poison. Resistance to DDT had been reported as early as 1946. The campaign managers knew that in mosquitoes, regular exposure to DDT tended to produce widespread resistance in four to seven years. Since it took three years to clear malaria from a human population, that didn't leave a lot of leeway for the eradication effort. As it turned out, the logistics simply couldn't be made to work in large, heavily infested areas with high human populations, poor housing and roads, and generally minimal infrastructure. In 1969, the campaign was abandoned. Today, DDT resistance is widespread in *Anopheles,* as is resistance to many more recent pesticides.

Undoubtedly, the campaign saved millions of lives, and it did clear malaria from some areas. But its broadest legacy has been of much more dubious value. It engendered the idea of DDT as a first resort against mosquitoes and it established the unstable dynamic of DDT resistance in *Anopheles* populations. In mosquitoes, the genetic mechanism that confers resistance to DDT does not usually come at any great competitive "cost"—that is, when no DDT is being sprayed, the resistant mosquitoes may do just about as well as nonresistant mosquitoes. So once a population acquires resistance, the trait is not likely to disappear even if DDT isn't used for years. If DDT is reapplied to such a population, widespread resistance will reappear very rapidly. The rule of thumb among entomologists is that you may get seven years of resistance-free use the first time around, but you only get about seven months the second time. Even that limited respite, however, is enough to make the chemical an attractive option as an emergency measure—or to keep it in the arsenals of bureaucracies committed to its use.

Malaria Taxes

In December 2000, the POPs Treaty negotiators convened in Johannesburg, South Africa, even though, by an unfortunate coincidence, South Africa had suffered a potentially embarrassing setback earlier that year in its own POPs policies. In 1996, South Africa had switched its mosquito control programs from DDT to a less persistent group of pesticides known as pyrethroids. The move seemed solid and supportable at the time, since years of DDT use had greatly reduced *Anopheles* populations and largely eliminated one of the most troublesome local vectors, the appropriately named *A. funestus* ("funestus" means deadly). South Africa seemed to have beaten the DDT habit: the chemical had been used to achieve a worthwhile objective; it had then been discarded. And the plan worked—until a year before the POPs summit, when malaria infections rose to 61,000 cases, a level not seen in decades. *A. funestus* reappeared as well, in KwaZulu-Natal, and in a form resistant to pyrethroids. In early 2000, DDT was reintroduced, in an indoor spraying program. (This is now a standard way of using DDT for mosquito control; the pesticide is usually applied only to walls, where mosquitoes alight to rest.) By the middle of the year, the number of infections had dropped by half.

Initially, the spraying program was criticized, but what reasonable alternative was there? This is said to be the African predicament, and yet the South African situation is hardly representative of sub-Saharan Africa as a whole.

Malaria is considered endemic in 105 countries throughout the tropics and warm temperate zones, but by far the worst region for the disease is sub-Saharan Africa. The deadliest of the four parasite species, *Plasmodium falciparum,* is widespread throughout this region, as is one of the world's most effective malaria vectors, *Anopheles gambiae.* Nearly half the population of sub-Saharan

Africa is at risk of infection, and in much of eastern and central Africa, and pockets of west Africa, it would be difficult to find anyone who has not been exposed to the parasites. Some 90 percent of the world's malaria infections and deaths occur in sub-Saharan Africa, and the disease now accounts for 30 percent of African childhood mortality. It is true that malaria is a grave problem in many parts of the world, but the African experience is misery on a very different order of magnitude. The average Tanzanian suffers more infective bites each *night* than the average Thai or Vietnamese does in a year.

As a broad social burden, malaria is thought to cost Africa between $3 billion and $12 billion annually. According to one economic analysis, if the disease had been eradicated in 1965, Africa's GDP would now be 35 percent higher than it currently is. Africa was also the gaping hole in the global eradication program: the WHO planners thought there was little they could do on the continent and limited efforts to Ethiopia, Zimbabwe, and South Africa, where eradication was thought to be feasible.

But even though the campaign largely passed Africa by, DDT has not. Many African countries have used DDT for mosquito control in indoor spraying programs, but the primary use of DDT on the continent has been as an agricultural insecticide. Consequently, in parts of west Africa especially, DDT resistance is now widespread in *A. gambiae*. But even if *A. gambiae* were not resistant, a full-bore campaign to suppress it would probably accomplish little, because this mosquito is so efficient at transmitting malaria. Unlike most *Anopheles* species, *A. gambiae* specializes in human blood, so even a small population would keep the disease in circulation. One way to get a sense for this problem is to consider the "transmission index"—the threshold number of mosquito bites necessary to perpetuate the disease. In Africa, the index overall is 1 bite per person per month. That's all that's necessary to keep malaria in circulation. In India, by comparison, the TI is 10 bites per person per month.

And yet Africa is not a lost cause—it's simply that the key to progress does not lie in the general suppression of mosquito populations. Instead of spraying, the most promising African programs rely primarily on "bednets"—mosquito netting that is treated with an insecticide, usually a pyrethroid, and that is suspended over a person's bed. Bednets can't eliminate malaria, but they can "deflect" much of the burden. Because *Anopheles* species generally feed in the evening and at night, a bednet can radically reduce the number of infective bites a person receives. Such a person would probably still be infected from time to time, but would usually be able to lead a normal life.

In effect, therefore, bednets can substantially reduce the disease. Trials in the use of bednets for children have shown a decline in malaria-induced mortality by 25 to 40 percent. Infection levels and the incidence of severe anemia also declined. In Kenya, a recent study has shown that pregnant women who use bednets tend to give birth to healthier babies. In parts of Chad, Mali, Burkina Faso, and Senegal, bednets are becoming standard household items. In the tiny west African nation of The Gambia, somewhere between 50 and 80 percent of the population has bednets.

Bednets are hardly a panacea. They have to be used properly and retreated with insecticide occasionally. And there is still the problem of insecticide resistance, although the nets themselves are hardly likely to be the main cause of it. (Pyrethroids are used extensively in agriculture as well.) Nevertheless, bednets can help transform malaria from a chronic disaster to a manageable public health problem—something a healthcare system can cope with.

So it's unfortunate that in much of central and southern Africa, the nets are a rarity. It's even more unfortunate that, in 28 African countries, they're taxed or subject to import tariffs. Most of the people in these countries would have trouble paying for a net even without the tax. This problem was addressed in the May 2000 "Abuja Declaration," a summit agreement on infectious diseases signed by 44 African countries. The Declaration included a pledge to do away with "malaria taxes." At last count, 13 countries have actually acted on the pledge, although in some cases only by reducing rather than eliminating the taxes. Since the Declaration was signed, an estimated 2 to 5 million Africans have died from malaria.

This failure to follow through with the Abuja Declaration casts the interest in DDT in a rather poor light. Of the 31 POPs treaty signatories that have reserved the right to use DDT, 21 are in Africa. Of those 21, 10 are apparently still taxing or imposing tariffs on bednets. (Among the African countries that have *not* signed the POPs treaty, some are almost certainly both using DDT and taxing bednets, but the exact number is difficult to ascertain because the status of DDT use is not always clear.) It is true that a case can be made for the use of DDT in situations like the one in South Africa in 1999—an infrequent flare-up in a context that lends itself to control. But the routine use of DDT against malaria is an exercise in toxic futility, especially when it's pursued at the expense of a superior and far more benign technology.

Learning to Live With the Mosquitoes

A group of French researchers recently announced some very encouraging results for a new anti-malarial drug known as G25. The drug was given to infected aotus monkeys, and it appears to have cleared the parasites from their systems. Although extensive testing will be necessary before it is known whether the drug can be safely given to people, these results have raised the hope of a cure for the disease.

Of course, it would be wonderful if G25, or some other new drug, lives up to that promise. But even in the absence of a cure, there are opportunities for progress that may one day make the current incidence of malaria look like some dark age horror. Many of these opportunities have been incorporated into an initiative that began in 1998, called the Roll Back Malaria (RBM) campaign, a collaborative effort between WHO, the World Bank, UNICEF, and the UNDP [United Nations Development Programme]. In contrast to the earlier WHO eradication program, RBM grew out of joint efforts between WHO and various African governments specifically to address African malaria. RBM focuses on household- and community-level intervention and it emphasizes apparently modest changes that could yield major progress. Below are four "operating principles" that are, in one way or another, implicit in RBM or likely to reinforce its progress.

1. Do away with all taxes and tariffs on bednets, on pesticides intended for treating bednets, and on antimalarial drugs. Failure to act on this front certainly undercuts claims for the necessity of DDT; it may also undercut claims for antimalaria foreign aid.

2. Emphasize appropriate technologies. Where, for example, the need for mud to replaster walls is creating lots of pothole sized cavities near houses—cavities that fill with water and then with mosquito larvae—it makes more sense to help people improve their housing maintenance than it does to set up a program for squirting pesticide into every pothole. To be "appropriate," a technology has to be both affordable and culturally acceptable. Improving home maintenance should pass this test; so should bednets. And of course there are many other possibilities. In Kenya, for example, a research institution called the International Center for Insect Physiology and Ecology has identified at least a dozen native east African plants that repel *Anopheles gambiae* in lab tests. Some of these plants could be important additions to household gardens.

3. Use existing networks whenever possible, instead of building new ones. In Tanzania, for example, an established healthcare program (UNICEF's Integrated Management of Childhood Illness Program) now dispenses antimalarial drugs—and instruction on how to use them. The UNICEF program was already operating, so it was simple and cheap to add the malaria component. Reported instances of severe malaria and anemia in infants have declined, apparently as a result. In Zambia, the government is planning to use health and prenatal clinics as the network for a coupon system that subsidizes bednets for the poor. Qualifying patients would pick up coupons at the clinics and redeem them at stores for the nets.

4. Assume that sound policy will involve action on many fronts. Malaria is not just a health problem—it's a social problem, an economic problem, an environmental problem, an agricultural problem, an urban planning problem. Health officials alone cannot possibly just make it go away. When the disease flares up, there is a strong and understandable temptation to strap on the spray equipment and douse the mosquitoes. But if this approach actually worked, we wouldn't be in this situation today. Arguably the biggest opportunity for progress against the disease lies, not in our capacity for chemical innovation, but in our capacity for *organizational innovation*—in our ability to build an awareness of the threat across a broad range of policy activities. For example, when government officials are considering loans to irrigation projects, they should be asking: has the potential for malaria been addressed? When foreign donors are designing antipoverty programs, they should be asking: do people need bednets? Routine inquiries of this sort could go a vast distance to reducing the disease.

Where is the DDT in all of this? There isn't any, and that's the point. We now have half a century of evidence that routine use of DDT simply will not prevail against the mosquitoes. Most countries have already absorbed this lesson, and banned the chemical or relegated it to emergency only status. Now the RBM campaign and associated efforts are showing that the frequency and intensity of those emergencies can be reduced through systematic attention to the chronic aspects of the disease. There is less and less justification for DDT, and the futility of using it as a matter of routine is becoming increasingly apparent: in order to control a disease, why should we poison our soils, our waters, and ourselves?

NO

Roger Bate

A Case of the DDTs

Militants from Greenpeace have been mounting protests in an effort to close down the only major DDT-production facility in the world, located in Cochin, India. The protesters won't be getting any support from Jocchonia Gumede, a domestic servant in Johannesburg, South Africa: In the past two years, six of Jocchonia's close relatives have died from malaria. His family lives in northern KwaZulu Natal, where malaria has always been endemic—Jocchonia himself contracted it twice while growing up—and DDT is simply the cheapest and most effective way to combat this dread disease.

Malaria is now on the increase, not just in Africa but in all tropical regions of the planet. It afflicted well over 300 million people last year, and killed over 1 million. Prof. Wen Kilama of the African Malaria Vaccine Testing Network in Tanzania characterizes the death toll as "equivalent to crashing seven jumbo jets filled with children every day."

That's a devastating human cost. And it has far-reaching consequences, beyond even the sad plight of the sufferers and the huge burden the disease imposes on health resources. In many countries, the disease is also clouding the long-term economic future: When people are unable to work effectively because of illness, productivity suffers—and this, in turn, scares away investors. Professor Jeffrey Sachs of the Harvard Center for International Development estimates that every year, malaria destroys around 1 percent of the wealth—not just income, but *total wealth*—of Africa.

Given such devastating human and economic costs, one might expect the "international community" to be fighting malaria with all its might. But the chief effort of the world's politicians has been to try force developing countries to abandon their best weapon in the fight against malaria—the pesticide dichlorodiphenyl-trichloroethane, known commonly as DDT. The United Nations is actually promoting a treaty that might ban the use of DDT globally—and on April 18 [2001], President Bush agreed to sign the treaty.

This is absurd, because DDT is the proven solution to malaria. Today malaria is a tropical disease, but until the 1920s it was endemic all over Europe and America. Epidemics were found as far north as Archangel in the Russian Arctic Circle, and occurred regularly in Holland and England. After World War II, Europe and North America eradicated it with DDT. The pesticide saved countless millions of lives by killing the malarial mosquito, but it never had complete success in some of the world's poorer countries, because their governments lacked the capacity to implement the necessary spraying programs and removal of mosquito breeding areas; without the appropriate medical and organizational ability, even the best sprays won't be effective in eradicating a disease. Then, in the late 1960s, environmentalists started to complain about DDT, and it was removed from the malaria-control program in many countries; some 20 countries—most in Africa—continued to use it.

According to Donald Roberts, a professor of tropical public health at the Uniformed Services University of the Health Sciences, the huge drop in the number of houses sprayed with DDT has had severe consequences: From the mid 1980s to the mid 1990s, Latin America experienced an annual increase of more than 1.8 million malaria cases (more than 4.8 per 1,000 people)—and the rate has continued to grow since 1996. Ecuador, however, continued to use DDT, and its malaria rate *fell* over the period 1988–97.

Other mosquito-borne diseases are also on the rise. Until the 1970s, DDT was used to eradicate the *Aedes aegypti* mosquito from most tropical regions of the Americas. A new invasion of *Aedes aegypti* has since brought devastating outbreaks of dengue fever and a renewed threat of urban yellow fever. Roberts says the international anti-DDT groups, with what he calls their "high-pressure tactics," bear some responsibility for this public-health disaster.

About 40 years ago, suspicions about DDT—inspired by Rachel Carson's book *Silent Spring*—sparked the first green crusade. When it was used in vast quantities in agriculture, DDT probably did harm reproduction in birds of prey. (This harm subsequently proved reversible.) But after decades of research, there is not one replicated study that shows any harm to human beings at all. Furthermore, DDT is now only sprayed inside houses. Dr. Amir Attaran, a researcher at Harvard's Center for International Development, estimates that the amount of DDT used to spray a few acres of cotton in America in the early 1960s would be enough to spray all the homes in Guyana of those at risk of malaria. Such indoor spraying, he concludes, would have "negligible impacts on the environment." Even Green presidential nominee Ralph Nader has come out in favor of DDT use.

Undeterred, Greenpeace continues to try to shut down the Indian DDT factory—and, in effect, to prevent some of the world's poorest countries from using the least expensive method of eradicating malaria. [Recently], the Indian government gave an assurance to Greenpeace that production would cease in 2005—but officials of that country's anti-malaria program, which has used DDT since 1953, objected to this commitment; the government may make an embarrassing but felicitous U-turn. (There is precedent for such a switch. In 1996, South Africa stopped using DDT, and the death rate from malaria rose by around 1,000 percent. In desperation, the country has returned to using the pesticide.)

There are other pesticides that work against malaria, but they are at least twice—and sometimes up to 20 times—as expensive, and none is as effective a repellent. These substitutes are less persistent than DDT—they don't linger in the environment as long—and this is what makes them so attractive to the greens. But when you are using pesticides indoors, a persistent material is better: It means you might have to spray only once a year—and to a poor country, this kind of cost consideration could make all the difference.

Despite these basic facts of science and economics, international-pesticide-treaty negotiators decided... to restrict the use of DDT. Their decision is not final; they will meet again in Sweden..., when environmental ministers are expected to sign a final text. It may get worse between now and then; the negotiators may yet decide to replace the restrictions with an outright ban. But even the restrictions demanded under the existing draft will be onerous for the poorest countries, some of which have health budgets of less than $5 per person per year (rich countries spend well over 400 times this amount on health care). Even worse, many countries have been coming under pressure from international health and environment agencies to give up DDT or face losing aid grants: Belize and Bolivia are on record admitting they gave in to pressure on this issue from the U.S. Agency for International Development.

South Africa has asked for an exemption from the DDT restrictions. Jocchonia Gumede reports that the number of mosquitoes is down since DDT was reintroduced in February 2000. He is cautiously optimistic that the situation will improve, but his quick smile belies the anguish that he feels. Jocchonia was not aware of the reason for the removal of DDT, nor is he aware of the debate surrounding whether to ban the substance outright. He is simply one of countless millions around the world whose health and prosperity depend, ominously, on sensible decision-making by global bureaucrats.

POSTSCRIPT

Should DDT Be Banned Worldwide?

Over and over again, the debates over environmental issues come down to which we should do first: Should we meet human needs regardless of whether or not species die and air and water are contaminated? Or should we protect species, air, water, and other aspects of the environment even if some human needs must go unmet? What if this means endangering the lives of children? In the debate over DDT, the human needs are clear, for insect-borne diseases have killed and continue to kill a great many people. Yet the environmental needs are also clear. The question is one of choosing priorities and balancing risks. See John Danley, "Balancing Risks: Mosquitoes, Malaria, Morality, and DDT," *Business and Society Review* (Spring 2002).

Mosquitoes can be controlled in various ways: Swamps can be drained (which carries its own environmental price), and other breeding opportunities can be eliminated. Fish can be introduced to eat mosquito larvae. And mosquito nets can be used to keep the insects away from people. But these (and other) alternatives do not mean that there does not remain a place for chemical pesticides. In "Pesticides and Public Health: Integrated Methods of Mosquito Management," *Emerging Infectious Diseases* (January–February 2001), Robert I. Rose, an arthropod biotechnologist with the Animal and Plant Health Inspection Service of the U.S. Department of Agriculture, says, "Pesticides have a role in public health as part of sustainable integrated mosquito management. Other components of such management include surveillance, source reduction or prevention, biological control, repellents, traps, and pesticide-resistance management."

Researchers have long sought a vaccine against malaria, but the parasite has demonstrated a persistent talent for evading all attempts to arm the immune system against it. The difficulties are covered by Thomas L. Richie and Allan Saul in "Progress and Challenges for Malaria Vaccines," *Nature* (February 7, 2002). In March 2002 a new vaccine was reported to be effective in animals; see Michael Greer, "Malaria Vaccine Based on Parasite Protein Effective in Animals," *Vaccine Weekly* (March 13 & 20, 2002). A newer approach is to develop genetically engineered (transgenic) mosquitoes that either cannot support the malaria parasite or cannot infect humans with it; see Jane Bradbury, "Transgenic Mosquitoes Bring Malarial Control Closer," *The Lancet* (May 25, 2002).

It is worth stressing that malaria is only one of several mosquito-borne diseases that pose threats to public health. Two others are yellow fever and dengue. A recent arrival to the United States is West Nile virus, which mosquitoes can transfer from birds to humans. However, West Nile virus is far less fatal than malaria, yellow fever, or dengue, and a vaccine is in development. See Dwight G. Smith, "A New Disease in the New World," *The World & I* (February 2002) and Michelle Mueller, "The Buzz on West Nile Virus," *Current Health 2* (April/May 2002).

It is also worth stressing that global warming means climate changes that may increase the geographic range of disease-carrying mosquitoes. Many climate researchers are concerned that malaria, yellow fever, and other now mostly tropical and subtropical diseases may return to temperate-zone nations and even spread into areas where they have never been known.

ISSUE 9

Is Genetic Engineering an Environmentally Sound Way to Increase Food Production?

YES: Royal Society of London et al., from "Transgenic Plants and World Agriculture," A Report Prepared Under the Auspices of the Royal Society of London, the U.S. National Academy of Sciences, the Brazilian Academy of Sciences, the Chinese Academy of Sciences, the Indian National Science Academy, the Mexican Academy of Sciences, and the Third World Academy of Sciences (July 2000)

NO: Brian Halweil, from "The Emperor's New Crops," *World Watch* (July/August 1999)

ISSUE SUMMARY

YES: The national academies of science of the United Kingdom, the United States, Brazil, China, India, Mexico, and the Third World argue that genetically modified crops hold the potential to feed the world during the twenty-first century while also protecting the environment.

NO: Brian Halweil, a researcher at the Worldwatch Institute, argues that the genetic modification of crops threatens to produce pesticide-resistant insect pests and herbicide-resistant weeds, will victimize poor farmers, and is unlikely to feed the world.

In the early 1970s scientists first discovered that it was technically possible to move genes—the biological material that determines a living organism's physical traits—from one organism to another and thus (in principle) to give bacteria, plants, and animals new features. Most researchers in molecular genetics were excited by the potentialities that suddenly seemed within their reach. However, a few researchers—as well as many people outside the field—were disturbed by the idea; they thought that genetic mix-and-match games might spawn new diseases, weeds, and pests. Some people even argued that genetic engineering should be banned at the outset, before unforeseeable horrors were unleashed. Researchers in support of genetic experimentation responded by declaring a moratorium on their own work until suitable safeguards (in the form of government regulations) could be devised.

A 1987 National Academy of Sciences report said that genetic engineering posed no unique hazards. And, despite continuing controversy, by 1989 the technology had developed tremendously: researchers could obtain patents for mice with artificially added genes ("transgenic" mice); firefly genes had been added to tobacco plants to make them glow (faintly) in the dark; and growth hormone produced by genetically engineered bacteria was being used to grow low-fat pork and increase milk production in cows. The growing biotechnology industry promised more productive crops that made their own fertilizer and pesticide. Proponents argued that genetic engineering was in no significant way different from traditional selective breeding. Critics argued that genetic engineering was unnatural and violated the rights of both plants and animals to their "species integrity"; that expensive, high-tech, tinkered animals gave the competitive advantage to big agricultural corporations and drove small farmers out of business; and that putting human genes into animals, plants, or bacteria was downright offensive. See Betsy Hanson and Dorothy Nelkin, "Public Responses to Genetic Engineering," *Society* (November/December 1989).

In 1992 the U.S. Office of Science and Technology issued guidelines to bar regulations that are based on the assumption that genetically engineered crops pose greater risks than similar crops produced by traditional breeding methods. The result was the rapid commercial introduction of crops that were genetically engineered to make the bacterial insecticide Bt and to resist herbicides, among other things. Between 1996 and 1998 the areas planted with genetically engineered crops jumped from 1.7 million hectares to 27.8 million hectares. Sales of genetically engineered crop products are expected to reach $25 billion by 2010.

Skepticism about the benefits remains, but agricultural genetic engineering has proceeded at a breakneck pace, largely because, as Robert Shapiro, CEO of the Monsanto Corporation, said in June 1998, it "represents a potentially sustainable solution to the issue of feeding people." Many people are not reassured. They see potential problems in nutrition, toxicity, allergies, and ecology. Europe has paid more attention to the critics than the United States has, and the growing and marketing of genetically engineered crops has been either banned or severely restricted across the continent.

The following selections illustrate the different current perspectives on the use of genetic engineering in agriculture. In the first selection, the national academies of science of the United Kingdom, the United States, Brazil, China, India, Mexico, and the Third World recognize that the use of genetically modified crops has some worrisome potentials that deserve further research, but they conclude that such crops hold the potential to feed the world during the twenty-first century while also protecting the environment. In the second selection, Brian Halweil argues that the genetic modification of crops threatens to produce pesticide-resistant insect pests and herbicide-resistant weeds, will victimize poor farmers, and is unlikely to feed the world.

155

Royal Society of London et al.  **YES**

Transgenic Plants and World Agriculture

During the 21st century, humankind will be confronted with an extraordinary set of challenges. By 2030, it is estimated that eight billion persons will populate the world—an increase of two billion people from today's population. Hunger and poverty around the globe must be addressed, while the life-support systems provided by the world's natural environment are maintained. Meeting these challenges will require new knowledge generated by continued scientific advances, the development of appropriate new technologies, and a broad dissemination of this knowledge and technology along with the capacity to use it throughout the world. It will also require that wise policies be implemented through informed decision-making on the part of national, state, and local governments in each nation.

Scientific advances require an open system of information exchange in which arguments are based on verifiable evidence. Although the primary goal of science is to increase our understanding of the world, knowledge created through science has had immense practical benefits. For example, through science, we have developed a more complete understanding of our natural environment, improved human health with new medicines, and discovered specific plant genes that control disease- or drought-resistance.

Biotechnology can be defined as the application of our knowledge and understanding of biology to meet practical needs. By this definition, biotechnology is as old as the growing of crops and the making of cheeses and wines. Today's biotechnology is largely identified with applications in medicine and agriculture based on our knowledge of the genetic code of life. Various terms have been used to describe this form of biotechnology including genetic engineering, genetic transformation, transgenic technology, recombinant DNA technology, and genetic modification technology. For the purposes of this report, which is focused on plants and products from plants, the term genetic modification technology, or GM technology is used.

GM technology was first developed in the 1970s. One of the most prominent developments, apart from the medical applications, has been the development of novel transgenic crop plant varieties. Many millions of hectares of commercially produced transgenic crops such as soybean, cotton, tobacco,

From Royal Society of London et al., "Transgenic Plants and World Agriculture," A Report Prepared Under the Auspices of the Royal Society of London, the U.S. National Academy of Sciences, the Brazilian Academy of Sciences, the Chinese Academy of Sciences, the Indian National Science Academy, the Mexican Academy of Sciences, and the Third World Academy of Sciences (July 2000).

potato and maize have been grown annually in a number of countries including the USA (28.7 million hectares in 1999), Canada (4 million), China (0.3 million), and Argentina (6.7 million) (James 1999). However, there has been much debate about the potential benefits and risks that may result from the use of such crops.

The many crucial decisions to be made in the area of biotechnology in the next century by private corporations, governments, and individuals will affect the future of humanity and the planet's natural resources. These decisions must be based on the best scientific information in order to allow effective choices for policy options. It is for this reason that representatives of seven of the world's academies of science have come together to provide recommendations to the developers and overseers of GM technology and to offer scientific perspectives to the ongoing public debate on the potential role of GM technology in world agriculture. . . .

The Need for GM Technology in World Agriculture

Today there are some 800 million people (18% of the population in the developing world) who do not have access to sufficient food to meet their needs (Pinstrup-Anderson and Pandya-Lorch 2000, Pinstrup-Anderson et al 1999), primarily because of poverty and unemployment. Malnutrition plays a significant role in half of the nearly 12 million deaths each year of children under five in developing countries (UNICEF 1998). In addition to lack of food, deficiencies in micro-nutrients (especially vitamin A, iodine and iron) are widespread. Furthermore, changes in the patterns of global climate and alterations in use of land will exacerbate the problems of regional production and demands for food. Dramatic advances are required in food production, distribution and access if we are going to address these needs. Some of these advances will occur from non-GM technologies, but others will come from the advantages offered by GM technologies.

Achieving the minimum necessary growth in total production of global staple crops—maize, rice, wheat, cassava, yams, sorghum, potatoes and sweet potatoes—without further increasing land under cultivation, will require substantial increases in yields per acre. Increases in production are also needed for other crops, such as legumes, millet, cotton, rape, bananas and plantains.

It is important to increase yield on land that is already intensively cultivated. However, increasing production is only one part of the equation. Income generation, particularly in low-income areas together with the more effective distribution of food stocks, are equally, if not more, important. GM technologies are relevant to both these elements of food security.

In developing countries, it is estimated that about 650 million of the poorest people live in rural areas where the local production of food is the main economic activity. Without successful agriculture, these people will have neither employment nor the resources they need for a better life. Farming the land, and in particular small-holder farming, is the engine of progress in the rural communities, particularly of less developed countries.

The domestication of plants for agricultural use was a long-term process with profound evolutionary consequences for many species. One of its most valuable results was the creation of a diversity of plants serving human needs. Using this stock of genetic variability through selection and breeding, the 'Green Revolution' produced many varieties that are used throughout the world. This work, carried out largely in publicly-supported research institutions, has resulted in our present high-yielding crop varieties. A good example of such selective breeding was the introduction of 'dwarf' genes into rice and wheat which, in conjunction with fertilizer applications, dramatically increased the yield of traditional food crops in the Indian sub-continent, China and elsewhere. Despite past successes, the rate of increase of food crop production has decreased recently (yield increase in the 1970s of 3% per annum has declined in the 1990s to approximately 1% per annum) (Conway *et al* 1999). There are still heavy losses of crops owing to biotic (e.g. pests and disease) and abiotic (e.g. salinity and drought) stresses. The genetic diversity of some crop plants has also decreased and there are species without wild relatives with which to cross-breed. There are fewer options available than previously to address current problems through traditional breeding techniques though it is recognised that these techniques will continue to be important in the future.

Increasing the amount of land available to cultivate crops, without having a serious impact on the environment and natural resources, is a limited option. Modern agriculture has increased production of food, but it has also introduced large-scale use of pesticides and fertilisers that are expensive and can potentially affect human health or damage the ecosystem. A major challenge faced by humankind today is how to increase world food production and people's access to food, which requires local and employment-intensive staples production, without further depleting non-renewable resources and causing environmental damage. In other words, how do we move towards sustainable agricultural practices that do not compromise the health and economic well-being of the current and future generations? In order to think in terms of sustainable agriculture, factors responsible for soil, water and environmental deterioration must be identified and corrective measures taken.

Research on transgenic crops, as with conventional plant breeding and selection by farmers, aims selectively to alter, add or remove a character of choice in a plant, bearing in mind the regional needs and opportunities. It offers the possibility of not only bringing in desirable characteristics from other varieties of the plant, but also of adding characteristics from other unrelated species. Thereafter the transgenic plant becomes a parent for use in traditional breeding. Modification of qualitative and quantitative characteristics such as the composition of protein, starch, fats or vitamins by modification of metabolic pathways has already been achieved in some species. Such modifications increase the nutritional status of the foods and may help to improve human health by addressing malnutrition and under-nutrition. GM technology has also shown its potential to address micro-nutrient deficiencies and thus reduce the national expenditure and resources required to implement the current supplementation programmes (Texas A&M University 1997). These nutritional improvements have rarely been achieved previously by traditional methods of plant breeding.

Transgenic plants with important traits such as pest and herbicide resistance are most necessary where no inherent resistance has been demonstrated within the local species. There is intense research on the development of resistance to viral, bacterial, and fungal diseases; modification of plant architecture (eg height) and development (eg early or late flowering or seed production); tolerance to abiotic stresses (eg salinity and drought); production of industrial chemicals (plant-based renewable resources); and the use of transgenic plant biomass for novel and sustainable sources of fuel. Other benefits from transgenic plants under study include increased flexibility in crop management, decreased dependency on chemical insecticides and soil disturbance, enhanced yields, easier harvesting and higher proportions of the crop available for trading. For the consumer this should lead to decreased cost of food and higher nutritive value.

A large proportion of developing world agriculture is in the hands of small-scale farmers whose interests must be taken into account. Concerns regarding GM technology range from its potential impact on human health and the environment to concerns about private sector monopolies of the technology. It is essential that such concerns are addressed if we are to reap the potential benefits of this new technology.

We conclude that steps must be taken to meet the urgent need for sustainable practices in world agriculture if the demands of an expanding world population are to be met without destroying the environment or natural resource base. In particular, GM technology, coupled with important developments in other areas, should be used to increase the production of main food staples, improve the efficiency of production, reduce the environmental impact of agriculture, and provide access to food for small-scale farmers.

Examples of GM Technology That Would Benefit World Agriculture

GM technology has been used to produce a variety of crop plants to date, primarily with 'market-led' traits, some of which have become commercially successful. Developments resulting in commercially produced varieties in countries such as the USA and Canada have centred on increasing shelf-life of fruits and vegetables, conferring resistance to insect pests or viruses, and producing tolerance to specific herbicides. While these traits have had benefits for farmers, it has been difficult for the consumers to see any benefit other than, in limited cases, a decreased price owing to reduced cost and increased ease of production (University of Illinois 1999; Falck-Zepeda et al 1999).

A possible exception is the development of GM technology that delays ripening of fruit and vegetables, thus allowing an increased length of storage. Farmers would benefit from this development by increased flexibility in production and harvest. Consumers would benefit by the availability of fruits and vegetables such as transgenic tomatoes modified to soften much more slowly than traditional varieties, resulting in improved shelf-life and decreased cost of production, higher quality and lower cost. It is possible that farmers in developing countries could benefit considerably from crops with delayed ripening or softening as this may allow them much greater flexibility in distribution than they have at present. In many cases small-scale farmers suffer heavy losses due to excessive or uncontrolled ripening or softening of fruit or vegetables.

The real potential of GM technology to help address some of the most serious concerns of world agriculture has only recently begun to be explored. The following examples show how GM technology can be applied to some of the specific problems of agriculture indicating the potential for benefits.

Pest Resistance

There is clearly a benefit to farmers if transgenic plants are developed that are resistant to a specific pest. For example, papaya-ringspot-virus-resistant Papaya has been commercialised and grown in Hawaii since 1996 (Gonsalves 1998). There may also be a benefit to the environment if the use of pesticides is reduced. Transgenic crops containing insect resistance genes from *Bacillus thuringiensis* have made it possible to reduce significantly the amount of insecticide applied on cotton in the USA. One analysis, for example, showed a reduction of five million acre-treatments (two-million-hectare-treatments) or about one million kilograms of chemicals insecticides in 1999 compared with 1998 (US National Research Council 2000). However, populations of pests and disease-causing organisms adapt readily and become resistant to pesticides, and there is no reason to suppose that this will not occur equally rapidly with transgenic plants. In addition, pest biotypes are different in various regions. For instance, insect resistant crops developed for use in the USA and Canada may be resistant to pests that are of no concern in developing countries, and this is true both for transgenic plants and those developed by conventional breeding techniques. Even where the same genes for insect or herbicide resistance are useful in different regions, typically these genes will need to be introduced into locally adapted cultivars. There is need, therefore, for more research on transgenic plants that have been made resistant to local pests to assess their sustainability in the face of increased selection pressures for ever more virulent pests.

Improved Yield

One of the major technologies that led to the 'Green Revolution' was the development of high-yielding semi-dwarf wheat varieties. The genes responsible for height reduction were the Japanese NORIN 10 genes introduced into Western wheats in the 1950s (Gibberellin-insensitive-dwarfing-genes). These genes had two benefits: they produced a shorter, stronger plant that could respond to more fertiliser without collapsing, and they increased yield directly by reducing cell elongation in the vegetative plant parts, thereby allowing the plant to invest more in the reproductive plant parts that are eaten. These genes have recently been isolated and demonstrated to act in exactly the same way when used to transform other crop plant species (Peng *et al* 1999, Worland et al 1999). This dwarfing technique can now potentially be used to increase productivity in any crop plant where the economic yield is in the reproductive rather than the vegetative parts.

Tolerance to Biotic and Abiotic Stresses

The development of crops that have an inbuilt resistance to biotic and abiotic stress would help to stabilise annual production. For example, Rice Yellow Mottle Virus (RYMV) devastates rice in Africa by destroying the majority of the crop directly, with a secondary effect on any surviving plants that makes them more susceptible to fungal infections. As a result this virus has seriously threatened rice production in Africa. Conventional approaches to the control of RYMV using traditional breeding methods have failed to introduce resistance from wild species to cultivated rice. Researchers have used a novel technique that mimics 'genetic immunisation' by creating transgenic rice plants that are resistant to RYMV (Pinto *et al* 1999). Resistant transgenic varieties are currently entering field trials to test the effectiveness of their resistance to RYMV. This could provide a solution to the threat of total crop failure in the sub-Saharan African rice growing regions.

Numerous other examples could be given to illustrate the range of current scientific research including transgenic plants modified to combat papaya ring spot virus (Souza *et al* 1999), blight resistant potatoes (Torres *et al* 1999) and rice bacterial leaf blight (Zhai *et al* 2000); or as an example of an abiotic stress, plants modified to overproduce citric acid in roots and provide better tolerance to aluminum in acid soils (de la Fuente *et al* 1997). These examples have clear commercial potential but it will be imperative to maintain publicly funded research in GM technology if their full benefits are to be realised. For example, while GM technology provides access to new gene pools for sources of resistance, it needs to be established that these sources of resistance will be more stable than the traditional intra-species sources.

Use of Marginalised Land

A vast landmass across the globe, both coastal as well as terrestrial has been marginalised because of excessive salinity and alkalinity. A salt tolerance gene from mangroves (*Avicennia marina*) has been identified, cloned and transferred to other plants. The transgenic plants were found to be tolerant to higher concentrations of salt. The gutD gene from *Escherichia coli* has also been used to generate salt-tolerant transgenic maize plants (Liu *et al* 1999). Such genes are a potential source for developing cropping systems for marginalised lands (MS Swaminathan, personal communication, 2000).

Nutritional Benefits

Vitamin A deficiency causes half a million children to become partially or totally blind each year (Conway and Toennissen 1999). Traditional breeding methods have been unsuccessful in producing crops containing a high vitamin A concentration and most national authorities rely on expensive and complicated supplementation programs to address the problem. Researchers have introduced three new genes into rice: two from daffodils and one from a micro-organism. The transgenic rice exhibits an increased production of beta-carotene as a precursor to vitamin A and the seed is yellow in colour (Ye *et al* 2000). Such yellow, or golden, rice may be a useful tool to help treat the problem of vitamin A deficiency in young children living in the tropics.

Iron fortification is required because cereal grains are deficient in essential micro-nutrients such as iron. Iron deficiency causes anaemia in pregnant women and young children. About 400 million women of childbearing age suffer as a result and they are more prone to stillborn or underweight children and to mortality at childbirth. Anaemia has been identified as a contributing factor in over 20% of maternal deaths (after giving birth) in Asia and Africa (Conway 1999). Transgenic rice with elevated levels of iron has been produced using genes involved in the production of an iron-binding protein and in the production of an enzyme that facilitates iron availability in the human diet (Goto *et al* 1999, Lucca *et al* (it). (1999). These plants contain 2 to 4 times the levels of iron normally found in non-transgenic rice, but the bioavailability of this iron will need to be ascertained by further study.

Reduced Environmental Impact

Water availability and efficient usage have become global issues. Soils subjected to extensive tillage (ploughing) for controlling weeds and preparing seed beds are prone to erosion, and there is a serious loss of water content. Low tillage systems have been used for many years in traditional communities. There is a need to develop crops that thrive under such conditions, including the introduction of resistance to root diseases currently controlled by tillage and to herbicides that can be used as a substitute for tillage (Cook 2000). Applications in more developed countries show that GM technology offers a useful tool for the introduction of root disease resistance for conditions of reduced tillage. However, a careful cost-benefit analysis would be needed to ensure that maximum advantage is achieved. Regional differences in agricultural systems and the potential impact of substituting a traditional crop with a new transgenic one would also need to be carefully evaluated.

Other Benefits of Transgenic Plants

First generation transgenic varieties have benefited many farmers in the form of reduced production costs, higher yields, or both. In many cases, they have also benefited the environment because of reduced pesticide usage or by providing the means to grow crops with less tillage. Insects are responsible for huge losses to crops in the field and to harvested products in transit or storage, but health concerns for consumers and for environmental impact have limited the registration of many promising chemical pesticides. Genes for pest resistance carefully deployed in crops to avoid selecting for future pest resistance, provides alternative opportunities to reduce the use of chemical pesticides in many important crops. In addition, lowering the contamination of our food supply by pathogens that cause food safety problems (eg mycotoxins) would be beneficial to farmers and consumers alike.

Pharmaceuticals and Vaccines From Transgenic Plants

Vaccines are available for many of the diseases that cause widespread death or human discomfort in developing countries, but they are often expensive both to produce and use. The majority must be stored under conditions of refrigeration and administered by trained specialists, all of which adds to the expense. Even the cost of needles to administer vaccines is prohibitive in some countries. As a result, the vaccines often do not reach those in most need. Researchers are currently investigating the potential for GM technology to produce vaccines and pharmaceuticals in plants. This could allow easier access, cheaper production, and an alternative way to generate income. Vaccines against infectious diseases of the gastro-intestinal tract have been produced in plants such as potato and bananas (Thanavala *et al* 1995). Another appropriate target would be cereal grains. An anti-cancer antibody has recently been expressed in rice and wheat seeds that recognises cells of lung, breast and colon cancer and hence could be useful in both diagnosis and therapy in the future (Stoger *et al* 2000). Such technologies are at a very early stage in development and obvious concerns about human health and environmental safety during production must be investigated before such plants can be approved as speciality crops. Nevertheless, the development of transgenic plants to produce therapeutic agents has immense potential to help in solving problems of disease in developing countries.

About one third of medicines used today are derived from plants, one of the most famous examples being aspirin (the acetylated form of a natural plant product, salicylic acid). It is believed that less that 10% of medicinal plants have been identified and characterised, and the potential exists to use GM technology in a way that increases yields of these medicinal substances once identified. For example, the valuable anti-cancer agents vinblastine and vincristine are the only approved drugs for treatment of Hodgkin's lymphoma. Both products are derived from the Madagascar Periwinkle, which produces them in minute concentrations along with 80 to 100 very similar chemicals. The therapeutic compounds are therefore extremely expensive to produce. Currently, there is intensive research in progress to investigate the potential of GM technology to increase the yields of active compounds, or to allow their production in other plants that are easier to manage than the Periwinkle.

We recommend that transgenic crop research and development should focus on plants that will (i) improve production stability; (ii) give nutritional benefits to the consumer; (iii) reduce the environmental impacts of intensive and extensive agriculture; and (iv) increase the availability of pharmaceuticals and vaccines; while (v) developing protocols and regulations that ensure that transgenic crops designed for purposes other than food, such as pharmaceuticals, industrial chemicals, etc. do not spread or mix with either transgenic or non-transgenic food crops.

Transgenic Plants and Human Health and Safety

Through classical plant breeding techniques, present day cultivated crops have become significantly different from their wild counterparts. Many of these crops were originally less productive and at times unsuitable for human consumption. Over the years, traditional plant breeding and selection of these crops have resulted in plants that are more productive and nutritious. The advent of GM technology has allowed further development. To date, over 30 million hectares of transgenic crops have been grown and no human health problems associated specifically with the ingestion of transgenic crops or their products have been identified. However numerous potential concerns have been raised since the development of GM technology in the early 1970s. Such concerns have focused on the potential for allergic reactions to food products, the possible introduction or increase in production of toxic compounds as a result of the GM technology, and the use of antibiotic resistance as markers in the transformation process.

Every effort should be made to avoid the introduction of known allergens into food crops. Information concerning potential allergens and natural plant toxins should be made available to researchers, industry, regulators, and the general public. In order to facilitate this effort, public databases should be developed which facilitate access of all interested parties to data.

Traditional plant breeding methods include wide crosses with closely related wild species, and may involve a long process of crossing back to the commercial parent to remove undesirable genes. A feature of GM technology is that it involves the introduction of one or at most, a few, well-defined genes rather than the introduction of whole genomes or parts of chromosomes as in traditional plant breeding. This makes toxicity testing for transgenic plants more straightforward than for conventionally produced plants with new traits, because it is much clearer what the new features are in the modified plant. On the other hand, GM technology can introduce genes from diverse organisms, some of which have little history in the food supply.

Decisions regarding safety should be based on the nature of the product, rather than on the method by which it was modified. It is important to bear in mind that many of the crop plants we use contain natural toxins and allergens. The potential for human toxicity or allergenicity should be kept under scrutiny for any novel proteins produced in plants with the potential to become part of food or feed. Health hazards from food, and how to reduce them, are an issue in all countries, quite apart from any concerns about GM technology.

Since the advent of GM technology, researchers have used antibiotic resistance genes as selective markers for the process of genetic modification. The concern has been raised that the widespread use of such genes in plants could increase the antibiotic resistance of human pathogens. Kanamycin, one of the most commonly used resistance markers for plant transformation is still used for the treatment of the following human infections: bone, respiratory tract, skin, soft-tissue, and abdominal infections, complicated urinary tract infections, endocarditis, septicaemia, and enterococcal infections. Scientists now have the means to remove these marker genes before a crop plant is developed for commercial use (Zubko *et al* 2000). Developers should continue to move rapidly to remove all such markers from transgenic plants and to utilise alternative markers for the selection of new varieties. No definitive evidence exists that these antibiotic resistance genes cause harm to humans, but because of public concerns, all those involved in the development of transgenic plants should move quickly to eliminate these markers.

Ultimately, no credible evidence from scientists or regulatory institutions will influence popular public opinion unless there is public confidence in the institutions and mechanisms that regulate such products.

We recommend: (i) public health regulatory systems need to be put in place in every country to identify and monitor any potential adverse human health effects of transgenic plants, as for any other new variety. Such systems must remain fully adaptable to rapid advances in scientific knowledge. The possibility of long-term adverse effects should be kept in view when setting up such systems. This will require coordinated efforts between nations, the sharing of experience, and the standardisation of some types of risk assessments specifically related to human health; (ii) Information should be made available to the public concerning how their food supply is regulated and its safety ensured.

Transgenic Plants and the Environment

Modern agriculture is intrinsically destructive of the environment. It is particularly destructive of biological diversity, notably when practised in a very resource-inefficient way, or when it applies technologies that are not adapted to environmental features (soils, slopes, climatic regions) of a particular area. This is true of both small-scale and large-scale agriculture. The widespread application of conventional agricultural technologies such as herbicides, pesticides, fertilisers and tillage has resulted in severe environmental damage in many parts of the world. Thus the environmental risks of new GM technologies need to be considered in the light of the risks of continuing to use conventional technologies and other commonly used farming techniques.

Some agricultural practices in parts of the developing world maintain biological diversity. This is achieved by simultaneously cultivating several varieties of a crop and mixing them with other secondary crops, thus maintaining a highly diverse community of plants (Toledo *et al* 1995; Nations *et al* 1980; Whitmore *et al* 1992).

Most of the environmental concerns about GM technology in plants have derived from the possibility of gene flow to close relatives of the transgenic plant, the possible undesirable effects of the exotic genes or traits (eg insect resistance or herbicide tolerance), and the possible effect on non-target organisms.

As with the development of any new technology, a careful approach is warranted before development of a commercial product. It must be shown that the potential impact of a transgenic plant has been carefully analysed and that if it is not neutral or innocuous, it is preferable to the impact of the conventional agricultural technologies that it is designed to replace (Campbell *et al* 1997; May 1999; Toledo *et al* 1995).

Given the limited use of transgenic plants world-wide and the relatively constrained geographic and ecological conditions of their release, concrete information about their actual effects on the environment and on biological diversity is still very sparse. As a consequence there is no consensus as to the seriousness, or even the existence, of any potential environmental harm from GM technology. There is therefore a need for a thorough risk assessment of likely consequences at an early stage in the development of all transgenic plant varieties, as well as for a monitoring system to evaluate these risks in subsequent field tests and releases.

Risk assessments need base-line information including the biology of the species, its ecology and the identification of related species, the new traits resulting from GM technology, and relevant ecological data about the site(s) in which the transgenic plant is intended to be released. This information can be very difficult to obtain in highly diverse environments. Centres of origin or diversity of cultivated plants should receive careful consideration because there will be many wild relatives to which the new traits could be transferred (Ellstrand *et al* 1999; Mikkelsen *et al* 1996; Scheffler 1993; Van Raamsdonk *et al* 1997). For special environments, transgenic plants can be developed using technologies that minimise the possibilities of gene flow via pollen and its effects on wild relatives, through the use of male sterility methods or maternal inheritance resulting from chloroplast transformation (Daniell 1999; Daniell *et al* 1998; Scott & Wilkinson 1999).

Studies of gene transfer from conventional and transgenic plants to wild relatives and other plants in the ecosystem have so far concentrated on species of economic importance such as wheat, oilseed rape and barley. A virtual absence of data, particularly for species like maize, imposes the need to carefully and continuously monitor any possible effects of novel transgenic plants in the field (Hokanson *et al* 1997; Daniell *et al* 1998). In addition there is a continued need for research on the rates of gene transfer from traditional crops to indigenous species (Ellstrand *et al* 1999).

When monitoring a small-scale pilot release of a transgenic crop the following issues should be considered in addition to any concerns specific to a particular local environment:

(a) Does the existence of a transgenic plant with resistance for a particular pest or disease exacerbate the emergence of new resistant pests or diseases, and is this problem worse than that with the traditional alternative? (Riddick & Barbosa 1998; Hillbeck *et al* 1998; Birch *et al* 1999).

(b) If traits (eg salt tolerance, disease resistance, etc) are transferred to wild varieties, is there an expansion in the niche of these species that may result in the suppression of biological diversity in the surrounding areas?

(c) Would the widespread adoption of stress-tolerant plants promote a considerable increase in the use of land where formerly agriculture could not be practised in a way that destroys valuable natural ecosystems?

The risk assessments performed should be standardized for plants new to an environment. Most nations already have procedures for the approval and local release of new varieties of crop plants. Although these assessments are based primarily on the agronomic performance of the new variety compared with existing varieties, this approval process could serve as the beginning or model for a more formal risk assessment process to investigate the potential environmental impact of the new varieties, including those with transgenes.

Historically, both poverty and structural change in rural areas have resulted in severe environmental deterioration. The adoption of modern biotechnology should not accelerate this deterioration. It should instead be used in a way that reduces poverty and its deleterious effects on the environment.

We recommend that: (i) coordinated efforts be undertaken to investigate the potential environmental effects, both positive and negative, of transgenic plant technologies in their specific applications; (ii) all environmental effects should be assessed against the background of effects from conventional agricultural practices currently in use in places for which the transgenic crop has been developed or grown; and (iii) *in situ* and *ex situ* conservation of genetic resources for agriculture should be promoted that will guarantee the widespread availability of both conventional and transgenic varieties as germplasm for future plant breeding.

Membership of Working Group and Methodology

The following individuals represented the Councils of the Brazilian Academy of Sciences, the Chinese Academy of Sciences, the Indian National Science Academy, the Mexican Academy of Sciences, the Royal Society (UK), the Third World Academy of Sciences and the National Academy of Sciences of the USA during the preparation of this report. The text of the report was produced following meetings at the Royal Society (Chairman Professor Brian Heap FRS, Secretary Dr Rebecca Bowden) in London in July 1999 and February 2000 at which the issues covered in this report were discussed in detail.

The Brazilian Academy of Sciences
On working group:

- Dr Ernesto Paterniani
- Dr Fernando Perez
- Professor Fernando Reinach Professor
- Jose Galizia Tundisi

The Chinese Academy of Sciences
On working group:

- Professor Zhihong Xu
- Professor Rongxiang Fang
- Professor Qian Yingqian

The Indian National Science Academy
On working group:

- Professor R P Sharma
- Professor S K Sopory

Reviewers on behalf of Council:

- Professor P N Tandon, Chairman
- Dr H K Jain
- Dr Manju Sharma
- Dr R S Paroda
- Dr Anupam Varma
- Ms Suman Sahai
- Dr J Thomas
- Professor K Muralidhar

The Mexican Academy of Sciences
On working group:

- Mr Jorge Larson
- Dr Jorge Nieto Sotelo
- Professor Josè Sarukhàn

The Royal Society of London
On working group:

- Sir Aaron Klug OM PRS
- Professor Michael Gale FRS
- Professor Michael Lipton

Reviewed and approved by the Council of the Royal Society.

The Third World Academy of Sciences
On working group:

- Professor Muhammad Akhtar FRS

The National Academy of Sciences of the USA
(Staff Officer to NAS Delegation—Mr John Campbell)

On working group:

- Professor Bruce Alberts
- Professor F Sherwood Rowland
- Professor Luis Sequiera
- Professor R James Cook
- Professor Alex McCalla

References

Birch et al. (1999). *Molecular Breeding* **5** 75–83

Campbell L H et al. (1997). The Joint Nature Conservation Committee, Report 227, UK.

Conway G. (1999). *Biotechnology, Food & Drought* in Proceedings of the World Commission on Water, Nov 1999.

Conway G. (1999). *The Doubly Green Revolution: Food for All in the 21st Century.* London: Penguin Books

Conway G et al. (1999). *Nature* 402 C55–58

Cook R J (2000) in *Agricultural Biotechnology and the Poor*, proceedings of an international conference, Washington DC, October 1999, 123–130 (G I Persley and M M Lantin, eds).

Daniell H et al (1998). *Nature Biotechnology* **16** 345–348.

Daniell H. (1999). *Trends in Plant Science* **4(12)** 467–469.

De la Fuente J M et al. (1997). *Science* **276 (5318)** 1566–1568.

Ellestrand N C *et al.* (1999). Annual Review of ecological Systems 30 539-563.

Falck-Zepeda B J et al. (1999). *International Service for the Acquisition of Agri-Biotech Applications* **14** 17.

Gonsalves D. (1998). *Annual Review of Phytopathology* **36** 415–437.

Goto F et al. (1999). *Nature Biotechnology* **17** 282–286.

Hillbeck A et al. (1998). *Environmental Entomology* **27** 480–487.

Hokanson et al. (1997). *Euphytica* **96** 397–403.

James C. (1999). *Global status of transgenic crops in* 1999. ISAAA: Ithaca, New York.

Liu Y et al. (1999). *Science in China (Series C)* **42** 90–95.

Lucca P et al (1999). In *Proceedings of General Meeting of the International Programme on Rice Biotechnology*, Sept 1999, Phuket, Thailand.

Mikkelsen T et al. (1996). *Nature* 380 31.

Nations J et al. (1980). *Journal of Anthropological Research* **36** 1–30.

Peng J R et al. (1999). *Nature* **400** 256–261.

Pinstrup-Anderson P et al. (1999). *World food prospects: critical issues for the early 21st Century.* International Food Policy Research Institute: Washington D C, USA.

Pinstrup-Anderson P *et al.* (2000). *Meeting food needs in the 21 st century: how many and who will be at risk?* Presented at AAAS Annual Meeting, Feb 2000, Washington D C, USA.

Pinto Y M et al. (1999) *Nature Biotechnology* **17** 702–707.

Riddick E et al. (1998). *Annals of the Entomology Society of America* **91** 303–307.

Scheffler J. (1993). *Transgenic Research* **2** 356–364.

Scott S E et al. (1999). *Nature Biotechnology* **17** 390–392.

Souza M T. (1999). *Analysis of the resistance in genetically engineered papaya against papaya ringspot potyvirus, partial characterisation of the PRSV.Brasil. Bahia isolate and development of transgenic papaya for Brazil.* PhD dissertation Cornell University, USA.

Stoger E *et al.* (2000). *Plant Molecular Biology* **42** 583–590.

Texas A & M University. (1997). *Report filed with USEPA for hearing on 21 May 1997.* Docket OPP-0478.

Thanavala Y et al. (1995). *Proceedings of the National Academy of Sciences of the USA* **92(8)** 3358–3361.

Toledo V M et al. (1995). *Interciencia* **20** 177–187.

Torres A C et al. (1999). Biotechnologia—*Ciencia & Disenvolvimento* **2** (**7**) 74–77.

UNICEF. (1998). *The state of the world's children 1998: focus on nutrition.* United Nations: New York, USA.

U S National Research Council. (2000). *Genetically modified pest-protected plants: science and regulation.* National Academy Press: Washington, D C, USA.

University of Illinois. (1999). *The economics and politics of genetically modified organisms in agriculture: implications for WTO 2000.* USA: University of Illinois Bulletin 809, November 1999

Van Raamsdonk L et al. (1997). *Acta Botanica Neerlandica* **48** 9–84.

Whitmore T M et al. (1992). *Annals of the Association of American Geographers* **82** 402–425.

Worland A J *et al.* (1999) Nature 400 256-261.

Xudong Ye et al. (2000). *Science* **287** 303–306.

Zhai W et al. (2000). *Science in China (Series C)* 43 361–368.

NO

Brian Halweil

The Emperor's New Crops

It's June 1998 and Robert Shapiro, CEO of Monsanto Corporation, is delivering a keynote speech at "BIO 98," the annual meeting of the Biotechnology Industry Organization. "Somehow," he says, "we're going to have to figure out how to meet a demand for a doubling of the world's food supply, when it's impossible to conceive of a doubling of the world's acreage under cultivation. And it is impossible, indeed, even to conceive of increases in productivity—using current technologies—that don't produce major issues for the sustainability of agriculture."

Those "major issues" preoccupy a growing number of economists, environmentalists, and other analysts concerned with agriculture. Given the widespread erosion of topsoil, the continued loss of genetic variety in the major crop species, the uncertain effects of long-term agrochemical use, and the chronic hunger that now haunts nearly 1 billion people, it would seem that a major paradigm shift in agriculture is long overdue. Yet Shapiro was anything but gloomy. Noting "the sense of excitement, energy, and confidence" that engulfed the room, he argued that "biotechnology represents a potentially sustainable solution to the issue of feeding people."

To its proponents, biotech is the key to that new agricultural paradigm. They envision crops genetically engineered to tolerate dry, low-nutrient, or salty soils—allowing some of the world's most degraded farmland to flourish once again. Crops that produce their own pesticides would reduce the need for toxic chemicals, and engineering for better nutrition would help the overfed as well as the hungry. In industry gatherings, biotech appears as some rare hybrid between corporate mega-opportunity and international social program.

The roots of this new paradigm were put down nearly 50 years ago, when James Watson and Francis Crick defined the structure of DNA, the giant molecule that makes up a cell's chromosomes. Once the structure of the genetic code was understood, researchers began looking for ways to isolate little snippets of DNA—particular genes—and manipulate them in various ways. In 1973, scientists managed to paste a gene from one microbe into another microbe of a different species; the result was the first artificial transfer of genetic information across the species boundary. In the early 1980s, several research teams—including one at Monsanto, then a multinational pesticide company—

succeeded in splicing a bacterium gene into a petunia. The first "transgenic" plant was born.

Such plants represented a quantum leap in crop breeding: the fact that a plant could not interbreed with a bacterium was no longer an obstacle to using the microbe's genes in crop design. Theoretically, at least, the world's entire store of genetic wealth became available to plant breeders, and the biotech labs were quick to test the new possibilities. Among the early creations was a tomato armed with a flounder gene to enhance frost resistance and with a rebuilt tomato gene to retard spoilage....

Transgenic crops are no longer just a laboratory phenomenon. Since 1986, 25,000 transgenic field trials have been conducted worldwide—a full 10,000 of these just in the last two years. More than 60 different crops—ranging from corn to strawberries, from apples to potatoes—have been engineered. From 2 million hectares in 1996, the global area planted in transgenics jumped to 27.8 million hectares in 1998. That's nearly a fifteenfold increase in just two years.

In 1992, China planted out a tobacco variety engineered to resist viruses and became the first nation to grow transgenic crops for commercial use. Farmers in the United States sowed their first commercial crop in 1994; their counterparts in Argentina, Australia, Canada, and Mexico followed suit in 1996. By 1998, nine nations were growing transgenics for market and that number is expected to reach 20 to 25 by 2000.

Ag biotech is now a global phenomenon, but it remains powerfully concentrated in several ways:

In terms of where transgenics are planted.   Three-quarters of transgenic cropland is in the United States. More than a third of the U.S. soybean crop last year was transgenic, as was nearly one-quarter of the corn and one-fifth of the cotton. The only other countries with a substantial transgenic harvest are Argentina and Canada: over half of the 1998 Argentine soybean crop was transgenic, as was over half of the Canadian canola crop. These three nations account for 99 percent of global transgenic crop area. (Most countries have been slow to adopt transgenics because of public concern over possible risks to ecological and human health.)

In terms of which crops are in production.   While many crops have been engineered, only a very few are cultivated in appreciable quantities. Soybeans account for 52 percent of global transgenic area, corn for another 30 percent. Cotton—almost entirely on U.S. soil—and canola in Canada cover most of the rest.

In terms of which traits are in commercial use.   Most of the transgenic harvest has been engineered for "input traits" intended to replace or accommodate the standard chemical "inputs" of large-scale agriculture, especially insecticides and herbicides. Worldwide, nearly 30 percent of transgenic cropland is planted in varieties designed to produce an insect-killing toxin, and almost all of the rest is in crops engineered to resist herbicides....

These two types of crops—the insecticidal and the herbicide-resistant varieties—are biotech's first large-scale commercial ventures. They provide the first real opportunity to test the industry's claims to be engineering a new agricultural paradigm.

The Bugs

The only insecticidal transgenics currently in commercial use are "Bt crops." Grown on nearly 8 million hectares worldwide in 1998, these plants have been equipped with a gene from the soil organism *Bacillus thuringiensis* (Bt), which produces a substance that is deadly to certain insects.

The idea behind Bt crops is to free conventional agriculture from the highly toxic synthetic pesticides that have defined pest control since World War II. Shapiro, for instance, speaks of Monsanto's Bt cotton as a way of substituting "information encoded in a gene in a cotton plant for airplanes flying over cotton fields and spraying toxic chemicals on them." . . . At least in the short term, Bt varieties have allowed farmers to cut their spraying of insecticide-intensive crops, like cotton and potato. In 1998, for instance, the typical Bt cotton grower in Mississippi sprayed only once for tobacco budworm and cotton bollworm—the insects targeted by Bt—while non-Bt growers averaged five sprayings.

Farmers are buying into this approach in a big way. Bt crops have had some of the highest adoption rates that the seed industry has ever seen for new varieties. In the United States, just a few years after commercialization, nearly 25 percent of the corn crop and 20 percent of the cotton crop is Bt. In some counties in the southeastern states, the adoption rate of Bt cotton has reached 70 percent. The big draw for farmers is a lowering of production costs from reduced insecticide spraying, although the savings is partly offset by the more expensive seed. Some farmers also report that Bt crops are doing a better job of pest control than conventional spraying, although the crops must still be sprayed for pests that are unaffected by Bt. (Bt is toxic primarily to members of the Lepidoptera, the butterfly and moth family, and the Coleoptera, the beetle family.)

Unfortunately, there is a systemic problem in the background that will almost certainly erode these gains: pesticide resistance. Modern pest management tends to be very narrowly focused; the idea, essentially, is that when faced with a problematic pest, you should look for a chemical to kill it. The result has been a continual toughening of the pests, which has rendered successive generations of chemicals useless. After more than 50 years of this evolutionary rivalry, there is abundant evidence that pests of all sorts—insects, weeds, or pathogens—will develop resistance to just about any chemical that humans throw at them.

The Bt transgenics basically just replace an insecticide that is sprayed on the crop with one that is packaged inside it. The technique may be more sophisticated but the strategy remains the same: aim the chemical at the pest. Some entomologists are predicting that, without comprehensive strategies to prevent it, pest resistance to Bt could appear in the field within three to five years of widespread use, rendering the crops ineffective. Widespread resistance to Bt would affect more than the transgenic crops, since Bt is also commonly used in conventional spraying. Farmers could find one of their most environmentally benign pesticides beginning to slip away.

In one respect, Bt crops are a throwback to the early days of synthetic pesticides, when farmers were encouraged to spray even if their crops didn't appear to need it. The Bt crops show a similar lack of discrimination: they are programmed to churn out toxin during the entire growing season, regardless of the level of infestation. This sort of prophylactic control greatly increases the likelihood of resistance because it tends to maximize exposure to the toxin—it's the plant equivalent of treating antibiotics like vitamins.

Agricultural entomologists now generally agree that Bt crops will have to be managed in a way that discourages resistance if the effectiveness of Bt is to be maintained. In the United States, the Environmental Protection Agency, which regulates the use of pesticides, now requires producers of Bt crops to develop "resistance management plans." This is a new step for the EPA, which has never required analogous plans from manufacturers of conventional pesticides.

The usual form of resistance management involves the creation of "refugia"—areas planted in a crop variety that isn't armed with the Bt gene. If the refugia are large enough, then a substantial proportion of the target pest population will never encounter the Bt toxin, and will not be under any selection pressure to develop resistance to it. Interbreeding between the refugia insects and the insects in the Bt fields should stall the development of resistance in the population as a whole, assuming the resistance gene is recessive.

The biotech companies themselves have been recommending that their customers plant refugia, although the recommendations generally fall short of what most resistance experts consider necessary. This is not surprising, of course, since there is an inherent inconsistency between the refugia idea and the inevitable interest on the part of the manufacturer in selling as much product as possible. An even greater obstacle may be the reactions of farmers themselves, since the refugia concept is counter-intuitive: farmers, who spend much of their lives trying to control pests, are being told that the best way to maintain a high yield is to leave substantial portions of their land vulnerable to pests. The impulse to plant smaller refugia—or to count someone else's land as part of one's own refugia—may prove irresistible. And the possibility of enforcing the planting of larger refugia seems remote, especially once Bt crops are deployed to hundreds of millions of small-scale farmers throughout the developing world....

According to Gary Barton, director of ag biotech communications at Monsanto, "products now in the pipeline which rely on different insecticidal toxins or multiple toxins could replace Bt crops in the event of widespread resistance."...

The result, according to Fred Gould, an entomologist at the University of North Carolina, would be "a crop with a series of silver bullet pest solutions." And each of these solutions, in Gould's view, would be highly vulnerable to pest resistance. This scenario does not differ essentially from the current one: in place of a pesticide treadmill, we would substitute a sort of gene treadmill. The arms race between farmers and pests would continue, but would include an additional biochemical dimension. Transgenic plants, designed to secrete increasingly potent combinations of pesticides, would vie with a host of increasingly resistant pests.

Figure 1

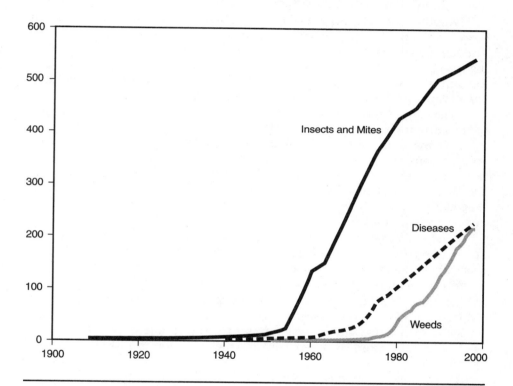

Reported Numbers of Pesticide-Resistant Species, 1908–98

Source: *Worldwatch Institute,* Vital Signs 1999

The Weeds

The global transgenic harvest is currently dominated, not by Bt crops, but by herbicide-resistant crops (HRCs), which occupy 20 million hectares worldwide. HRCs are sold as part of a "technology package" comprised of HRC seed and the herbicide the crop is designed to resist. The two principal product lines are currently Monsanto's "Roundup Ready" crops—so-named because they tolerate Monsanto's best-selling herbicide, "Roundup" (glyphosate)—and AgrEvo's "Liberty Link" crops, which tolerate that company's "Liberty" herbicide (glufosinate).

It may sound contradictory, but one ostensible objective of HRCs is to reduce herbicide use. By designing crops that tolerate fairly high levels of exposure to a broad-spectrum herbicide (a chemical that is toxic to a wide range of plants), the companies are giving farmers the option of using a heavy, once-in-the-growing-season dousing with that herbicide, instead of the standard practice, which calls for a series of applications of several different compounds. It's not yet clear whether this new herbicide regime actually reduces the amount of material used, but its simplicity is attracting many farmers into the package.

Another potential benefit of HRCs is that they may allow for more "conservation tillage," farming techniques that reduce the need for plowing or even —under "no till" cultivation—eliminate it entirely. A primary reason for plowing is to break up the weeds, but because it exposes bare earth, plowing causes top soil erosion. Top soil is the capital upon which agriculture is built, so conserving soil is one of agriculture's primary responsibilities....

The bigger problem is that HRCs, like Bt crops, are really just an extension of the current pesticide paradigm. HRCs may permit a reduction in herbicide use over the short term, but obviously their widespread adoption would encourage herbicide dependency. In many parts of the developing world, where herbicides are not now common, the herbicide habit could mean substantial additional environmental stresses: herbicides are toxic to many soil organisms, they can pollute groundwater, and they may have long-term effects on both people and wildlife.

And of course, resistance will occur. Bob Hartzler, a weed scientist at Iowa State University, warns that if HRCs encourage reliance on just a few broad-spectrum herbicides, then resistance is likely to develop faster—and agriculture is likely to be more vulnerable to it....

In the U.S. Midwest, heavy use of Roundup on Roundup Ready soybeans is already encouraging weed species, like waterhemp, that are naturally resistant to that herbicide. (As Roundup suppresses the susceptible weeds, the resistant ones have more room to grow.) Thus far, the evolution of resistance in weed species that are susceptible to Roundup has been relatively rare, despite decades of use. The first reported case involved wild ryegrass in Australia, in 1995. But with increasing use, more such cases are all but inevitable—especially since Monsanto is on the verge of releasing Roundup Ready corn. Corn and soybeans are the classic crop rotation in the U.S. Midwest—corn is planted in one year, soy in the next. Roundup Ready varieties of both crops could subject vast areas of the U.S. "breadbasket" to an unremitting rain of that herbicide. As with the Bt crops, the early promise of HRCs is liable to be undercut by the very mentality that inspired them: the single-minded chemical pursuit of the pest.

Transgenes on the Loose

In 1997, just one year after its first commercial planting in Canada, a farmer reported—and DNA testing confirmed—that Roundup Ready canola had cross-pollinated with a related weed species growing in the field's margins, and produced an herbicide-tolerant descendant....

If a transgenic crop is capable of sexual reproduction (and they generally are), the leaking of "transgenes" is to some degree inevitable, if any close relatives are growing in the vicinity. This type of genetic pollution is not likely to be common in the industrialized countries, where most major crops have relatively few close relatives. But in the developing world—especially in regions where a major crop originated—the picture is very different. Such places are the "hot spots" of agricultural diversity: the cultivation of the ancient, traditional varieties—whether it's corn in Mexico or soybeans in China—often involves a subtle genetic interplay between cultivated forms of a species, wild forms, and related species that aren't cultivated at all. The possibilities for genetic pollution in such contexts are substantial.

Ordinary breeding creates some degree of genetic pollution too. But according to Allison Snow, an Ohio State University plant ecologist who studies transgene flow, biotech could amplify the process considerably because of the far more diverse array of genes it can press into service. Any traits that confer a substantial competitive advantage in the wild could be expected to spread widely. The Bt gene would presumably be an excellent candidate for this process, since its toxin affects so many insect species.

There's no way to predict what would happen if the Bt gene were to escape into a wild flora, but there's good reason to be concerned. John Losey, an entomologist at Cornell University, has been experimenting with Monarch butterflies, by raising their caterpillars on milkweed dusted with Bt-corn pollen. Losey found that nearly half of the insects raised on this fare died and the rest were stunted. (Caterpillars raised on milkweed dusted with ordinary corn pollen did fine.) According to Losey, "these levels of mortality are comparable to those you find with especially toxic insecticides." If the gene were to work a change that dramatic in a wild plant's toxicity, then it could trigger a cascade of second- and third-order ecological effects.

The potential for this kind of trouble is likely to grow, since a major interest in biotech product development is "trait-stacking"—combining several engineered genes in a single variety, as with the attempts to develop corn with multiple toxins. Monsanto's "stacked cotton"—Roundup Ready and Bt-producing—is already on the market in the United States....

In the agricultural hot spots, there is an important practical reason to be concerned about any resulting genetic pollution. Plant breeders depend on the genetic wealth of the hot spots to maintain the vigor of the major crops—and there's no realistic possibility of biotech rendering this natural wealth "obsolete." But it certainly is possible that foreign genes could upset the relationships between the local varieties and their wild relatives....

Toward a New Feudalism

The advent of transgenic crops raise serious social questions as well—beginning with ownership. All transgenic seed is patented, as are most nontransgenic commercial varieties. But beginning in the 1980s, the tendency in industrialized countries and in international law has been to permit increasingly broad agricultural patents—and not just on varieties but even on specific genes. Under the earlier, more limited patents, farmers could buy seed and use it in their own breeding; they could grow it out and save some of the resulting seed for the next year; they could even trade it for other seed. About the only thing they couldn't do was sell it outright. But under the broader patents, all of those activities are illegal; the purchaser is essentially just paying for one-time use of the germplasm.

The right to own genes is a relatively new phenomenon in world history and its effects on agriculture—and life in general—are still very uncertain. The biotech companies argue that ownership is essential for driving their industry: without exclusive rights to a product that costs hundreds of millions of dollars to develop, how will it be possible to attract investors? ... Val Giddings of the Biotechnology Industry Organization makes this case: "intellectual property rights allow us to harness genetic resources for commercial use, making biodiversity concretely more valuable." ...

Patents are clearly an important ingredient in the industry's expansion. Global sales of transgenic crop products grew from $75 million in 1995 to $1.5 billion in 1998—a 20-fold increase. Sales are expected to hit $25 billion by 2010. And as the market has expanded, so has the scramble for patents. Recently, for example, the German agrochemical firm AgrEvo, the maker of "Liberty" herbicide, bought a Dutch biotech company called Plant Genetic Systems (PGS), which owned numerous wheat and corn patents. The patents were so highly valued that AgrEvo was willing to pay $730 million for the acquisition—$700 million more than PGS's annual sales. . . .

This patent frenzy is contributing to an intense wave of consolidation within the industry. . . . Hoechst recently merged with one of its French counterparts, Rhône-Poulenc, to form Aventis, which is now the world's largest agrochemical firm and a major player in the biotech industry. On the other side of the Atlantic, Monsanto has spent nearly $8 billion since 1996 to purchase various seed companies. DuPont, a major competitor, has bought the world's largest seed company, Pioneer Hi-Bred. DuPont and Monsanto were minor players in the seed industry just a decade ago, but are now respectively the largest and second-largest seed companies in the world.

. . . Of the 56 transgenic products approved for commercial planting in 1998, 33 belonged to just four corporations: Monsanto, Aventis, Novartis, and DuPont. The first three of these companies control the transgenic seed market in the United States, which amounts to three-fourths of the global market. . . .

But there is far more at stake here than the fortunes of the industry itself: patents and similar legal mechanisms may be giving companies additional control over farmers. As a way of securing their patent rights, biotech companies are requiring farmers to sign "seed contracts" when they purchase transgenic seed —a wholly new phenomenon in agriculture. The contracts may stipulate what brand of pesticides the farmer must use on the crop—a kind of legal cement for those crop-herbicide "technology packages." ...

The most troubling aspect of these contracts is the possible effect on seed saving—the ancient practice of reserving a certain amount of harvested seed for the next planting. In the developing world, some 1.4 billion farmers still rely almost exclusively on seed saving for their planting needs. As a widespread, low-tech form of breeding, seed saving is also critical to the husbandry of crop diversity, since farmers generally save seed from plants that have done best under local conditions. The contracts have little immediate relevance to seed saving in the developing world, since the practice there is employed largely by farmers who could not afford transgenic seed in the first place. But even in industrialized countries, seed saving is still common in certain areas and for certain crops, and Monsanto has already taken legal action against over 300 farmers for replanting proprietary seeds.

... The substitution of commercial for farm-saved seed has been a primary reason for the loss of genetic diversity in the agricultural hot spots. Hope Shand, research director for the Rural Advancement Foundation International (RAFI), a farmer advocacy group based in Winnipeg, Canada, regards the extension of patents in general as a means of reducing farmers to "bioserfs," who provide little more than land and labor to agribusiness.

... [B]eyond these control issues, there remains the basic question of biotech's potential for feeding the world's billions. Here too, the current trends are not very encouraging. At present, the industry has funneled its immense pool of investment into a limited range of products for which there are large, secured markets within the capital-intensive production systems of the First World. There is very little connection between that kind of research and the lives of the world's hungry. HRCs, for example, are not helpful to poor farmers who rely on manual labor to pull weeds because they couldn't possibly afford herbicides. (The immediate opportunities for biotech in the developing world are not the subsistence farmers, of course, but the larger operations, which are often producing for export rather than for local consumption.)

Just to get a sense of proportion on this subject, consider this comparison. The entire annual budget of the Consultative Group for International Agricultural Research (CGIAR), a consortium of international research centers that form the world's largest public-sector crop breeding effort, amounts to $400 million. The amount that Monsanto spent to develop Roundup Ready soybeans alone is estimated at $500 million. In such numbers, one can see a kind of financial disconnect. Per Pinstrup-Andersen, director of the International Food Policy Research Institute, the CGIAR's policy arm, puts it flatly: "the private sector will not develop crops to solve poor people's problems, because there is not enough money in it." The very nature of their affliction—poverty—makes hungry people poor customers for expensive technologies.

In addition to the financial obstacle, there is a biological obstacle that may limit the role of biotech as agricultural savior. The crop traits that would be most useful to subsistence farmers tend to be very complex. Miguel Altieri, an entomologist at the University of California at Berkeley, identifies the kind of products that would make sense in a subsistence context: "crop varieties responsive to low levels of soil fertility, crops tolerant of saline or drought conditions and other stresses of marginal lands, improved varieties that are not dependent on agro-chemical inputs for increased yields, varieties that are compatible with small, diverse, capital-poor farm settings." In HRCs and Bt crops, the engineering involves the insertion of a single gene. Most of the traits Altieri is talking about are probably governed by many genes, and for the present at least, that kind of complexity is far beyond the technology's reach.

Beyond the Techno-Fix

... On a 300-acre farm in Boone, Iowa—the heart of the U.S. corn belt—Dick Thompson rotates corn, soybeans, oats, wheat interplanted with clover, and a hay combination that includes an assortment of grasses and legumes. The pests that plague neighboring farmers—including the corn borer targeted by Bt corn—are generally a minor part of the picture on Thompson's farm. High crop diversity tends to reduce insect populations because insect pests are usually "specialists" on one particular crop. In a very diverse setting, no single pest is likely to be able to get the upper hand. Diversity also tends to shut out weeds, because complex cropping uses resources more efficiently than mono-cultures, so there's less left over for the weeds to consume.... Even without herbicides, Thompson's farm has been on conservation tillage for the last three decades.... [C]attle, a hog operation, and the nitrogen-fixing legumes provide the soil nutrients that most U.S. farmers buy in a bag. The soil organic matter content—the sentinel indicator of soil health—registers at 6 percent on Thompson's land, which is more than twice that of his neighbors.... Thompson's soybean and corn yields are well above the county average and even as the U.S. government continues to bail out indebted farmers, Thompson is making money. He profits both from his healthy soil and crops, and from the fact that his "input" costs—for chemical fertilizer, pesticides, and so forth—are almost nil.

In the activities of people like Herren and Thompson it is possible to see a very different kind of agricultural paradigm, which could move farming beyond the techno-fix approach that currently prevails. Known as agroecology, this paradigm recognizes the farm as an ecosystem—an agroecosystem—and employs ecological principles to improve productivity and build stability. The emphasis is on adapting farm design and practice to the ecological processes actually occurring in the fields and in the landscape that surrounds them. Agroecology aims to substitute detailed (and usually local) ecological knowledge for off-the-shelf and off-the-farm "magic bullet" solutions. The point is to treat the disease, rather than just the symptoms. Instead of engineering a corn variety that is toxic to corn rootworm, for example, an agroecologist would ask why there's a rootworm problem in the first place.

Where would biotech fit within such a paradigm? In the industry's current form, at least, it doesn't appear to fit very well at all. Biotech's first agricultural products are "derivative technologies," to use a term favored by Frederick Buttel, a rural sociologist at the University of Wisconsin. Buttel sees those products as "grafted onto an established trajectory, rather than defining or crystallizing a new one."

There is no question that biotech contains some real potential for agriculture, for instance as a supplement to conventional breeding or as a means of studying crop pathogens. But if the industry continues to follow its current trajectory, then biotech's likely contribution will be marginal at best and at worst, given the additional dimensions of ecological and social unpredictabil-ity—who knows? In any case, the biggest hope for agriculture is not something biochemists are going to find in a test tube. The biggest opportunities will be found in what farmers already know, or in what they can readily discover on their farms.

POSTSCRIPT

Is Genetic Engineering an Environmentally Sound Way to Increase Food Production?

The full report of the Royal Society of London et al. includes sections that deal with funding issues, capacity building, and intellectual property. The report, complete with references, can be found on the Internet at http://www.royalsoc.ac.uk/policy/index.html.

Kathleen Hart, in *Eating in the Dark* (Pantheon, 2002), expresses horror at the fact that "Frankenfood" is not labeled and that U.S. consumers are not as alarmed by genetically modified foods as European consumers are. The worries—and the scientific evidence to support them—are summarized by Kathryn Brown, in "Seeds of Concern," and Karen Hopkin, in "The Risks on the Table," *Scientific American* (April 2001). In the same issue, Sasha Nemecek poses the question "Does the World Need GM Foods?" to two prominent figures in the debate: Robert B. Horsch, vice president of the Monsanto Corporation and recipient of the 1998 National Medal of Technology for his work on modifying plant genes, says yes; Margaret Mellon, of the Union of Concerned Scientists, says no, adding that much more work needs to be done with regard to safety. The May 2002 U.S. General Accounting Office Report to Congressional Requesters, "Genetically Modified Foods: Experts View Regimen of Safety Tests as Adequate, but FDA's Evaluation Process Could Be Enhanced," urges more attention to verifying safety testing performed by biotechnology companies. Carl F. Jordan, in "Genetic Engineering, the Farm Crisis, and World Hunger," *Bioscience* (June 2002), says that a major problem is already apparent in the way agricultural biotechnology is widening the gap between the rich and the poor.

Charles Mann, in "Biotech Goes Wild," *Technology Review* (July/August 1999), discusses the continuing "lack of a rigorous regulatory framework to sort out the risks inherent in agricultural biotech." Margaret Kriz, in "Global Food Fight," *National Journal* (March 4, 2000), describes the January 2000 Montreal meeting, in which representatives of 130 countries reached "an agreement that requires biotechnology companies to ask permission before importing genetically altered seeds [and] forces food companies to clearly identify all commodity shipments that may contain genetically altered grain." Also see "Environmental Effects of Transgenic Plants: The Scope and Adequacy of Regulation," a report of the National Research Council's Committee on Environmental Impacts Associated With Commercialization of Transgenic Crops (National Academy Press, 2002).

ISSUE 10

Do Environmental Hormone Mimics Pose a Potentially Serious Health Threat?

YES: Sheldon Krimsky, from "Hormone Disruptors: A Clue to Understanding the Environmental Causes of Disease," *Environment* (June 2001)

NO: Stephen H. Safe, from "Environmental and Dietary Estrogens and Human Health: Is There a Problem?" *Environmental Health Perspectives* (April 1995)

ISSUE SUMMARY

YES: Professor of urban and environmental policy Sheldon Krimsky summarizes the evidence indicating that many chemicals released to the environment affect the endocrine systems of animals and humans and may threaten human health with cancers, reproductive anomalies, and neurological effects.

NO: Toxicologist Stephen H. Safe argues that the suggestion that industrial estrogenic compounds contribute to increased cancer incidence and reproductive problems in humans is not plausible.

Following World War II there was an exponential growth in the industrial use and marketing of synthetic chemicals. These chemicals, known as "xenobiotics," were used in numerous products, including solvents, pesticides, refrigerants, coolants, and raw materials for plastics. This resulted in increasing environmental contamination. Many of these chemicals, such as DDT, PCBs, and dioxins, proved to be highly resistant to degradation in the environment; they accumulated in wildlife and were serious contaminants of lakes and estuaries. Carried by winds and ocean currents, these chemicals were soon detected in samples taken from the most remote regions of the planet, far from their points of introduction into the ecosphere.

Until very recently most efforts to assess the potential toxicity of synthetic chemicals to bio-organisms, including human beings, focused almost exclusively on their possible role as carcinogens. This was because of legitimate public concern about rising cancer rates and the belief that cancer causation was the most likely outcome of exposure to low levels of synthetic chemicals.

Some environmental scientists urged public health officials to give serious consideration to other possible health effects of xenobiotics. They were generally ignored because of limited funding and the common belief that toxic effects other than cancer required larger exposures than usually resulted from environmental contamination.

In the late 1980s Theo Colborn, a research scientist for the World Wildlife Fund who was then working on a study of pollution in the Great Lakes, began linking together the results of a growing series of isolated studies. Researchers in the Great Lakes region, as well as in Florida, the West Coast, and Northern Europe, had observed widespread evidence of serious and frequently lethal physiological problems involving abnormal reproductive development, unusual sexual behavior, and neurological problems exhibited by a diverse group of animal species, including fish, reptiles, amphibians, birds, and marine mammals. Through Colborn's insights, communications among these researchers, and further studies, a hypothesis was developed that all of these wildlife problems were manifestations of abnormal estrogenic activity. The causative agents were identified as more than 50 synthetic chemical compounds that have been shown in laboratory studies to either mimic the action or disrupt the normal function of the powerful estrogenic hormones responsible for female sexual development and many other biological functions.

Concern that human exposure to these ubiquitous environmental contaminants may have serious health repercussions was heightened by a widely publicized European research study, which concluded that male sperm counts had decreased by 50 percent over the past several decades (a result that is disputed by other researchers) and that testicular cancer rates have tripled. Some scientists have also proposed a link between breast cancer and estrogen disrupters.

In response to the mounting scientific evidence that environmental estrogens may be a serious health threat, the U.S. Congress passed legislation requiring that all pesticides be screened for estrogenic activity and that the Environmental Protection Agency (EPA) develop procedures for detecting environmental estrogenic contaminants in drinking water supplies; see the EPA's Endocrine Disruptor Screening Program Web site at http://www.epa.gov/scipoly/oscpendo/index.htm. Government-sponsored studies of synthetic endocrine disrupters and other hormone mimics are also under way in the United Kingdom and in Germany.

In the following selection, Sheldon Krimsky summarizes the evidence indicating that many chemicals released to the environment affect the endocrine systems of animals and humans and may threaten human health with cancers, reproductive anomalies, and neurological effects. He cautions that the regulatory machinery is likely to move very slowly and says that if we wish to take a precautionary approach, we cannot wait for scientific certainty about the hazards before we act. Stephen H. Safe's opposing article presumes that the reader knows more biochemistry than most college students, but it remains the most often cited opposition to the contention that there is a causative link between environmental estrogens and human health problems.

Sheldon Krimsky  **YES**

Hormone Disruptors: A Clue to Understanding the Environmental Causes of Disease

An increasing body of evidence indicates that certain chemicals in the environment—known as hormone or endocrine disruptors—cause developmental and reproductive abnormalities in humans and animals. Studies of wildlife show associations between hormone-disrupting chemicals in the environment and declining populations, thinning eggshells, morphologic abnormalities, and impaired viability of offspring. Scientists also have postulated a relationship between these chemicals and abnormalities and diseases in humans, including declining sperm counts; breast, testicular, and prostate cancers; and neurological disorders, including cognitive and neurobehavioral effects. While continuing to learn about the effects and interactions these chemicals have on humans and wildlife, scientists recently began working with policy makers to develop a testing program and identify chemicals that require further regulation.

One individual in particular has helped to amass and interpret the scientific evidence on endocrine disruptors, thereby stimulating the interest in and funding for endocrine disruptor research that continues today. Her name is Theo Colborn.

[Recently], Colborn, a senior scientist at the World Wildlife Fund (WWF) in Washington, DC, received one of the most honored environmental awards—the Blue Planet Prize—an award established by Japan's Asahi Glass Foundation at the 1992 Earth Summit in Rio de Janeiro. The foundation recognized Colborn for her "systematic research that certain types of chemical compounds pose a danger as disruptors of endocrine systems to the development of wild organisms and people." Colburn is credited for her synthesis of scientific evidence and her facilitation of multidisciplinary symposia that have focused the attention of many scientists and policy makers worldwide on the potential effects of scores of agricultural and industrial chemicals on the body's endocrine system.

Through her careful study of the patterns of wildlife abnormalities in the Great Lakes, Colborn found that some diseases associated with high levels of chemical pollutants, resulting in mortality and gross birth defects, abated after the first generation of pollution controls were established. Furthermore, her

From Sheldon Krimsky, "Hormone Disruptors: A Clue to Understanding the Environmental Causes of Disease," *Environment*, vol. 43, no. 5 (June 2001). Copyright © 2001 by Sheldon Krimsky. Reprinted by permission of the author. Notes omitted.

astute investigations into the scientific literature revealed patterns of a second category of diseases affecting wildlife offspring. Colborn hypothesized that these diseases arose because certain chemicals interfere with the delicate and intricate signaling pathways of organisms that are particularly vulnerable during fetal development. The brain and reproductive organs, as well as sperm- and egg-producing cells, depend on a proper balance of hormonal signals for their normal development. If the developing fetus is exposed to a cocktail of industrial chemicals which, singly or in combination, interfere with the functions of the body's natural hormones, developmental and reproductive abnormalities can occur. According to some laboratory studies on animals, observable effects of an endocrine-disrupting chemical on fetal development are detectable at concentrations as low as parts per trillion.

A Chemical With a Clue

While Colborn was researching the patterns and etiology of disease in wildlife in the Great Lakes during the late 1980s, scientists already had identified a chemical that provided important clues about the way a substance foreign to an organism (a xenobiotic) interferes with the body's natural signaling pathways. The chemical is a drug called diethylstilbestrol, or DES, as it is commonly known. Synthesized in 1938, DES was the first in a new generation of synthetic hormones prescribed to promote healthy pregnancies; slow the growth of tall, pre-adolescent girls; and ameliorate post-menopausal problems. In 1971, medical researchers published evidence that DES administered to seven pregnant women was associated with a rare form of vaginal cancer in their young adult daughters. This synthetic estrogen did not harm the mothers who took the drug, but most likely affected cells in the developing fetus that caused the cancer in their daughters several decades later.

DES was not a typical pharmaceutical. Not only was it widely used in clinical practice, it also was approved for use as a growth promoter in cows and poultry. The cattle industry continued its use of DES until 1979, when the courts finally upheld the regulatory ban of the veterinary drug in animals used for human consumption.

As a result of the well-documented cases of human disease caused by DES, funding became available to study how this synthetic estrogen could affect the developing fetus. John McLachlan, then chief of the Laboratory of Reproductive and Developmental Toxicology at the National Institute of Environmental Health Sciences, showed that the effects of DES on mice could be used as a model to explain how other chemicals, such as the insecticide 1,1,1-trichloro-2,2-*bis*-(p-chlorophenyl)ethane (also known as dichlorodiphenyltrichloroethane or DDT), can interfere with the body's normal hormone signaling system. McLachlan organized a series of scientific meetings, beginning in 1979 and continuing through the 1980s and 1990s, that explored the subject of environmental estrogens, particularly the estrogenic properties of selected agricultural and industrial chemicals and the molecular mechanisms through which chemical agents induce hormonal effects.

Bringing It All Together

In 1991, Colborn, in collaboration with John Peterson Myers, director of the W. Alton Jones Foundation, which is based in Charlottesville, Virginia, organized a conference that brought together wildlife biologists, human toxicologists, and endocrinologists to share their knowledge about the hormonal effects of environmental chemicals. They dubbed the meeting the "Wingspread Work Session," named after its location at the Wingspread Conference Center in Racine, Wisconsin. This watershed gathering brought the issues surrounding endocrine-disrupting chemicals to the attention of scientists. Wildlife biologists displayed the data they had amassed on reproductive and developmental abnormalities of birds and marine species and the links to environmental toxicants. McLachlan shared the progress he and his colleagues had made toward understanding the biochemical and genetic mechanisms through which chemicals can interfere with or mimic estrogen in the body. McLachlan recently had returned from a meeting in Europe at which Danish scientists postulated a link between in utero chemical exposure and sperm abnormalities in humans. Therefore, he was able to share with work session participants information on two of the three research paths that eventually converged into a single framework for explaining the hormonal effects of environmental chemicals. These independent research programs include the study of the relationship between fetal exposure to DES and transgenerational carcinogenesis; wildlife reproductive and developmental abnormalities; and male reproductive problems, including sperm count decline, testicular cancer, and hypospadia (abnormal urethral openings in newborn human males)....

The 1991 work session consensus statement, a part of which reads, "We estimate with confidence that ... experimental results are being seen at the low end of current environmental conditions," has been further supported by recently published research in the March 2001 issue of *Environmental Health Perspectives.* The study indicates that the doses at which arsenic may increase cancer through its endocrine-disrupting effects are lower than the doses found to cause DNA damage and mutation. These findings suggest that the hormone-disrupting effects of chemicals may be a more sensitive indicator of their potential harm to humans and fetuses than traditional animal carcinogen studies and cell mutagen studies.

Defining "Endocrine Disruptor"

The image that many people have of toxic chemicals is that they destroy cells and organs, impair the immune system, and cause mutations in DNA. Another widely held, and generally correct, belief is that a chemical can be hazardous at a high dose and benign at a low dose, often described by toxicologists with the aphorism, "the dose makes the poison." Endocrine disruptors, however, do not behave like any garden-variety chemical. These are compounds that can trick the organism's body into believing that they are supposed to play a role in the body's functions. To use an analogy, one can foul up a computer's operation by physically damaging the hardware or by releasing a computer virus that fools the computer's software into thinking the virus belongs there. Endocrine disruptors, like computer viruses, engage with the body's mechanism for regulating growth and development, while sabotaging its normal functions.

For example, the body's natural hormones send chemical messages to cells that result in the production of protein. The surface and inside of cells contain receptor molecules that bind to the body's natural hormones and transmit the hormone's instructions to the cell's DNA. It is now understood that foreign chemicals can bind to the receptor molecules dedicated to one of the body's hormone messengers, either mimicking or obstructing the role of the natural hormones. The foreign chemical and the natural hormone may both bind to the receptor molecules, but they do not function equivalently in the body.

There are also other ways that foreign chemicals can interfere with hormone-signaling pathways that do not involve receptor molecules. As one study notes, "some [endocrine disruptors] relay molecular messages through a complex array of cellular proteins, hormone and nonhormone response elements that indirectly turn genes on and alter cell growth and division." Thus, no single mechanism can completely describe the patterns of interference of these chemicals, which complicates the effort to discover cause-effect relationships.

Currently, scientists cannot use chemical structure to predict with any degree of certainty which chemicals will interfere with an organism's endocrine system. Therefore, chemicals have to be tested to evaluate their endocrine effects on the organism. To define an endocrine disruptor as "any chemical that may interfere with the normal function of the hormonal systems of humans and animals" is too imprecise for a hazard assessment. A foreign substance may exhibit hormone-like activity in one species, but not in others, and it may affect, but not interfere with, the body's normal function. Further, as stated in a 1999 U.S. National Academy of Sciences (NAS) panel report, "A single chemical can have multiple effects on an organism that act through several mechanisms, not all of which involve hormone receptors."

Because of this uncertainty, defining an "endocrine disruptor" has become a politicized issue. Some scientists object to the terms "hormone disruptor" and "endocrine disruptor," because these terms imply that chemicals that may play a role in hormonal pathways also cause an adverse effect. There are natural plant estrogens (phytoestrogens) that are hormonally active but are not known to cause adverse effects in normal adults. To preserve the distinction between chemicals that interact with hormone receptors or other hormone-mediated pathways and chemicals that cause adverse physiological effects on an organism, NAS chose the term "hormonally active agents" or HAAs. This term refers to substances that interact with the body's hormone systems, regardless of whether they are linked to a disease or abnormality. What evidence exists that HAAs or endocrine disruptors cause adverse effects? The evidence is generally categorized into four types: wildlife, laboratory animals, human cells and tissue, and human epidemiological studies.

Wildlife Evidence

A significant body of field and laboratory studies confirm that wildlife exposure to certain chemicals has produced reproductive and developmental abnormalities. When certain organisms, such as mollusks, were exposed to marine antifouling paints—usually containing tributyl tin (TBT)—the organisms were observed to have a condition called "imposex," in which the females develop male sexual organs. It is believed that TBT interferes with the biosynthesis of sex steroid hormones (hormones produced by the testes and ovaries, such as estrogens, progestins, and androgens).

Other cases of imposex were discovered in the United Kingdom among certain river fish living in the vicinity of sewage treatment plants. The effluents from the treatment plants suspected of causing these developmental pathologies were found to be comprised of synthetic estrogenic substances, such as nonylphenol (a by-product of detergents) and ethinylestradiol (the main ingredient of the contraceptive pill), as well as natural estradiol from human wastes.

Alligator populations are declining in Florida lakes contaminated with hormonally active pesticides, such as DDT, dicofol, and toxaphene. Genital abnormalities and low egg production have been observed in these alligators. These pesticides or their breakdown products (metabolites) are known to reduce plasma testosterone (a male hormone and one of the androgens) levels in alligators and are considered antiandrogenic. For example, DDT breaks down to a chemical called 1,1-dichloro-2,2-*bis*-(p-chlorophenyl)ethylene or DDE, which has been shown to be a powerful antiandrogen.

Environmental pollutants also have been linked to effects on the reproductive health of several species of birds in the Great Lakes regions and the skewed sex ratios of western gull populations. For more than 40 years, wildlife toxicologists have known that DDT, through its active metabolite DDE, was responsible for eggshell thinning of avian species.

The primary evidence of the effects of endocrine-disrupting chemicals on wildlife is not disputed, although there is still much uncertainty about the mechanisms and dose-responses of the agents in specific species. . . .

The human evidence used in support of the theory that certain chemicals can cause developmental and reproductive abnormalities in organisms has been much more controversial, in regard to the interpretation of evidence and the policy responses to that evidence.

Human Evidence

Many human diseases have unknown etiologies. These include most breast, prostate, and ovarian cancers; infertility; many thyroid-related abnormalities; hypospadia; and abnormal testicular development. Some developmental abnormalities are caused by disruption to the endocrine system or are mediated by hormonal messages. The endocrine disruptor framework, according to which foreign chemicals can interfere with the body's natural hormones, offers a new approach for investigating possible chemical causes or contributing factors of reproductive disorders and developmental abnormalities for which the etiologies remain poorly understood.

Scientists have postulated a link between endocrine-disrupting chemicals and three areas of human abnormalities and diseases in published studies of sperm count declines, cancer (mainly breast, testicular, and prostate), and neurological disorders, including cognitive and neurobehavioral effects.

During the past 50 years, there appears to be an overall downward trend in the density and quality of human sperm within major industrialized regions of the world. Because the patterns are not consistent within each region, some scientists dispute the conclusion that endocrine-disrupting chemicals are the cause of declining sperm counts....

[But] for many reproductive toxicologists, the link between endocrine disruptors and sperm abnormalities in humans remains a working hypothesis with supporting evidence from laboratory studies of animals.

There is ample reason to suspect xenobiotic estrogens as a primary or contributing cause of cancer. Estrogens are known to activate the growth of certain classes of cancer cells. Women with a longer lifetime exposure to estrogens (who experience early menarche and later menopause) are at a higher risk for breast cancer. On the other hand, women who have had their ovaries removed, significantly lowering their production of estrogen, are at a reduced risk of breast cancer, with the lowest risk in those who had the procedure as children.

A series of epidemiological studies conducted on women with breast cancer has not found a consistent link between levels of suspected agents (organochlorines in breast tissue and blood serum, for example) and the risk of breast cancer. One of the largest and best designed of these studies shows no association. Linking endocrine-disrupting chemicals to breast cancer is complicated by many cofounding factors. Women who breastfeed their children mobilize the toxic chemicals that have taken residence in their body and transfer them to their infant. This makes it difficult to get an accurate measure of the amount and the timing of a woman's exposure to endocrine disruptors, because measurements of chemical residues taken after breast-feeding may underestimate the woman's actual exposure. The hypothesis that endocrine-disrupting chemicals may cause breast cancer is further complicated because there may be a long latency period between the exposure of cells to endocrine-disrupting chemicals and the development of cancer decades later. Further, many of the studies testing the link between endocrine disruptors and breast cancer have focused on DDE, which is not estrogenic, and PCBs [polychlorinated biphyenyls], the most persistent of which are antiestrogenic. Thus, these studies have not examined the link between purely estrogenic compounds and breast cancer in the human population a difficult hypothesis to test. Ongoing studies in communities with elevated breast cancer rates, such as Cape Cod, Massachusetts, and Long Island, New York, are still exploring the possible connection between environmental estrogens and cancer.

The strongest evidence linking nonoccupational exposure of endocrine-disrupting chemicals with adverse human health consequences exists in the areas of neurobehavioral and neurodevelopmental effects. Most of the evidence for these effects comes from studies of three groups of halogenated compounds including PCBs, polychlorinated dibenzo-p-dioxins (PCDDs or dioxins), and polychlorinated dibenzofurans (PCDFs). Several studies included children with high in utero PCB exposure from mothers who ate two to three fish meals a month (26 pounds over 6 years) of Great Lakes fish. These children exhibited a variety of neurological and neurodevelopmental abnormalities, including significantly poorer performances on intelligence tests and delayed neuromuscular development compared to children with lower exposures. Infants breast-fed with milk that contained elevated levels of dioxins and PCBs showed signs of thyroid disturbances. Thyroid hormones play an important role in prenatal brain growth, and even small changes at critical times during the first trimester of pregnancy can result in impairments in cognitive development.

In its 1999 panel report, NAS provided a cautious review of the subject, stating,

> In humans, results of cognitive and neurobehavioral studies of mother-infant cohorts accidentally exposed to high concentrations of PCBs and PCDFs and of mother-infant cohorts eating contaminated fish and other food products containing mixtures of PCBs, dioxin, and pesticides, such as DDE, dieldrin, and lindane, provide evidence that prenatal exposure to those HAAs can affect the developing nervous system.

Phthalates, a class of compounds found in many types of chlorine-based plastics, have also been identified as endocrine disruptors. Used as softening agents in plastic, phthalates typically are not tightly bound to the resin and can leach out of food packaging or medical devices. One of the most widely used phthalates, di-(2-ethylhexyl)phthalate (DEHP), has been linked to reproductive abnormalities in women who have been occupationally exposed; it also has been shown to depress hormone levels, including estradiol, in test animals. In September 2000, the U.S. Centers for Disease Control (CDC) reported widespread phthalate contamination among people in the United States, with the highest contamination in women of childbearing age. CDC suggests that perfumes and cosmetics may be the source of contamination. Several phthalate compounds are used as the volatile components of perfumes, nail polishes, and hair sprays.

The published science suggests that there is solid evidence that some chemical agents in sufficient concentrations will affect human reproduction through hormone-mediated mechanisms. The debates over whether there is an observable human effect at lower concentrations center around the substance, dose, and age of development at the time of exposure. Because test animals metabolize phthalates differently than humans and most human exposures are from traces in food, cosmetics, and medical devices, some policy makers are hesitant to use animals studies as a basis for regulating most human exposure. A paradox for setting policy exists because the data most relevant to humans at common exposures are weak, and the most definitive data are irrelevant to humans. Human data are obtained directly from unusual cases of high exposures or indirectly from epidemiological studies. The relevance of using the best animal data to evaluate human risks, even at doses approximating human exposure, is questionable because of species dissimilarities. The challenge for policy making is to find a path between the best available data, with their uncertainties, and the protection of human health.

Current Policies and Programs

In 1996, the U.S. Congress passed the Food Quality Protection Act (FQPA), the new pesticide law, which represents a long sought after compromise between business groups and environmentalists. FQPA granted the business community its wish to insure that the 1958 Delaney clause of the Food and Drug Amendments would not apply to pesticide residues on food. The Delaney clause had prohibited the use of any additive in processed food shown to be carcinogenic at any dose in animals or humans. However, it was not applied to carcinogenic pesticide residues in processed foods that came from the pesticides sprayed on crops. In recent years, in response to a lawsuit filed by the Natural Resources Defense Council, the courts overturned the practice of not applying the Delaney clause to carcinogenic pesticide residues from crops in processed foods. That decision threatened pesticide manufacturers, large agricultural businesses, and processed food manufacturers, whose products would be in violation of the Delaney clause even if they contained minute quantities of pesticides carcinogenic in test animals at high doses.

As a compromise, FQPA eliminated the Delaney clause for pesticide residues but requires higher safety standards for evaluating pesticide health risks and gives more consideration to exposed children. FQPA requires the U.S. Environmental Protection Agency (EPA) to take into account infants' and children's potentially greater exposure and sensitivity to pesticide residues and provide an additional tenfold safety factor in tolerance limits to ensure levels that are safe for infants and children.

During the debates over FQPA, now former Senator Alphonse D'Amato (R-NY), prompted by Long Island breast cancer activists, introduced a hormone disruptor amendment to the bill, which mandated that EPA test agricultural and industrial chemicals for their potential endocrine-disrupting effects. FQPA required EPA to complete its plan for this testing by 1998 and to begin implementing testing by 1999. Congress also amended the Safe Drinking Water Act (SDWA) in 1996, giving EPA the authority to conduct testing for estrogenic or other substances found in drinking water for which a "substantial population may be exposed."

Although SDWA and FQPA address the human health aspects of endocrine disruptors, the Endocrine Disruptor Screening and Testing Advisory Committee (EDSTAC)—an EPA committee of scientists and stakeholders—recommended, and EPA approved, a policy to consider the human and wildlife effects of endocrine-disrupting chemicals. This policy gave rise to EPA's Endocrine Disruptor Screening Program, which is a program designed to identify and evaluate the hazard potential of endocrine-disrupting chemicals. Beginning with an inventory of about 87,000 chemicals in current use, EDSTAC established a priority target of 15,000 high-volume chemicals for screening and testing, including pesticides, commercial chemicals, and environmental contaminants. After EPA validates the assay and testing protocols, industry will conduct much of the chemical testing.

EPA submitted its screening and testing plan to Congress in August 1998 and a year later reported to Congress that implementation had begun on the development of standardized and validated tests. Many of the tests are still on the drawing boards, and until the in vitro assays and animal tests have been approved by the agency, companies will not begin to test chemical substances or submit data to EPA.

The testing program was recently challenged by People for the Ethical Treatment of Animals (PETA), an animal advocacy group opposed to the use of animals for testing chemicals. Largely in response to the PETA campaign, EPA received more than 50,000 post cards objecting to the use of animals for its endocrine disruptor testing program. The agency, meanwhile, is committed to minimizing the number of animals used in the program and to using alternative methods when available.

PETA's legal challenges and the stringent federal requirements for obligating money and selecting groups of contractees to carry out different components of the screening and testing validation studies have slowed the progress of implementation. In contrast to the broad mandate of EDSTAC, which deliberated from 1996 through 1998 and set the general principles for a screening and testing program, the validation studies will concentrate on focused technical issues such as standardization, consistency, reliability, and replicability of tests in different laboratories, as well as dose-response relationships and measurement protocols.

Challenges to the Testing Program

U.S. legislation on endocrine disruptors is directed solely at developing and implementing this new screening and testing program. If a chemical tests positively as an endocrine disruptor, the legislation offers no mandated responses on the part of a regulatory agency. Further, the battery of tests done under the screening program will be vulnerable to all sorts of criticisms. For example, critics could argue that animal and human endocrine systems will respond differently to xenobiotics. Also, inevitably there will be some inconsistencies in the data. A chemical may test positively for endocrine effects in some tests and negatively in others. A substance also may show up as a weak estrogen mimic in one species and as a stronger one in another. EPA will compare foreign chemicals exhibiting estrogen-like effects to natural human and plant estrogens to examine how the chemicals are metabolized and the physiological responses the chemicals induce. Policy makers will have to decide which tests are most salient for understanding potential human and ecological risks.

Two chemicals in combination may produce twice or more than twice the estrogenic effects of either one, and under FQPA, EPA must take into account the additive, cumulative, and synergistic effects of endocrine disruptors. However, because humans are typically exposed to cocktails of chemicals, it will be a formidable task to test chemicals in combinations. Testing even two at a time from the group of 15,000 may exceed the program's capacity.

EPA has targeted 15 tests for standardization and validation and expects these tests will be operational by 2005, a timetable that some insiders view as overly optimistic. The screening program's tests include the use of rodent assays, for which conducting a single mammalian test can take 2–3 years. Moreover, any regulatory action against the use of a highly profitable chemical likely will be contested by litigation, resulting in long regulatory delays. One study reported that approximately 60 percent of the herbicides (by weight) used in the United States are endocrine disruptors. However, the history of chemical regulations indicates that it may take years before suspect substances are prohibited for commercial use. DDT, lead, mercury, DES, and ethylene dibromide (EDB) are just a few of the chemicals that remained in commercial use years after the evidence of their hazards were known by scientists and after regulatory actions began.

A New Regulatory Approach

When EPA completes the process of standardizing and validating the tests, many political and value judgments will come into play. Among the decisions to be made are the amount of evidence that will trigger a regulatory action against a chemical, how potency factors and species effects will be used to determine acceptable levels, whether chemicals already in use will be subject to weaker regulations, and whether EPA will follow the historical tradition of regulating one chemical at a time.

The snail's pace of pesticide testing was eloquently expressed by former congressman Mike Synar (D-OK) at a 1993 hearing. He said, "Almost 20,000 pesticide products have been under review since 1972 and only 31 have been reregistered. At this rate, it will take us to the year 15,520 AD to complete. I believe in flexibility. I believe in good science. What I don't believe in is geological time."

Whether our system of production will be weaned from endocrine-disrupting chemicals will depend, in part, on how many such chemicals are discovered, the availability of substitutes at reasonable costs, whether the evidence of human and ecological effects grows and shows consistent patterns, and whether the regulatory bodies are given stronger mandates to act. The regulation of toxic substances has been dominated by the search for cancer-causing chemicals. The endocrine-disruption theory of disease and developmental abnormalities offers a new framework requiring greater subtlety in testing that introduces new challenges in assessing additive and interactive effects of chemicals. Moreover, policymaking is further complicated by the recent scientific consensus that estrogenic chemicals can cause biological effects at very low doses that have long been considered safe. Thus far, the policy requirements have outpaced the science, and the science is outpacing regulatory actions.

A precautionary approach (also known as the precautionary principle) requires that society take bold steps to accelerate the identification of problematic endocrine-disrupting chemicals and to remove them from the environment, even while scientists and policy makers are filling in the knowledge gaps. This approach was highlighted in a recent report issued by the Royal Society of the United Kingdom, which states, "Regulation cannot be put on hold until all the evidence has been collected." The precautionary approach was framed during the Second International Conference on the Protection of the North Sea in 1987, where agreement was reached to control a group of suspect chemicals even before a definitive causal link had been established between the chemicals and human health or environmental effects.

According to this principle, strong circumstantial, but scientifically inconclusive, evidence can justify policy responses if the consequences of not acting represent unacceptable risks: the mistake of overregulating a "safe chemical" is of secondary moral significance to the mistake of approving or continuing the use of a potentially dangerous chemical. EPA's current response to endocrine-disrupting chemicals is a first step toward a precautionary approach.

In her commemorative lecture for the Blue Planet Prize, Theo Colborn called for a change in the way nations of the world manage industrial chemicals, imploring, "Are we going to wait until every child is affected? . . . The individual costs and societal costs are just too high not to change the system."

NO

Stephen H. Safe

Environmental and Dietary Estrogens and Human Health: Is There a Problem?

Recent reports have suggested that background levels of industrial chemicals and other environmental pollutants may play a role in development of breast cancer in women and decreased male reproductive success as well as the reproductive failures of some wildlife species. These suggestions have been supported by articles in the popular and scientific press and by a television documentary which have described the perils of exposure to endocrine-disrupting chemicals such as estrogenic organochlorine pesticides and pollutants. During the past two decades, environmental regulations regarding the manufacture, use, and disposal of chemicals have resulted in significantly reduced emissions of most industrial compounds and their by-products. Levels of the more environmentally stable organochlorine pesticides and pollutants are decreasing in most ecosystems including the industrialized areas around the Great Lakes in North America. Decreased levels of organochlorine compounds correlates with the improved reproductive success of highly susceptible fish-eating water birds in the Great Lakes region. This article reviews key papers that have been used to support the hypotheses that environmental estrogens play a role in the increased incidence of breast cancer in women and decreased sperm counts in males. Environmental/dietary estrogens and antiestrogens are identified and intakes of "estrogen equivalents" are estimated to compare the relative dietary impacts of various classes of estrogenic chemicals.

Role of Estrogens in Breast Cancer and Male Reproductive Problems

Concerns regarding the role of environmental and dietary estrogens as possible contributors to the increased incidence of breast cancer were fueled by several reports that showed elevated levels of organochlorine compounds in breast cancer patients.... Polychlorinated biphenyls (PCBs) and 1, 1-dichloro-2, 2-bis(*p*-chlorophenyl)ethylene (DDE) are the two most abundant organochlorine

From Stephen H. Safe, "Environmental and Dietary Estrogens and Human Health: Is There a Problem?" *Environmental Health Perspectives,* vol. 103, no. 4 (April 1995). References omitted.

pollutants identified in all human tissues with high frequencies. In one Scandinavian study, levels of DDE or PCBs in adipose tissue from breast samples were not significantly different in breast cancer patients compared to controls. In another study in Finland, β-hexachlorocyclohexane levels were elevated in breast cancer patients; however, this compound was not detected in adipose tissue of some individuals in the patient and control groups and has a relatively low frequency of detection in human tissue samples. Falck and co-workers reported that PCB levels were elevated in mammary adipose tissue samples from breast cancer patients in Connecticut. In contrast, serum levels of DDE (but not PCBs) were significantly elevated in breast cancer patients enrolled in the New York University Women's Health Study. DDE (but not PCB) levels were also elevated in estrogen receptor (ER)-positive but not ER-negative breast cancer patients from Quebec compared to levels in women with benign breast disease. It was initially concluded by Wolff and co-workers that "these findings suggest that environmental chemical contamination with organochlorine residues may be an important etiologic factor in breast cancer." The correlations reported in the two U.S. studies heightened public and scientific concern regarding the potential role of these compounds in development of breast cancer. These observations undoubtedly reinforced advocacy by some groups for a ban on the use of all chlorine-containing chemicals. However, the proposed linkage between PCBs and/or DDE and breast cancer is questionable for the following reasons:

- Most studies with PCBs indicate that these mixtures are not estrogenic, and the weak estrogenic activity observed for lower chlorinated PCB mixtures may be due to their derived hydroxylated metabolites;
- p,p'-DDE, the dominant persistent metabolite of 1, 1, 1-trichloro-2, 2-bis(p-chlorophenyl)ethane (p,p'-DDT), is not estrogenic, and levels of $o,p' =$ DDT, the estrogenic member of the DDT family, are low to nondetectable in most environmental samples;
- Epidemiology studies of individuals occupationally exposed to relatively high levels of DDT or PCBs do not show a higher incidence of breast cancer; and
- No single class of organochlorine compounds was elevated in all studies, suggesting that other factors may be critical for development of breast cancer.

Krieger and co-workers recently reported results from a nested case–control study of women from the San Francisco area which showed that there were no differences in serum DDE or PCB levels between breast cancer patients and control subjects. The authors concluded that "the data do not support the hypothesis that exposure to DDE and PCBs increases risk of breast cancer." This was duly noted in *Time* magazine by a three-line statement in "The Good News" section. Moreover, combined analysis of the 6 studies which report PCB and DDE levels in 301 breast cancer patients and 412 control patients showed that there were no significant increases in either DDE or PCB levels in breast cancer patients versus controls.

The second major link between environmental/dietary estrogens and human disease was precipitated by an article published in the *Lancet*, in which Sharpe and Skakkebaek hypothesized that increased estrogen exposure may be responsible for falling sperm counts and disorders of the male reproductive tract. Unlike the proposed link between environmental estrogens and breast cancer, this hypothesis was not based on experimentally derived measurements of increased levels of any estrogenic compounds in males. Previous studies with diethylstilbestrol, a highly potent estrogenic drug, showed that *in utero* exposure results in adverse effects in male offspring, and the authors' hypothesized that *in utero* exposure to environmental/dietary estrogens may also result in adverse effects in male offspring. A critical experimental component supporting the authors' hypothesis was their analysis of data from several studies which indicated that male sperm counts had decreased by over 40% during the past 50 years. These observations, coupled with the hypothesis that environmental estrogens including organochlorine chemicals were possible etiologic agents, were reported with alarm in the popular and scientific press and in a BBC television program entitled "Assault on the Male: a Horizon Special." Subsequent and prior scientific studies have cast serious doubts on both the hypothesis and the observed decrease in male sperm counts. In 1979, Macleod and Wang reported that there had been no decline in sperm counts, and reanalysis of the data presented by Carlsen and co-workers showed that sperm counts had not decreased from 1960 to 1990. Thus, during the time in which environmental levels of organochlorine compounds were maximal, there was not a corresponding decrease in sperm counts. Moreover, a reevaluation of the sperm concentration data was recently reported by Brownwich et al. in the *British Medical Journal*, and their analysis suggested that the decline in sperm values in males was a function of the choice of the normal or reference value for sperm concentrations. The authors contend that their analysis of the data does "not support the hypothesis that the sperm count declined significantly between 1940 and 1990."

These results suggest that the increasing incidence of human breast cancer is not related to organochlorine environmental contaminants and that decreases in sperm counts is highly debatable. Nevertheless, human populations are continually exposed to a wide variety of environmental and dietary estrogens, and these compounds clearly fit into the category of "endocrine disrupters." The remainder of this article briefly describes the different structural classes of both environmental and dietary estrogens and quantitates human exposures to these compounds.

Synthetic Industrial Chemicals With Estrogenic Activity

The estrogenic activities of different structural classes of industrial chemicals were reported by several research groups in the late 1960s and 1970s in which *o,p'*-DDT and other diphenylmethane analogs and the insecticide kepone were characterized as estrogens. Subsequent studies have confirmed the estrogenic activity of *o,p'*-DDT and related compounds whereas the *p,p'*-substituted analogs were relatively inactive. In addition, *p,p'*-methoxychlor and its hydroxylated metabolites elicit estrogenic responses. Ecobichon and Comeau investigated the estrogenic activities of commercial PCB mixtures (Aroclors) and individual congeners in the female rat uterus and reported estrogenic responses for some Aroclors and individual congeners. Studies in this laboratory showed that a number of commercial PCBs did not significantly increase secretion of procathepsin D, an estrogen-regulated gene product, in MCF-7 human breast cancer cells. It should be noted that several hydroxylated PCBs bind to the ER, and it is possible that *para*-hydroxylated PCB metabolites may be the active estrogenic compounds associated with lower chlorinated PCBs. A recent study reported that several additional organochlorine pesticides including endosulfan, toxaphene, and dieldrin exhibit estrogenlike activity and induce proliferation of MCF-7 human breast cancer cells.

Other industrial chemicals or intermediates that have been identified as estrogenic compounds include bisphenol-A, a chemical used in the manufacture of polycarbonate-derived products; phenol red, a pH indicator used in cell culture media; and alkyl phenols and their derivatives, which are extensively used for preparation of polyethoxylates in detergents.

Natural Estrogenic Compounds

Human exposure to estrogenic chemicals is not confined to xenoestrogens derived from industrial compounds. Several different structural classes of naturally occurring estrogens have been identified, including plant bioflavonoids and various mycotoxins including zearalenone and related compounds. The plant bioflavonoids include different structural classes of compounds which contain a flavonoid backbone: flavones, flavanones, flavonols, isoflavones, and related condensation products (e.g., coumesterol). The estrogenic activities of diverse phytoestrogenic bioflavonoids and mycotoxins have been extensively investigated in *in vivo* models, *in vitro* cell culture systems, and in ER binding assays, and most of these compounds elicit multiple estrogenic responses in these assays. In addition, a number of plant foodstuffs contain 17 β-estradiol (E_2) and estrone.

Environmental and Dietary Antiestrogens

Several different structural classes of chemicals found in the human diet also exhibit antiestrogenic activity. 2, 3, 7, 8-Tetrachlorodibenzo-*p*-dioxin (TCDD) and related halogenated aromatics including polychlorinated dibenzo-*p*-dioxins (PCDDs), dibenzofurans (PCDFs), and PCBs are also an important class of organochlorine pollutants that elicit a diverse spectrum of biochemical and toxic responses. These chemicals act through the aryl hydrocarbon receptor (AhR)-mediated signal transduction pathway, which is thought to play a role in most of the responses elicited by these compounds. AhR agonists such as TCDD have been characterized as antiestrogens using rodent and cell models similar to those used for determining the estrogenic activity of dietary and environmental chemicals. In the rodent model, TCDD and related compounds inhibit several estrogen-induced uterine responses including increased uterine wet weight, peroxidase activity, cytosolic and nuclear progesterone receptor (PR) and ER binding, epidermal growth factor (EGF) receptor binding, EGF receptor mRNA, and c-*fos* mRNA levels. In parallel studies, the antiestrogenic activities of TCDD and related compounds have also been investigated in several human breast cancer cell lines. For example, structurally diverse AhR agonists inhibit the following E_2-induced responses in MCF-7 human breast cancer cells: post-confluent focus production, secretion of tissue plasminogen activator activity, procathepsin D (52-kDa protein), cathepsin D (34-kDa protein), a 160-kDa protein, PR binding sites, glucose-to-lactate metabolism, pS2 protein levels, and PR, cathepsin D, ER, and pS2 gene expression. Moreover, TCDD inhibits formation and/or growth of mammary tumors in athymic nude mice and female Sprague-Dawley rats after long-term feeding studies or initiation with 7,12-dimethylbenzanthracene. A recent epidemiology study on women exposed to TCDD after an industrial accident in Seveso reported that breast cancer incidence was decreased in areas with high levels of TCDD contamination (particularly in the age class 45 to 74) and among women living longest in an area of low TCDD contamination. Endometrial cancer showed a remarkable decrease, particularly in areas with medium and low TCDD contamination. Thus, TCDD and related compounds exhibit a broad spectrum of antiestrogenic activities and, not surprisingly, so do other AhR agonists such as the polynuclear aromatic hydrocarbons (PAHs), indole-3-carbinol (IC), and related compounds found in relatively high levels in foodstuffs. PAHs are found in cooked foods and are ubiquitous environmental contaminants. IC is a major component of cruciferous vegetables (e.g., brussels sprouts, cauliflower) and exhibits antiestrogenic and anticancer (mammary) activities.

Bioflavonoids have been extensively characterized as weak estrogens and therefore may also be active as antiestrogens at lower concentrations. The interaction between estrogenic bioflavonoids and E_2 depends on their relative doses or concentrations, the experimental model, and the specific estrogen-induced endpoint. Markaverich and co-workers reported that the estrogenic bioflavonoids quercetin and luteolin inhibited E_2-induced proliferation of MCF-7 human breast cancer cells and E_2-induced uterine wet weight increase in 21-day-old female rats. Similar results were also observed in this laboratory for quercetin, resperetin, and naringenin. For example, the bioflavonoid naringenin inhibited estrogen-induced uterine hyptertrophy in female rats and estrogen-induced luciferase activity in MCF-7 cells transfected with an E_2-responsive plasmid construct containing the 5′-promoter region of the pS2 gene and a luciferase reporter gene (unpublished results). In contrast, a recent study reported that coumestrol, genistein, and zearalenone were not antiestrogenic in human breast cancer cells. The antiestrogenic activities of weak dietary and environmental estrogens require further investigation; however, it is clear that at subestrogenic doses, some of these compounds exhibit antiestrogenic activities in both *in vivo* and *in vitro* models.

Mass/Potency Balance

The uptake of environmental or dietary chemicals that elicit common biochemical/toxic responses can be estimated by using an equivalency factor approach in which estrogen equivalents (EQs) in any mixture are equal to the sum of the concentration of the individual compounds (EC_i) times their potency (EP_i) relative to an assigned standard such as diethylstilbestrol (DES) or E_2. The total EQs in a mixture would be:

$$EQ = \Sigma \, ([EC_i] \times EP_i)$$

A similar approach is being used to determine the TCDD equivalents (TEQs) of various mixtures containing halogenated hydrocarbons. Verdeal and Ryan have previously used this approach with DES equivalents assuming that the oral potency of E_2 is 15% that of DES. Winter has estimated the dietary intake of pesticides based on FDA's total diet study, which includes estimates of food intakes and pesticide residue levels in these foods. The results presented in Table 1 summarize the estimated exposure of different groups to estrogenic pesticides. For example, 14- to 16-year-old males were exposed to a total of 0.0416 [μ]g/kg/day of the estrogenic pesticides, DDT, dieldrin, endosulfan, and p,p'-methoxychlor (note: the DDT value represents p,p'-DDE and related metabolites, which are primarily nonestrogenic). Thus, the overall dietary intake of these compounds by this age group was 2.5[μ]g/day.

Table 1

Estimated Dietary Intake of Estrogenic Pesticides by Different Age Groups Based on Food Intakes and Pesticide Levels in These Foodstuffs

	Estimated exposure (µg/kg/day)		
Pesticide	6–11 months	14–16[a] years	60–65 years
DDT (total)	0.077	0.0260	0.0103
Dieldrin	0.0014	0.0016	0.0016
Endosulfan	0.0274	0.0135	0.0210
p,p'-Methoxychlor	0.0005	0.0005	0.0001

[a] Maximum exposure: 60 x 0.0416 = 2.5 µg/day.

The relative potencies of dietary and xenoestrogens are highly variable. The results of *in vitro* cell culture studies suggest that estrogenic potencies of bioflavonoids relative to E_2 are 0.001 to 0.0001 whereas Soto and co-workers have assigned an estrogen potency factor of 0.000001 for the estrogenic pesticides. These relative estrogen potency factors for bioflavonoids and pesticides may be lower when derived from *in vivo* studies since pharmacokinetic factors and metabolism may decrease bioavailability. Thus, a more accurate assessment of dietary/environmental EQs requires further data from dietary feeding studies that evaluate these compounds using the same experimental protocols.

The results in Table 2 summarize human exposure to dietary and environmental estrogens and the estimated daily dose in terms of EQs. The relative estrogenic intakes for various hormonal drug therapies were previously estimated by Verdeal and Ryan; the average estimated daily intake of all flavonoids in food products was 1020 and 1070 mg/day, (winter and summer, respectively). The results show that the estimated dietary EQ levels of estrogenic pesticides are 0.0000025 [µ]g/day, whereas the corresponding dietary EQ levels for the bioflavonoids are 102 [µ]g/day. Thus, the EQ values for the dietary intake of flavonoids was 4×10 times higher than the daily EQ intake of estrogenic pesticides, illustrating the minimal potential of these industrial estrogens to cause an adverse endocrine-related response in humans.

Table 2

Estimated Mass Balance of Human Exposures to Environmental and Dietary Estrogens and Antiestrogens

Source	Estrogen equivalents (µg/day)
Estrogens	
Morning after pill	333,500
Birth control pill	16,675
Post-menopausal therapy	3,350
Flavonoids in foods (1,020 mg/day x 0.0001)	102
Environmental organochlorine estrogens	
(2.5 x 0.000001)	0.0000025
Antiestrogens	TCDD antiestrogen equivalents (µg/day)
TCDD and organochlorines (80–120 pg/day)	0.000080–0.000120[a]
PAHs in food (1.2–5.0 x 106 pg/day;	
relative potency ~ 0.001)	0.001200–0.0050[b]
Indolo[3,2-*b*]carbazole in 100 g brussels sprouts	0.000250–0.00128[b]
(0.256–1.28 x 106 pg/day; relative potency	
~ 0.001)	

[a] In most studies, 1 nM TCDD inhibits 50–100% of 1 nM E_2-induced responses in MCF-7 cells; therefore, 1 estrogen equivalent $\cong$ 1 antiestrogen equivalent.
[b] The antiestrogenic potencies of PAHs and indolo[3,2-*b*]carbazole compared to E_2 were approximately 0.001.

Previous studies have also shown that AhR agonists, such as TCDD and related compounds, PAHs and IC and its most active derivative, indolo[3,2-*b*]carbazole (ICZ) all inhibit E_2-induced responses in MCF-7 cells. At a concentration of 10^{-9} M, TCDD inhibits 50–100% of most E_2-induced responses *in vitro* in which the concentration of E_2 is 10^{-9} M. Therefore, 1 TEQ is approximately equal to 1 EQ. The estimated daily intakes of TCDD and related compounds, PAHs, and ICZ (in 100 g brussels sprouts) are summarized in Table 2. The relative potencies of PAHs and ICZ as antiestrogens compared to TCDD are approximately 0.001 in MCF-7 cells. Thus, the TEQs or antiestrogen TEQs can be calculated for the dietary intakes of TCDD and related organochlorines and PAHs (in all foods). The antiestrogen TEQs for the three classes of dietary AhR agonists are orders of magnitude higher than the estimated dietary intakes of estrogenic pesticide EQs. Thus, the major human intake of endocrine disrupters associated with the estrogen-induced response pathways are naturally occurring estrogens found in foods. Relatively high serum levels of estrogenic bioflavonoids have also been detected in a Japanese male population, whereas lower levels were observed in a Finnish group, and this is consistent with their dietary intakes of these estrogenic compounds, *p,p'*-DDE is present in human serum; however, the estrogenic *o,p'*-DDE and *o,p'*-DDT analogs and other weakly estrogenic organochlorine compounds are not routinely detected in serum samples. A recent study identified several hydroxylated PCB congeners in human serum. All of the hydroxylated compounds were also substituted with chlorine groups at both adjacent meta positions. Based on results of previous structure-activity studies (43) for hydroxylated PCBs, these compounds would exhibit minimal estrogenic activity; however, further studies on the activity of hydroxylated PCBs are warranted.

Summary

The hypothesized linkage between dietary/environmental estrogens and the increased incidence of breast cancer is unproven; there is a lack of correlation between higher organochlorine levels in breast cancer patients compared to controls and the low levels of organochlorine EQs in the diet (Table 2). Higher levels of bioflavonoids are unlikely to contribute to increased breast cancer incidence because these compounds and the foods they are associated with tend to exhibit anticarcinogenic activity. The hypothesis that male reproductive problems and decreased sperm counts are related to increased exposure to environmental and dietary estrogens is also unproven. As noted above, dietary exposure to xenoestrogens derived from industrial chemical residues in foods is minimal compared to the daily intake of EQs from naturally occurring bioflavonoids. Moreover, there are serious questions regarding the decreased sperm counts reported by Carlsen and co-workers. Reanalysis of Carlsen et al.'s data suggests that there has not been a decrease in sperm counts in males over the past 30 years and possibly over the past 50 years. Thus, in response to articles in the popular and scientific press such as "The Estrogen Complex" and "Ecocancers: Do Environmental Factors Underlie a Breast Cancer Epidemic?", the results would suggest that the linkage between dietary or environmental estrogenic compounds and breast cancer has not been made, and further research is required to determine the factors associated with the increasing incidence of this disease.

Note Added in Proof: A recent study reported a 2.1% decrease in sperm concentrations in France from 1973 to 1979.

POSTSCRIPT

Do Environmental Hormone Mimics Pose a Potentially Serious Health Threat?

The statement by Safe that is most often quoted by authors of popular articles about the environmental estrogen controversy is "The suggestion that industrial estrogenic chemicals contribute to an increased incidence of breast cancer in women and male reproductive problems is not plausible." This quote is taken from the abstract that was published along with his article in *Environmental Health Perspectives*. Note that in the article he draws the much more cautious conclusion that a link between these consequences and estrogenic compounds is "unproven." Drawing conclusions about estrogenic potency based upon the average person's low dietary exposure to a few common pesticides is suspect for several reasons. Individuals living near highly contaminated bodies of water and consuming fish and other foods from them may take in far higher levels of synthetic estrogens. The complex and variable manner by which different compounds with estrogenic properties affect organisms makes comparisons risky. Some researchers have suggested that humans and other animals may have evolved tolerances to naturally occurring estrogenic compounds but not to the synthetics.

Krimsky credits Theo Colborn, a senior scientist with the World Wildlife Fund, with first drawing public attention to the potential problems of environmental estrogens with the book *Our Stolen Future* (Dutton, 1996), coauthored by Dianne Dumanoski and John Peterson Myers. Colborn clearly believes that the problem is real; she finds the evidence that extensive damage is being done to wildlife by synthetic estrogenic chemicals convincing and thinks it likely that humans are experiencing similar health problems. Recent data reinforce her and Krimsky's points; see Rebecca Renner, "Human Estrogens Linked to Endocrine Disruption," *Environmental Science and Technology* (January 1, 1998) and Ted Schettler et al., *Generations at Risk: Reproductive Health and the Environment* (MIT Press, 1999). In 1999 the National Research Council published *Hormonally Active Agents in the Environment* (National Academy Press), in which the council's Committee on Hormonally Active Agents in the Environment reports on its evaluation of the scientific evidence pertaining to endocrine disruptors. The National Environmental Health Association has called for more research and product testing; see Ginger L. Gist, "National Environmental Health Association Position on Endocrine Disrupters," *Journal of Environmental Health* (January–February 1998). And in 2002 legislation was introduced in Congress to provide funds for expanded research efforts; see Neil Franz, "Bill Floated to Boost Endocrine Funding," *Chemical Week* (May 22, 2002).

Elisabete Silva, Nissanka Rajapakse, and Andreas Kortenkamp, in "Something From 'Nothing'—Eight Weak Estrogenic Chemicals Combined at Concentrations Below NOECs Produce Significant Mixture Effects," *Environmental Science and Technology* (April 2002), find synergistic effects of exactly the kind mentioned by Krimsky. Also, after reviewing the evidence, the U.S. National Toxicology Program found that low-dose effects had indeed been demonstrated in animals (see Ronald Melnick et al., "Summary of the National Toxicology Program's Report of the Endocrine Disruptors Low-Dose Peer Review," *Environmental Health Perspectives* [April 2002]).

The evidence seems to favor the view that environmental hormone mimics have potentially serious effects, despite Safe's cautious interpretation of the data that were available in 1995. If the testing of chemicals and the development of suitable regulations takes as long as Krimsky suggests, one might find useful Michele L. Trankina's article "The Hazards of Environmental Estrogens," *The World & I* (October 2001), in which the author recommends several measures one can take to minimize exposure to environmental estrogens.

ISSUE 11

Is the Environmental Protection Agency's Decision to Tighten Air Quality Standards for Ozone and Particulates Justified?

YES: Carol M. Browner, from Statement Before the Subcommittee on Clean Air, Wetlands, Private Property and Nuclear Safety, Committee on Environment and Public Works, U.S. Senate (February 12, 1997)

NO: Daniel B. Menzel, from Statement Before the Subcommittee on Clean Air, Wetlands, Private Property and Nuclear Safety, Committee on Environment and Public Works, U.S. Senate (February 5, 1997)

ISSUE SUMMARY

YES: Carol M. Browner, administrator for the Environmental Protection Agency (EPA), summarizes the evidence and arguments that were the basis for the EPA's proposal for more stringent standards for ozone and particulates. She contends that the present acceptable levels for these pollutants may be inadequate and that the new regulations are both necessary and sufficient to protect public health.

NO: Daniel B. Menzel, a professor of environmental medicine and a researcher on air pollution toxicology, agrees that ozone and particulates are serious health hazards. He argues, however, that adequate research has not been done to demonstrate that the new standards will result in the additional public health benefits that would justify the difficulty and expense associated with their implementation.

The fouling of air due to the burning of fuels has long plagued the inhabitants of populated areas. After the industrial revolution, the emissions from factory smokestacks were added to the pollution resulting from cooking and household heating. The increased use of coal combined with local meteorological conditions in London produced the dense, smoky, foggy condition first referred to as "smog" by Dr. H. A. Des Vouex in the early 1900s.

Dr. Des Vouex organized British smoke abatement societies. Despite the efforts of these organizations, the problem grew worse and spread to other industrial centers during the first half of the twentieth century. In December 1930 a dense smog in Belgium's Meuse Valley resulted in 60 deaths and an estimated 6,000 illnesses due to the combined effect of the dust and sulfur oxides from coal combustion. This event, and a similar one in Donora, Pennsylvania, received public attention but little response. Serious smog control efforts did not begin until a disastrous "killer smog" in London in December 1952 resulted in approximately 4,000 deaths.

The first major air pollution control victory was the regulation of high-sulfur coal burning in populated areas. The last life-threatening London smog incidents were recorded in the mid-1960s. Unfortunately, a new smog problem, linked to the rush hour traffic in large cities such as Los Angeles, was fast developing. This highly irritating, smelly, smoky haze—now referred to as photochemical smog—is caused by sunlight acting on air laden with nitrogen oxides, unburned hydrocarbons from automotive exhausts, and other sources. The resulting chemical reactions produce a variety of exotic chemicals that are very irritating to lungs and nasal passages—even at very low levels.

The Clean Air Act of 1963 was the first comprehensive U.S. legislation aimed at controlling air pollution from both stationary sources (factories and power plants) and mobile sources (cars and trucks). Under this law and its early amendments, regulations have been issued to establish maximum levels for both ambient air concentrations and emissions from tailpipes and smokestacks for common pollutants, such as sulfur dioxide, suspended particulates (dust), carbon monoxide, nitrogen dioxide, ozone, and airborne lead.

After years of debate Congress finally passed the 1990 Clean Air Act amendments. This complex piece of legislation includes more stringent standards and extends controls to many additional sources of air emissions. It was designed to improve air quality by the end of the twentieth century and to reduce the incidence of acid precipitation that occurs when nitrogen- and sulfur-oxide emissions return to earth hundreds of miles from their sources in the form of rain, snow, or fog laden with sulfuric and nitric acids. The 1990 amendments did not respond to concerns, expressed by many environmental health experts, that the standards for the maximum allowed ambient air levels of ground-level ozone and suspended particulates are too high.

Based on a controversial series of research reports, demands have been made for these levels to be made more stringent to protect the health of sensitive segments of the population, including children, asthmatics, and the elderly. In July 1997, with the support of the Clinton administration, the EPA announced its decision to significantly lower permissible ozone levels and to phase in a new, more restrictive standard for suspended particulate matter aimed at controlling smaller, presumably more toxic particles.

The following selections are excerpts from statements by Carol M. Browner and Daniel B. Menzel, which were included in extensive, contentious testimony at the series of congressional hearings held prior to the implementation of the new standards.

Statement of Carol M. Browner

$\mathbf{M}$r. Chairman, members of the Committee, I want to thank you for inviting me to discuss the Environmental Protection Agency's [EPA] proposed revisions to the national ambient air quality standards for particulate matter and ozone.

On these two pollutants, over the past three and a half years, EPA has conducted one of its most thorough and extensive scientific reviews ever. That review is the basis for the new, more stringent standards for particulate matter and ozone that we have proposed in order to fulfill the mandate of the Clean Air Act.

On average, an adult breathes in about 13,000 liters of air each day. Children breathe in 50 percent more air per pound of body weight than do adults.

For 26 years, the Clean Air Act has promised American adults and American children that they will be protected from the harmful effects of dirty air—based on the best available science. Thus far, when you consider how the country has grown since the Act was first passed, it has been a tremendous success. Since 1970, while the U.S. population is up 28 percent, vehicle miles travelled are up 116 percent and the gross domestic product has expanded by 99 percent, emissions of the six major pollutants or their precursors have *dropped* by 29 percent.

The Clinton Administration views protecting public health and the environment as one of its highest priorities. We have prided ourselves on protecting the most vulnerable among us—especially our children—from the harmful effects of pollution. When it comes to the Clean Air Act, I take very seriously the responsibility the Congress gave me to set air quality standards that "protect public health with an adequate margin of safety"—based on the best science available.

Mr. Chairman, the best available, current science tells me that the current standards for particulate matter and ozone are not adequate, and I have therefore proposed new standards that I believe, based on our assessment of the science, are required to protect the health of the American people.

The standard-setting process includes extensive scientific peer review from experts outside of EPA and the Federal Government. Under the law, we are not to take costs into consideration when setting these standards. This has been the case through six Presidential administrations and 14 Congresses, and has

From U.S. Senate. Committee on Environment and Public Works. Subcommittee on Clean Air, Wetlands, Private Property and Nuclear Safety. *Clean Air Act: Ozone and Particulate Matter Standards.* Hearing, February 12, 1997. Washington, D.C.: Government Printing Office, 1997. (S. Hrg. 105-50, Part 1.)

been reviewed by the courts. We believe that approach remains appropriate. However, once we revise any given air quality standard, it is both appropriate and, indeed, critical that we work with states, local governments, industry and others to develop the most cost-effective, common-sense strategies and programs possible to meet those new standards....

Background

The Clean Air Act directs EPA to identify and set national standards for certain air pollutants that cause adverse effects to public health and the environment. EPA has set national air quality standards for six common air pollutants —ground-level ozone (smog), particulate matter (measured as PM_{10}, or particles 10 micrometers or smaller in size), carbon monoxide, lead, sulfur dioxide, and nitrogen dioxide.

For each of these pollutants, EPA sets what are known as "primary standards" to protect public health and "secondary standards" to protect the public welfare, including the environment, crops, vegetation, wildlife, buildings and monuments, visibility, etc.

Under the Clean Air Act, Congress directs EPA to review these standards for each of the six pollutants every 5 years. The purpose of these reviews is to determine whether the scientific research available since the last review of a standard indicates a need to revise that standard. The ultimate purpose is to ensure that we are continuing to provide adequate protection of public health and the environment. Since EPA originally set the national air quality standards (most were set in 1971), only two of EPA's reviews of these standards have resulted in revised primary standards—in 1979, EPA revised the ozone standard to be less stringent; and in 1987, EPA revised the particulate matter standard to focus on smaller particles (those less than 10 micrometers in diameter), instead of all sizes of suspended particles.

By the early 1990's, thousands of new studies had been published on the effects of ozone and there was an emerging body of epidemiological studies showing significant health effects associated with particulate matter. EPA was sued by the American Lung Association to review and make decisions on both the ozone and particulate matter standards. I directed my staff to conduct accelerated reviews of both standards....

Rationale for EPA's Proposed Revision of the Ozone Standards

Since the mid-1980's, there have been more than 3,000 scientific studies published that are relevant to our understanding of the health and environmental effects associated with ground-level ozone. These peer-reviewed studies were published in independent scientific journals and included controlled human exposure studies, epidemiological field studies involving millions of people (including studies tracking children in summer camps), and animal toxicological studies. Taken as a whole, the evidence indicates that, at levels below the current standard, ozone affects not only people with impaired respiratory systems, such as asthmatics, but healthy children and adults as well. Indeed, one of the groups most exposed to ozone are children who play outdoors during the summer ozone season.

Certain key studies, for example, showed that some moderately exercising individuals exposed for 6 to 8 hours at levels as low as 0.08 parts per million (ppm) (the current ozone standard is set at 0.12 ppm and focuses on 1-hour exposures) experienced serious health effects such as decreased lung function, respiratory symptoms, and lung inflammation. Other recent studies also provide evidence of an association between elevated ozone levels and increases in hospital admissions. Animal studies demonstrate impairment of lung defense mechanisms and suggest that repeated exposure to ozone over time might lead to permanent structural damage in the lungs, though these effects have not been corroborated in humans.

As a result of these and other studies, EPA's staff paper recommended that the current ozone standard be revised from the current 1-hour form (that focuses on the highest "peak" hour in a given day) to an 8-hour standard (that focuses on the highest 8 hours in a given day). It also recommended setting an 8-hour standard in the range of 0.07 ppm to 0.09 ppm, with multiple exceedances (between one and five per year).

The CASAC [Clean Air Scientific Advisory Committee] panel reviewed the scientific evidence and the EPA staff paper and was unanimous in its support of eliminating the 1-hour standard and replacing it with an 8-hour standard. While I do not base my decisions on the views of any individual CASAC member (as a group they bring a range of expertise to the process), it is instructive to note the views of the individual members on these matters. While ten of the 16 CASAC members who reviewed the ozone staff paper expressed their preferences as to the level of the standard, all believe it is ultimately a policy decision for EPA to make. All ten favored a multiple exceedance form. Three favored a level of 0.08 ppm; one favored a level of either 0.08 or 0.09 ppm; three favored the upper end of the range (0.09 ppm); one favored a 0.09–0.10 range with health advisories when a 0.07 level was forecast to be exceeded; and two just endorsed the range presented by EPA as appropriate.

Consistent with the advice of the CASAC scientists and the EPA staff paper, we proposed a new eight-hour standard at 0.08 ppm, with a form that allows for multiple exceedances, by taking the third highest reading each year and averaging those readings over three years. We are asking for comments on a number of alternative options, ranging from eight-hour levels of 0.07 to 0.09 ppm to an option that would retain the existing standard. Just as a point of reference, based on our most recent analysis of children outdoors, when measuring the exposures and risks of concern, as well as the number of areas of the country that would be in "nonattainment" status, the current 1-hour ozone standard of 0.12 ppm is roughly equivalent to a 0.09 ppm 8-hour standard with approximately two to three exceedances.

We considered a number of complex public health factors in reaching the decision on the level and form proposed. The quantitative risk assessments that we performed indicated differences in risk to the public among the various levels within the recommended ranges, but they did not by themselves provide a clear break point for a decision.[1] The risk assessments did, however, point to clear differences among the various standard levels under consideration. These differences indicate that hundreds of thousands of children are not protected under the current standard but would be under EPA's ozone proposal.

Also, consistent with EPA's prior decisions over the years, it was my view that setting an appropriate air quality standard for a pollutant for which there is no discernible threshold means that factors such as the nature and severity of the health effects involved, and the nature and size of the sensitive populations exposed are very important. As a result, I paid particular attention to the health-based concerns reflected in the independent scientific advice and gave great weight to the advice of the health professionals on the CASAC. To me, this is particularly important given the fact that one of the key sensitive populations being protected would be children. The decision to propose at the 0.08 ppm level reflects this, because, though it is in the middle of the range recommended for consideration by CASAC and the EPA staff paper, as a policy choice it reflects the lowest level recommended by individual CASAC panel members and it is the lowest level tested and shown to cause effects in controlled human-exposure health studies.

Finally, air quality comparisons have indicated that meeting a 0.08 ppm, third highest concentration, 8-hour standard (as proposed by EPA) would also likely result in nearly all areas not experiencing days with peak 8-hour concentrations above the upper end of the range (0.09 ppm) referred to in the CASAC and the EPA staff paper. Given the uncertainties associated with this kind of complex health decision, we believe that an appropriate goal is to reduce the number of people exposed to ozone concentrations that are above the highest level recommended by any of the members of the CASAC panel. The form of the standard we proposed (third highest daily maximum 8-hour average) appears to do the best job of meeting that goal, while staying consistent with the advice of the CASAC as a group, as well as the personal views of individual members.

It is also important to note that ozone causes damage to vegetation including:

- interfering with the ability of plants to produce and store food, so that growth, reproduction and overall plant growth are compromised;
- weakening sensitive vegetation, making plants more susceptible to disease, pests, and environmental stresses; and
- reducing yields of economically important crops like soybeans, kidney beans, wheat and cotton.

Nitrogen oxides is one of the key pollutants that causes ozone. Controlling these pollutants also reduces the formation of nitrates that contributes to fish kills and algae blooms in sensitive waterways, such as the Chesapeake Bay.

As part of its review of the ozone science, the CASAC panel unanimously advised that EPA set a secondary standard more stringent than the current standard in order to protect vegetation from the effects of ozone. However, agreement on the level and form of the secondary standard was not reached.

Rationale for EPA's Proposed Revision to the Particulate Matter Standards

For particulate matter standard review, EPA assessed hundreds of peer reviewed scientific research studies, including numerous community-based epidemiological studies. Many of these community-based health studies show associations between particulate matter (known as PM) and serious health effects. These include premature death of tens of thousands of elderly people or others with heart and/or respiratory problems each year. Other health effects associated with exposure to particles include aggravation of respiratory and cardiovascular disease, including more frequent and serious attacks of asthma in children. The results of these health effects have been significantly increased numbers of missed work and school days, as well as increased hospital visits, illnesses, and other respiratory problems.

The recent health studies and a large body of atmospheric chemistry and exposure data have focused attention on the need to address the two major subfractions of PM_{10}—"fine" and "coarse" fraction particles—with separate programs to protect public health. The health studies have indicated a need to continue to stay focused on the relatively larger particles or "coarse" fraction that are a significant component of PM_{10} and are controlled under the current standards. We continue to see adverse health effects from exposures to such coarse particles above the levels of the current standards. As a result, CASAC scientists were unanimous that existing PM_{10} standards be maintained for the purpose of continuing to control the effects of exposure to coarse particles.

However, a number of the new health and atmospheric science studies have highlighted significant health concerns with regard to the smaller "fine" particles, those at or below 2.5 micrometers in diameter. These particles are so small that several thousand of them could fit on the type-written period at the end of a sentence. In the simplest of terms, fine particles are of health concern because they can remain in the air for long periods both indoors and outdoors contributing to exposures and can easily penetrate and be absorbed in the deepest recesses of the lungs. These fine particles can be formed in the air from sulfur or nitrogen gases that result from fuel combustion and can be transported many hundreds of miles. They can also be emitted directly into the air from sources such as diesel buses and some industrial processes. These fine particles not only cause serious health effects, but they also are a major reason for visibility impairment in the United States in places such as national parks that are valued for their scenic views and recreational opportunities. For example, visibility in the eastern United States should naturally be about 90 miles, but has been reduced to under 25 miles.

EPA analyzed peer-reviewed studies involving more than five and a half million people that directly related effects of "fine" particle concentrations to human health. For example, one study of premature mortality tracked almost 300,000 people over the age of 30 in 50 U.S. cities.

Based on the health evidence reviewed, the EPA staff paper recommended that EPA consider adding "fine particle" or $PM_{2.5}$ standards, measured both annually and over 24 hours. The staff paper also recommended maintaining the current annual and/or 24-hour PM_{10} standards to protect against coarse fraction exposures, but in a more stable form for the 24-hour standard. This more stable form would be less sensitive to extreme weather conditions.

When CASAC reviewed the staff paper, 19 out of 21 panel members recommended establishment of new standards (daily and/or annual) for $PM_{2.5}$. They also agreed with the retention of the current annual PM_{10} standards and consideration of retention of the 24-hour PM_{10} standard in a more stable form.

Regarding the appropriate levels for $PM_{2.5}$, staff recommended consideration of a range for the 24-hour standard of between 20 and 65 micrograms per cubic meter (ug/m3) and an annual standard to range from 12.5 to 20 ug/m^3. Individual members of CASAC expressed a range of opinions about the levels and averaging times for the standards based on a variety of reasons. Four panel members supported specific ranges or levels within or toward the lower end of the ranges recommended in the EPA staff paper. Seven panel members recommended ranges or levels near, at or above the upper end of the ranges specified in the EPA staff paper. Eight other panel members declined to select a specific range or level.

Consistent with the advice of the EPA staff paper and CASAC scientists, in November [1996] I proposed adding new standards for $PM_{2.5}$. Specifically, based on public health considerations, I proposed an annual standard of 15 ug/m^3 and a 24-hour standard of 50 ug/m^3. In terms of the relative protection afforded, this proposal is approximately in the lower portion of the ranges or options recommended by those CASAC panel members who chose to express their opinions on specific levels. However, taking into account the form of the standard proposed by EPA, we understand that the proposal would fall into the lower to middle portion of the ranges or options. In order to ensure the broadest possible consideration of alternatives, I also asked for comment on options both more and less protective than the levels I proposed.

Also consistent with the advice of the EPA staff paper and CASAC scientists, I proposed to retain the current annual PM_{10} standard and to retain the current 24-hour PM_{10} standard, but with a more stable form. I also requested comment on whether the addition of a fine particle standard and the maintenance of an annual PM_{10} standard means that we should revoke the current 24-hour PM_{10} standard.

As has been the case throughout the 25-year history of environmental standard setting, uncertainty has played an important role in decisionmaking on the particulate matter standards. Specifically, the uncertainty about the exact mechanism causing the observed health effects has led some to argue that not enough is known to set new or revised standards. In this case, however, because of the strong consistency and coherence across the large number of epidemiological studies conducted in many different locations, the seriousness and magnitude of the health risks, and/or the fundamental differences between "fine" and "coarse" fraction particles, the CASAC scientists and the experts in my Agency clearly believed that "no action" was an inappropriate response. The question then became one of how best to deal with uncertainty—that is, how best to balance the uncertainties with the need to protect public health.

Given the nature and severity of the adverse health effects, I chose to meet the Congressional requirement of providing the public with an "adequate margin of safety," by proposing $PM_{2.5}$ standards within the ranges recommended in the EPA staff paper and commented upon in the CASAC closure letter. I believe the levels chosen reflect the independent, scientific advice given me about the relationship between the observed adverse health effects and high levels of fine particle pollution. That advice led to a proposed decision toward the lower end of the range of levels for the annual standard which is designed to address widespread exposures and toward the middle of the range for the 24-hour standard, which would serve as a backstop for seasonal or localized effects.

One final note on particulate matter. Some have suggested we need more research before decisions are made about these standards. I strongly support the need for continued scientific research on this and other air pollutants as a high priority. However, as we pursue this research, we must simultaneously take all appropriate steps to protect public health. We believe that tens of thousands of people each year are at risk from fine particles and I believe we need to move ahead with strategies to control these pollutants.

Finding Common Sense, Cost-Effective Strategies for Implementing a Revised Ozone or PM Standard

Throughout the 25-year history of the Clean Air Act and air quality management in the United States, national ambient air quality standards have been established based on an assessment of the science concerning the effects of air pollution on public health and welfare. Costs of meeting the standards and related factors have never been considered in setting the national ambient air quality standards themselves. As you can see from the description of the process I went through to choose a proposed level on ozone and particulate matter, the focus has been entirely on health, risk, exposure and damage to the environment.

I continue to believe that this is entirely appropriate. Sensitive populations like children, the elderly and asthmatics deserve to be protected from the harmful effects of air pollution. And the American public deserves to know whether the air in its cities and counties is unsafe or not; that question should never be confused with the separate issues of how long it may take or how much it may cost to reduce pollution to safe levels. Indeed, to allow costs and related factors to influence the determination of what levels protect public health would be to mislead the American public in a very fundamental way.

While cost-benefit analysis is a tool that can be helpful in developing strategies to *implement* our nation's air quality standards, we believe it is inappropriate for use to *set* the standards themselves. In many cases, cost-benefit analysis has overstated costs. In addition, many kinds of benefits are virtually impossible to quantify—how do I put a dollar value on reductions in a child's lung function or the premature aging of lungs or increased susceptibility to respiratory infection? Very often I cannot set a value and these types of health benefits are, in effect, counted as zero.

At the same time, both EPA and industry have historically tended to overstate costs of air pollution control programs. In many cases, industry finds cheaper, more innovative ways of meeting standards than anything EPA estimates. For example, during the 1990 debates on the Clean Air Act's acid rain program, industry initially projected the costs of an emission allowance (the authorization to emit one ton of sulfur dioxide) to be approximately $1,500, while EPA projected those same costs to be $450 to $600. Today those allowances are selling for less than $100....

On the other hand, the Clean Air Act has always allowed that costs and feasibility of meeting standards be taken into account in devising effective emission control strategies and in setting deadlines for cities and counties to comply with air quality standards. This is certainly the case for any revision we might make to either the ozone or the particulate matter standards. This process has worked well. In fact, our preliminary studies indicate that from 1970 to 1990 implementation of the Act's requirements has resulted in significant monetizable benefits many times the direct costs for that same period.

If we ultimately determine that public health is better served by revising one or both of these standards, the Clean Air Act gives us the responsibility to devise new strategies and deadlines for attaining the revised standards. In doing so, we are determined to develop the most cost-effective, innovative implementation strategies possible, and to ensure a smooth transition from current efforts.

To meet this goal, we have used the Federal Advisory Committee Act to establish a Subcommittee for Ozone, Particulate Matter and Regional Haze Implementation Programs. It is composed of almost sixty members of state and local agencies, industry, small business, environmental groups, other Federal agencies and other groups and includes five working groups comprised of another 100 or so members of these same kinds of organizations.

The Subcommittee and the various workgroups have been meeting regularly for well over a year working to hammer out innovative strategies for EPA to consider in implementing any revised standards. Members from industry, state governments and others are putting forward position papers advocating innovative ways to meet air quality standards. It is our belief that results from this Subcommittee process will lead us to propose innovative approaches for implementing any new standards. The Subcommittee will continue to meet over the next year to help develop cost-effective, common-sense implementation programs.

The issues being addressed by the Subcommittee include:

- What will be the new deadlines for meeting any new standards? [If EPA tightens a standard, it has the authority to establish deadlines of up to ten years—with the possibility of additional extensions—beyond the date an area is designated "nonattainment."]
- What will be the size of the area considered "nonattainment"? If it revises an air quality standard, EPA has the ability to change the size

of the affected nonattainment areas and focus control efforts on those areas that are causing the pollution problems, not just the downwind areas that are monitoring unhealthy air.

- How do we address the problem of the pollutants that form ozone and/ or fine particles being transported hundreds of miles and contributing to nonattainment problems in downwind areas?
- What kinds of control strategies are appropriate for various nonattainment areas? Can we use the experience of the past several years to target those control strategies that are the most cost-effective?
- How can we promote innovative, market-based air pollution control strategies?

The implementation of these new standards is likely to focus on sources like trucks, buses, power plants and cleaner fuels. In some areas, as with the current standards, our analysis shows that reaching the standards will present substantial challenges. All of the air pollution control programs we are pursuing to meet the current ozone and particulate matter standards, as well as programs to implement other sections of the Clean Air Act, will help meet any revised standards. For example, the sulfur dioxide reductions achieved by the acid rain program will greatly help reduce levels of fine particles, particularly in the eastern United States. Cleaner technology in power plants would also greatly reduce the nitrogen oxides that help form ozone across the eastern United States. In fact, we believe that under certain comprehensive control strategies, more than 70 percent of the counties that could become nonattainment areas under a new ozone standard would be brought back into attainment as a result of a program to reduce nitrogen oxides from power plants and a large number of other sources. Programs underway to reduce emissions from cars, trucks, and buses will also help meet a revised particulate matter or ozone standard.

I intend to announce our proposals on implementation of the proposed new standards in phases that correspond to the Federal Advisory Committee Act Subcommittee's schedule for deliberating on various aspects of the program. I expect to propose the first phase of that program at the same time that I announce our final decision on revisions to the ozone and particulate matter standards.

In announcing the proposed ozone and particulate matter standards last November, I directed my Office of Air and Radiation to further expand the membership of the Federal Advisory Subcommittee to include more representation from small business and local governments. Also, in conjunction with the Small Business Administration and the Office of Management and Budget, we are holding meetings with representatives of small businesses and small governments to obtain their input and views on our proposed standards.

There is one last point I would like to make on this matter. Critics of the proposals have been saying that meeting these proposed standards means widespread carpooling and the elimination of backyard barbecues, among other lifestyle changes. The broad national strategy is being developed by EPA, as I have described, with extensive input from industry, small business, state and local governments and others. While the ultimate decisions as to what programs are needed to meet air quality standards are up to the state and local governments, I would like to state categorically that there will *not* be any new federal mandates eliminating backyard barbecues or requiring carpooling. These kinds of claims are merely scare tactics designed to shift the debate away from the critical, complex public health issues we are attempting to address.

Conclusions

Mr. Chairman, I commend you for holding these hearings. The issues we are discussing today are critical to the state of the Nation's public health and environment. It is imperative that the American public understand these important issues. In that regard, I am disappointed that some have chosen to distort this important discussion by raising distracting and misleading pseudo issues like "junk science" and "banning backyard barbecues." I am hopeful that this and other hearings and public forums will help focus the national debate on the real health and environmental policy implications of these national air quality standards.

In the Clean Air Act, the Congress has given me the responsibility to review every 5 years the most recent science to determine whether revisions to national air quality standards are warranted. In doing so, the law tells me to protect the public health with an adequate margin of safety.

We are constantly reviewing the science associated with these standards, but we do not often propose revisions to them. I have done so in the case of ozone and particulate matter because of compelling new scientific evidence. For the past three and a half years we have targeted our resources to conduct a thorough, intensive review of this scientific evidence. The scope and depth of this review process has been based on unprecedented external peer review activities.

Given the sensitive populations affected by these pollutants—children, asthmatics, the elderly—as well as possible effects on outdoor workers and other healthy adults, it was my judgment that it was appropriate to propose standards that tended to fall in the lower end of the range of protection supported by my independent science advisors and recommended by experts in my technical offices. Based on the record before the Agency at the time of proposal, including the advice and recommendations of the CASAC panels, I concluded—subject to further consideration based on public comments—that the proposed standards were both necessary and sufficient to protect the public health, including sensitive populations, with an adequate margin of safety.

At the same time, I recognize that the proposed standards involve issues of great complexity and I look forward to receiving a broad range of comments from all affected and interested parties. As I have described, we have gone to unprecedented lengths to provide the public with opportunities to express their views on the proposed standards. We have also expressly requested comments on options (including alternative levels and forms of the standards) that are both more protective and less protective than the levels we proposed.

Note

1. CASAC itself agreed that there are a continuum of effects—even down to back-ground—and that there is no "bright line" distinguishing any of the proposed standards as being significantly more protective of public health.

NO

Daniel B. Menzel

Statement of Daniel B. Menzel

My name is Daniel B. Menzel. I am Professor and Chair of the Department of Community and Environmental Medicine, University of California at Irvine, Irvine, California. I have had more than 30 years' experience in research in air pollution and toxicology. My expertise centers in two areas: mechanisms of air pollution toxicity and mathematical modeling of toxicology, particularly deposition of air pollutants in the respiratory tract. I have served as a senior author on multiple EPA Criteria Documents and recently as a Consultant to the Clean Air Scientific Advisory Committee examining the Particulate Matter Criteria Document and proposed standard.

The Committee has requested that I provide my views on the ozone and particulate matter standards, which EPA has published in the Federal Register and intends to implement under the Clean Air Act. I am pleased to do that and would also like to extend my testimony to include the research effort of EPA because it directly affects the standard-setting process. I understand that the two standards present different problems in terms of the form of the standard, the scientific data supporting each standard and the process by which the standard was promulgated. In my view, however, there are similarities between the two standards that reflect a major deficiency in EPA's efforts. The common deficiency is the lack of solid scientific data. EPA is a grossly underfunded Agency given the scope of its responsibilities. EPA has not done well with its resources by not sustaining research to meet the long-term goals of the Agency. Thus, I hope that the committee will allow me to express my concerns about the research planning at EPA.

Air Pollution Is a Major Long Term Public Health Problem

Air pollution is a worldwide problem. In the United States air pollution is of such public health importance that it is critical that a national debate be undertaken on the future directions of air pollution research and regulation. This committee is providing a very valuable forum to the people so that they may learn more about the scientific controversy surrounding these two air pollutants and the alternative views that exist concerning the future of air pollution

From U.S. Senate. Committee on Environment and Public Works. Subcommittee on Clean Air, Wetlands, Private Property and Nuclear Safety. *Clean Air Act: Ozone and Particulate Matter Standards.* Hearing, February 5, 1997. Washington, D.C.: Government Printing Office, 1997. (S. Hrg. 105-50, Part 1.)

remediation efforts. I am at the moment writing a review of the toxicology of ozone.[1] This will be the third review of ozone that I have written for the scientific literature. Almost ten years have elapsed since my last effort, and I was surprised and saddened to note on examining the literature that questions which we raised in the review in 1988 still remain unresolved. Much new human data has become available on ozone supporting a lower standard and shorter averaging time, but the book is far from closed on ozone. I also wrote the first part of the health section of the SO_x (sulfur oxides) Particulate Matter Criteria Document for EPA in 1980. Many of the questions raised in that document also remain unanswered. As a consultant to the Clean Air Scientific Advisory Committee I assisted in the review of the current Particulate Matter Criteria Document. Not only were the fundamental questions raised in the original SO_x Particulate Matter Criteria Document still existent, but new important questions arose for which we have no answer. All of these experiences suggest to me that a greatly enhanced and invigorated research effort in air pollution is needed if we are to make sound, reasonable and rational decisions on the implementation of clean air standards. If anything, air pollution research is now more important to the national public health than ever before.

Both the ozone and particulate matter standards have vast implications for the quality of life and the economy of the United States. It is my opinion that the vast majority of Americans support improving and enhancing the quality of their life by eliminating or decreasing air pollution. Americans are quite willing to shoulder the burden of cleaner air, cleaner water, and cleaner food if they can understand clearly the benefits to be gained by these activities. The confidence of the American people in the decisions being made on environmental issues is critical to the ability of this government to govern and implement these decisions. If ever the public loses confidence in the environmental strategies promulgated by the Federal Government then it will be impossible to carry out large national programs designed to eliminate or at least ameliorate the adverse effects of air pollution. I am very concerned that the Environmental Protection Agency and the Congress maintain the confidence of the U.S. public and demonstrate to the public their vigorous support for a better quality of life and clean air. Scientific truth is the only lasting commodity upon which decisions can be based.

Generic Issues

From my view the difficulties that we face with both the ozone and the particulate matter standard stem from generic issues in toxicology which must be addressed in a sound scientific manner. The first of these generic issues is a plausible biological mechanism of action for the particular pollutant. The second is the nature of the dose-response relationship. I will address each of these and give examples of how they impinge upon the two standards that we are discussing today.

Plausible Biological Mechanisms

What is a plausible mechanism? We have learned a great deal about the quantitative nature of toxic reactions in the last 40 years. It is now possible to divide biological reactions to toxicants into several categories under which plausible mechanisms have been elucidated. A plausible mechanism of action for a toxin places the toxin within the context of our knowledge of disease processes. Having a plausible mechanism of action increases our confidence that health effects observed in animals will occur in humans. Understanding a mechanism of action also makes experiments more meaningful and relevant. In this forum it is not possible for me to elaborate in greater technical detail on how a plausible mechanism influences the experimental design and interpretation of the results of experiments. Experimental design and the concept of plausible mechanism of action are dealt with in standard textbooks of toxicology, such as "Casserett and Doull's Fundamentals of Toxicology."

A plausible mechanism of action is critically essential to controlled human exposure studies. The extrapolation from animal experiments to human exposures as they occur in nature, that is with free-living people, depends upon an intermediate link of controlled exposures of human volunteers to the toxin. We must have a clear idea of a plausible mechanism so that human studies can be developed with due care that no harm will ever result to the volunteers who courageously commit themselves to these kinds of experiments. In air pollution many of the human studies have been very limited because of the lack of a clear understanding of a plausible mechanism. Investigators have been very reluctant to engage in high level exposures of human subjects because they fear that some long-term harm will result from their experiments. Clearly, we cannot and will not tolerate human experimental studies that result in harm to the volunteer. This is simply not ethically acceptable.

Plausible Mechanism of Ozone Toxicity

One plausible mechanism of action of ozone is the production of free radicals by the reaction of ozone with cellular constituents. The free radical theory is that which we proposed in 1971.[2] It is now clear that this mechanism of action is too naive and simplistic and clearly does not explain the consequences of chronic exposure to ozone. Studies with experimental animals clearly show that the results of a continuous or intermittent lifetime exposure to ozone are highly complex and are not predictable from the free radical hypothesis alone. Further experiments are needed with life-term exposures of experimental animals using the most modern molecular biology techniques. The complex pattern of lifetime ozone exposure must involve multiple signal transduction pathways. Simply put, the adverse health effects of chronic exposure to ozone are complex and beyond the free radical theory which we now recognize as accounting for the brief initial contact of ozone with the lung.

Chronic exposure is the critical issue in ozone exposure. EPA initiated and was carrying out an excellently conceived and implemented research program on the chronic effects of ozone in support of the current ozone standard. But this research has stopped and support for ozone research by other Federal agencies has stalled. Basic research support for ozone by the National Institutes of Health and particularly the National Institute of Environmental Health Sciences (NIEHS), has fallen away. The scientific community is in error in allowing this to have happened.

Very compelling controlled human exposure experiments suggest that the current ozone standard (0.12 ppm) may be toxic. The short term exposures under which humans can be safely exposed do not allow us to study the chronic effects of ozone exposure. Epidemiologic studies are underway in the South Coast Air Basin, particularly those by Professor John Peters of the University of Southern California but this study is hampered because no quantitative biomarker of ozone health effects has been developed.

We would not be... engaging in this discussion if EPA's chronic ozone study in experimental animals had been carried out. Nor would we still have doubts about the ozone standard if ozone research had received a high priority in research support by the other Federal research agencies such as NIH and NSF.

In summary, there is a preliminary biologically plausible mechanism of action for ozone. The free radical theory is not comprehensive and does not explain all of the effects of chronic exposure to ozone. Much additional work is needed to understand the chronic effects of ozone.

Particulate Matter

In contrast to the ozone problem, no plausible biological mechanism of action has so far been proposed for particulate matter. It has been very difficult to demonstrate toxicity for particulate matter in experimental animals. In my laboratory and that of my colleagues at UCI we have not been able to show major toxicity with particulate matter at potencies approaching the levels reported from epidemologic studies.[3,4]

To place this problem in a more global context, urban particulate matter is a universal problem. Particulate matter seems to be a common result of human concentration in urban areas. To eliminate all of the particulate matter in our cities would, in my view, be only possible by the elimination of all human activity. Clearly this Draconian approach is not reasonable.

The studies of Schwartz and his colleagues[3,4] have challenged our conclusions from experimental animal studies. These studies indicate that all particles regardless of their geographic origin have the same toxicity. It is well known that the chemical composition of the urban particles differ widely between geographic areas. For example, in the western US, especially in the South Coast Air Basin of Los Angeles and its environs, the chemical processes responsible for the formation of particulate matter depend on photochemical reactions. Nitric acid is the dominant end product. There are very few oxides of sulfur present because of the nature of the fossil fuels used in California. On the other hand, in the East Coast Corridor the consumption of sulfur-containing fuels is much greater, and the chemistry of the reactions leading to the formation of particulate matter is not as dependent upon photochemistry as it is upon chemical reactions. Sulfuric acid, not nitric acid, is the dominant end product present in particulate matter. The chemical nature of the particles formed in California are quite different [from] those of the East Coast Corridor. Yet the health effects measured by epidemiologic techniques suggest that all particles have the same effect despite the differences in chemical composition. This is a very troublesome problem. One of the basic tenets of toxicology is that the toxicity occurs via chemical reaction. How then can the same effect result from very different kinds of chemistries? We must conclude that there is no plausible mechanism now available for particulate matter which can account for the reported results.

Particle Size and Site of Action of Respirable Urban Particles

The toxicity of particles also depends on the site within the respiratory tract where they are deposited. A major advance has been the recognition of the dependence of toxicity on the site of deposition. The site of deposition in the respiratory tract depends, in turn, on the physical size of the particle. By measuring the amount of particles within the size range which can be deposited in the human lung, EPA adopted a biologically based criterion for its standard setting. This concept of defining particulate air pollution in terms of the size of particles most likely to be responsible for the adverse health effects is referred to as PM_{10} where 10 refers to particles of 10 micrometers aerodynamic mass median diameter or less. PM_{10} is a fairly good surrogate measurement for the amount of material that would actually be inhaled and deposited in the human respiratory tract. Schwartz and his colleagues extrapolated from measured PM_{10} values. PM_{10} is a major advance in public health policy pioneered by EPA. The PM_{10} concept shifts emphasis to particles of that size which are likely to be the most harmful to people. A network of PM_{10} monitors has been constructed in the US and large amounts of data have been accumulated.

Schwartz and his colleagues went beyond PM_{10} and extrapolated from a very limited set of measurements of $PM_{2.5}$ and PM_{10} to estimate $PM_{2.5}$ values and to relate mortality and morbidity to particulate matter exposure smaller than PM_{10} or particles less than 2.5 micrometers mass median aerodynamic diameter. Only a few data exist on the $PM_{2.5}$ exposure in our major cities. By shifting from PM_{10} to $PM_{2.5}$ values, a major difference in the regional deposition within the lung of these particles is suggested as the site of action. The smaller the particle the more deposition occurs in the deeper parts of the lung. By assigning toxicity to particles in the $PM_{2.5}$ range the site of action is also assigned to the thoracic region of the lung. Because these $PM_{2.5}$ values are calculated and not measured, it is very difficult to place the heavy weight of evidence on this ultrafine particle range as EPA has done in its criteria document. Even with a shift in attention to particles of this size range, there is still is no plausible mechanism for toxicity. Further, some of the CASAC members questioned the potency of the particles calculated from the mortality and morbidity data. All of this underscores the importance of the research program reviewed by CASAC as part of the particulate matter standard setting process.

Dose-Response Relationship

The dose-response relationship is a curve that relates the number of individuals responding with an adverse reaction (mortality, morbidity or the like) to a certain exposure concentration of the chemical. The shape of the dose-response curve is important when setting standards. All theories of the dose-response relationship so far indicate that these curves will be non-linear; that is, there will be a point at which the probability that a response would occur is very unlikely. To put it another way, all theories suggest that there is a concentration at which nothing will occur while above that concentration adverse effects will occur. The point at which there is nothing detectable is the threshold. The dose-response relationship is at the heart of the risk assessment. In both the particulate matter and ozone standard the dose-response relationship is only poorly understood. Consequently, estimates of risk are also uncertain. Examples for ozone and particulate matter follow.

The Particulate Matter Dose-Response Curve Is Linear Not Curved

The current assumption of epidemiologic studies is that the mortality or morbidity is a linear function passing through zero at zero concentration of particles. The dose-response function has no point at which no adverse effects occur. The linear dose-response curve is in opposition to all of the theories and experimental data derived for a host of chemicals acting by a variety of different mechanisms of action.

The epidemiologic basis for a linear relationship between effect and dose is very poor. The data are not supported by any kind of a generalized theory and are in many cases a default assumption coming about because the epidemiologic data are weak. It is very difficult for epidemiologists to relate exposure to effect. The methodologies of epidemiology at present are insensitive to the concentration or exposure effect. This is especially true in ecological studies where indirect evidence is used for adverse health effect.

For example, the epidemiologic studies of particulate matter health effects depend upon death certificates and the coincidence of an increase in death with an increase in particulate matter exposure. These studies again provide no indication of how a person might have died from the exposure to particulate matter. The studies only associate the death with the exposure to particulate matter. Nonetheless, the increases in mortality associated with particulate matter are troublesome. If the magnitude of mortality suggested by these studies is correct, then we are faced with a major public health problem that demands immediate attention.

Time and Intensity Relationships in Ozone Health Effects

EPA initiated a time and intensity study in cooperation with the USSR. This program was well thought out and attacked the question of which variable is most important in determining the health effects of ozone. From the data that were generated by this study it appears that the intensity is the most critical factor rather than the duration of exposure for ozone toxicity. These studies of the time and concentration effects on ozone toxicity led to the current hypothesis upon which the proposed ozone standard is based. If it is correct that the magnitude of the exposure is more important, then extremes of exposure should be reduced. One strategy to reduce exposure to extreme concentrations of ozone is to change the averaging time for the standard, making implementation plans stricter for short-term excursions. The US-USSR research program to study the time and concentration dependency of ozone adverse health effects was very productive and was progressing along a track which would, if continued, have provided us a great deal of information at this time. Unfortunately, EPA chose to reduce and essentially eliminate this line of study. Extramural support for the program lagged and ozone in general has become an unpopular topic for support by other government agencies such as NIEHS.

Based on the fragmentary information that we have available, I feel that it is appropriate to support the EPA proposal of changing the averaging time for the ozone standard so that large excursions over short time periods will be eliminated or reduced. However, one should recognize that changing the averaging time will have a major impact on State implementation plans and will have major economic consequences. Clearly, understanding the nature of the dose-response relationship is very important and affects which alternatives we choose to reduce ozone health effects.

Time and Intensity Relationship for Particulate Matter Health Effects Are Unknown

As stated above, most time and intensity (dose and dose-rate) relationships for chemicals follow a simple relationship that the product of the dose rate and the time of exposure form a constant. This constant is arbitrary and unique for each chemical. Epidemiologic studies of the increases in mortality associated with increases in particulate matter are strictly linear with the amount of particulate matter. One reason why this assumption occurs is that a lag period has been assumed. The lag period means that the increases in mortality occurring 2 to 3 days after an exposure are related to the exposure to particulate matter, not earlier or later. The underlying hypothesis is that particulate matter toxicity is not immediately evident but occurs after this lag period. This very short acting time raises the question as to what happens when people are exposed to concentrations of particulate matter over the long term. We really have no data on the chronic effects in humans of exposure to particulate matter. Chronic exposure studies are very difficult to achieve using epidemiologic data.

To my knowledge there are no experimental animal data or controlled human studies which relate this kind of lag time to exposure to the toxicity of particulate matter. In my laboratory and that of my colleagues at UCI we have found that experimental animals such as the rat are very insensitive to particulate matter exposures. We have never observed potencies equivalent to that proposed for humans based on the epidemiologic data. This again raises the question of a plausible biological mechanism of action....

Conclusions

The Proposed Ozone Standard

It is my opinion that we will have achieved only marginal effects by decreasing the current ambient air quality standards for ozone from 120 parts per billion to 90 parts per billion. The nature of the dose-response relationship is such that it may still be at a linear range and thus reduction to much lower levels may be necessary to result in the abolition of detectable health effects from ozone. My colleague, Robert Wolpert, and I published a simple analysis of different kinds of dose-response relationships for ozone looking toward this very issue. How much would one have to reduce the ozone concentration in the air in order to be able to find a detectable advance in public health? Because the data are so sparse, a multitude of different kinds of theoretical treatments are possible. None of them, however, are sufficiently sensitive that one could lead to a clear prediction of a health benefit. On the other hand, as I mentioned above, a change in the time constant alone is going to have a great benefit. I endorse EPA's analysis of the time constant and think that EPA's proposal to a change in the averaging time for ozone is likely to be of benefit to the public health.

Still, I think that translating these changes into new State implementation plans may be very difficult. To translate both a change in the concentration, that is the amount of ozone that is permissible in the air and the duration over which it is permissible, will be a very difficult task indeed to implement.

Continued research into the health effects of ozone are urgently needed. Further reductions in the ozone standard may be indicated in the near future. Because of the economic impact of ozone standards and strategies, the highest quality research is needed.

Particulate Matter Standard

As I have said previously, I do not doubt that the particulate matter problem is a very serious problem indeed. We need to place a very strong active and progressive research program into place in order for us to cope with this problem. It is my view that too little is known. In the report of the Clean Air Scientific Advisory Committee to Administrator Carol Browner, the committee pointed out that one of the areas in which additional research should be undertaken is chronic exposure.

I am not in favor of the use of a $PM_{2.5}$ standard. A viable network of monitoring instruments and sound research supports the PM_{10} standard. The $PM_{2.5}$ standard has no background. There is no existing research quality $PM_{2.5}$ network. Without a research quality $PM_{2.5}$ network it is not likely that we will make much progress towards the goal of a new particulate matter standard. We lack information on the actual $PM_{2.5}$ in the atmosphere of our cities. We do not know the duration of exposure of people to $PM_{2.5}$. The chemical nature of the $PM_{2.5}$ fraction is poorly known. We lack a plausible biological mechanism for particulate matter. We do not know if regulation of $PM_{2.5}$ will be of benefit. A strong aggressive long-term research program is essential to address the current data deficiencies if we are to convince people that this is a major problem.

Notes

1. Shoaf, C.R. and Menzel, D.B. Oxidative damage and toxicity of environmental pollutants. In: Cellular Antioxidant Defense Mechanisms. (ed., C. K. Chow) CRC Press, Inc, Vol. 1:197–213, 1988.

2. Roehm, J.N., Hadley, J.G. and Menzel, D.B. Oxidation of Unsaturated Fatty Acids by Ozone and Nitrogen Dioxide: A Common Mechanism of Action. *Arch. Environ. Health* 23:142–148, 1971.

3. Saldiva, P. H., Pope, C. A., Schwartz, J., Dockery, D. W., Lichtenfels, A. J., Salge, J. M., Barone, I. & Bohm, G. M. (1995) Air pollution and mortality in elderly people: a time-series study in Sao Paulo, Brazil. *Arch. Environ. Health* 50:159–163.

4. Schwartz, J. (1995) Short term fluctuations in air pollution and hospital admissions of the elderly for respiratory disease. *Thorax* 50: 531–538.

POSTSCRIPT

Is the Environmental Protection Agency's Decision to Tighten Air Quality Standards for Ozone and Particulates Justified?

Analysis of the options that are available for reducing the health effects of air pollution is complex, and, like most technological problems related to environmental protection, it requires the use of many unprovable assumptions. The National Research Council's Committee on Research Priorities for Airborne Particulate Matter, asked by Congress to assess research in this area, published *Research Priorities for Airborne Particulate Matter I: Immediate Priorities and a Long-Range Research Portfolio* (National Academy Press, 1998); *Research Priorities for Airborne Particulate Matter II: Evaluating Research Progress and Updating the Portfolio* (National Academy Press, 1999); and *Research Priorities for Airborne Particulate Matter III: Early Research Progress* (National Academy Press, 2001). A fourth volume updating the assessment is planned. A specific proposal for a new, costly set of regulatory standards requires a judgment about whether or not the analysis is based on an adequate set of valid research results and whether or not the interpretation of the results provides sufficient confidence that the benefits achieved will justify the expense. It is on this point—rather than the question of whether or not suspended particulate matter and ozone are serious health threats—that Browner, Menzel, and many other experts disagree. Menzel points out one serious problem with the proposal to base particulate matter standards on particles smaller than 2.5 microns rather than 10 microns: no system for monitoring the smaller particles exists. In announcing the actual schedule for implementing the new standards, the EPA acknowledged this problem and established a timetable associated with the development of the needed monitoring network.

One of the key research efforts that the EPA used in formulating the new particulate standard is the Harvard School of Public Health's "Six Cities" study. For a report on epidemiologist Joel Schwartz's role in directing that research, see Renée Skelton's article "Clearing the Air" in the Summer 1997 issue of *The Amicus Journal*. Another debate about the new standards, featuring Lester B. Lave and Robert W. Crandall, was published in the Summer 1997 issue of *The Brookings Review* under the title "EPA's Proposed Air Quality Standards." For a lengthy denunciation of the new particulate and ozone standards and the scientific research on which they are based, see "Polluted Science," by Michael Fumento, *Reason* (August/September 1997). For a summary of the Health Effects

Institute study released in July 2000, see Jocelyn Kaiser, "Evidence Mounts That Tiny Particles Can Kill," *Science* (July 7, 2000).

In "Who Will Be Protected by EPA's New Ozone and Particulate Matter Standards?" *Environmental Science and Technology* (January 1, 1998), Feng Liu reports that some moderation would occur under the new regulations in the disproportionate effect of air pollution on Hispanics, Asians/Pacific Islanders, and African Americans. In the June 1, 1998, issue of the same journal, see Allen S. Lefohn, Douglas S. Shadwick, and Stephen D. Ziman, "The Difficult Challenge of Attaining EPA's New Ozone Standard."

On October 15, 1997, the EPA issued a report entitled *The Benefits and Costs of the Clean Air Act, 1970 to 1990*. On November 15, 1999, a second report was issued entitled *The Benefits and Costs of the Clean Air Act, 1990 to 2010*. Both concluded that the benefits of the programs and standards required by the 1990 Clean Air Act Amendments significantly exceed costs. See http://www.epa.gov/airlinks. However, on October 29, 1999, the U.S. Court of Appeals for the District of Columbia upheld a May 1999 court decision calling the EPA's 1997 National Ambient Air Quality Standards (NAAQS) for ozone and particulates unconstitutional. See April Reese, "Bad Air Days," *E: The Environmental Magazine* (November–December 1999) and Richard J. Pierce, Jr., "The Inherent Limits on Judicial Control of Agency Discretion: The D.C. Circuit and the Nondelegation Doctrine," *Administrative Law Review* (Winter 2000) for criticism of the decision. In January 2000 New Jersey and Massachusetts asked the U.S. Supreme Court to review that decision. In February 2001 the Court "unanimously upheld the constitutionality of the Clean Air Act as EPA had interpreted it in setting health-protective air quality standards for ground-level ozone and particulates." In March 2002 the D.C. Circuit Court rejected all remaining challenges to the EPA's 1997 particulate and ozone standards, and the EPA declared its intent to "move forward with programs to protect Americans from the wide variety of health problems that these air pollutants can cause, such as respiratory illnesses and premature death." Unfortunately, in June 2002 the discovery of a small error in the calculations of the risks of fine particulates threatened to reawaken the debate; see Jocelyn Kaiser, "Software Glitch Threw Off Mortality Estimates," *Science* (June 14, 2002).

Much of the debate has centered on air quality standards, especially for particulates. Regulations to control downwind air pollution (mentioned by Browner) have also come under fire. Eastern states have demanded that the EPA enforce agreements that would reduce the amount of air pollutants emitted by Midwestern power plants, which travel on the wind and add to air quality problems in the East. Industry groups and states such as Michigan filed suit to block such reductions, but in June 2000 the U.S. Court of Appeals for the District of Columbia chose to allow the EPA to implement its plans. However, in June 2002 the EPA proposed to relax rules requiring old, high-emissions, coal-burning power plants to improve their emissions control systems when they undergo major upgrades. The announced rationale was to give utilities greater flexibility and to keep consumer electric bills down. Environmental groups immediately objected that the savings would not be worth the increased incidence of health problems.

ISSUE 12

Do Human Activities Threaten to Change the Global Climate?

YES: Intergovernmental Panel on Climate Change, from "Climate Change 2001: The Scientific Basis," A Report of Working Group I of the Intergovernmental Panel on Climate Change (2001)

NO: Kevin A. Shapiro, from "Too Darn Hot?" *Commentary* (June 2001)

ISSUE SUMMARY

YES: The Intergovernmental Panel on Climate Change states that global warming appears to be real, with strong effects on sea level, ice cover, and rainfall patterns to come, and that human activities—particularly emissions of carbon dioxide—are to blame.

NO: Neuroscience researcher Kevin A. Shapiro argues that past global warming predictions have been wrong and that the data do not support calls for immediate action to reduce emissions of carbon dioxide.

Scientists have known for more than a century that carbon doxide and other "greenhouse gases" (including water vapor, methane, and chlorofluorocarbons) help prevent heat from escaping the earth's atmosphere. In fact, it is this "greenhouse effect" that keeps the earth warm enough to support life. Yet there can be too much of a good thing. Ever since the dawn of the industrial age, humans have been burning vast quantities of fossil fuels, releasing the carbon they contain as carbon dioxide. Because of this, some estimate that by the year 2050, the amount of carbon dioxide in the air will be double what it was in 1850. By 1982 an increase was apparent. Less than a decade later, many researchers were saying that the climate had already begun to warm. Now there is a strong consensus that the global climate is warming and will continue to warm. There is less agreement on just how much it will warm or what the impact of the warming will be on human (and other) life. See Spencer R. Weart, "The Discovery of the Risk of Global Warming," *Physics Today* (January 1997).

The debate has been heated. The June 1992 issue of *The Bulletin of the Atomic Scientists* carries two articles on the possible consequences of the greenhouse effect. In "Global Warming: The Worst Case," Jeremy Leggett says that although there are enormous uncertainties, a warmer climate will release more carbon dioxide, which will warm the climate even further. As a result, soil will grow drier, forest fires will occur more frequently, plant pests will thrive, and methane trapped in the world's seabeds will be released and will increase global warming much further—in effect, there will be a "runaway greenhouse effect." Leggett also hints at the possibility that polar ice caps will melt and raise sea levels by hundreds of feet.

Taking the opposing view, in "Warming Theories Need Warning Label," S. Fred Singer emphasizes the uncertainties in the projections of global warming and their dependence on the accuracy of the computer models that generate them, and he argues that improvements in the models have consistently shrunk the size of the predicted change. There will be no catastrophe, he argues, and money spent to ward off the climate warming would be better spent on "so many pressing—and real—problems in need of resources."

Global warming, says the UN Environment Programme, will do some $300 billion in damage each year to the world economy by 2050. In March 2001 President George W. Bush announced that the United States would not take steps to reduce greenhouse emissions—called for by the international treaty negotiated in 1997 in Kyoto, Japan—because such reductions would harm the American economy (the U.S. Senate has not ratified the Kyoto treaty). Since the Intergovernmental Panel on Climate Change (IPCC) had just released its third report saying that past forecasts were, in essence, too conservative, Bush's stance provoked immense outcry.

The analysis of data and computer simulations described by the IPCC in the following selection indicates that global warming is a genuine problem. According to the IPCC, climate warming is already apparent and will get worse than previous forecasts had suggested. Sea level will rise, ice cover will shrink, rainfall patterns will change, and human activities—particularly emissions of carbon dioxide—are to blame. The report excerpt reprinted here does not suggest that anything in particular should be done, but other writers, such as Stephen H. Schneider and Kristin Kuntz-Duriseti ("Facing Global Warming," *The World & I* [June 2001]), pull no punches: "Nearly all knowledgeable scientists agree that some global warming is inevitable, that major warming is quite possible, and that for the bulk of humanity the net effects are more likely to be negative than positive. This will hold true particularly if global warming is allowed to increase beyond a few degrees, which is likely to occur by the middle of this century if no policies are undertaken to mitigate emissions."

Kevin A. Shapiro is more optimistic. In the second selection, he argues that past global warming predictions have been wrong and that there is too much room for error in the data and computer simulations to support calls for immediate action to reduce emissions of carbon dioxide.

 **YES**

Summary for Policymakers

The Third Assessment Report of Working Group I of the Intergovernmental Panel on Climate Change (IPCC) builds upon past assessments and incorporates new results from... five years of research on climate change. Many hundreds of scientists from many countries participated in its preparation and review.

This Summary for Policymakers (SPM), which was approved by IPCC member governments in Shanghai in January 2001, describes the current state of understanding of the climate system and provides estimates of its projected future evolution and their uncertainties....

An increasing body of observations gives a collective picture of a warming world and other changes in the climate system.

Since the release of the Second Assessment Report (SAR), additional data from new studies of current and palaeoclimates, improved analysis of data sets, more rigorous evaluation of their quality, and comparisons among data from different sources have led to greater understanding of climate change.

The global average surface temperature has increased over the 20th century by about 0.6°C.

- The global average surface temperature (the average of near surface air temperature over land, and sea surface temperature) has increased since 1861. Over the 20th century the increase has been 0.6 ± 0.2°C. This value is about 0.15°C larger than that estimated by the SAR for the period up to 1994, owing to the relatively high temperatures of the additional years (1995 to 2000) and improved methods of processing the data. These numbers take into account various adjustments, including urban heat island effects. The record shows a great deal of variability; for example, most of the warming occurred during the 20th century, during two periods, 1910 to 1945 and 1976 to 2000.
- Globally, it is very likely that the 1990s was the warmest decade and 1998 the warmest year in the instrumental record, since 1861.

- New analyses of proxy data for the Northern Hemisphere indicate that the increase in temperature in the 20th century is likely to have been the largest of any century during the past 1,000 years. It is also likely that, in the Northern Hemisphere, the 1990s was the warmest decade and 1998 the warmest year. Because less data are available, less is known about annual averages prior to 1,000 years before present and for conditions prevailing in most of the Southern Hemisphere prior to 1861.
- On average, between 1950 and 1993, night-time daily minimum air temperatures over land increased by about 0.2°C per decade. This is about twice the rate of increase in daytime daily maximum air temperatures (0.1°C per decade). This has lengthened the freeze-free season in many mid- and high latitude regions. The increase in sea surface temperature over this period is about half that of the mean land surface air temperature.

Temperatures have risen during the past four decades in the lowest 8 kilometres of the atmosphere.

- Since the late 1950s (the period of adequate observations from weather balloons), the overall global temperature increases in the lowest 8 kilometres of the atmosphere and in surface temperature have been similar at 0.1°C per decade.
- Since the start of the satellite record in 1979, both satellite and weather balloon measurements show that the global average temperature of the lowest 8 kilometres of the atmosphere has changed by $+0.05 \pm 0.10$°C per decade, but the global average surface temperature has increased significantly by $+0.15 \pm 0.05$°C per decade. The difference in the warming rates is statistically significant. This difference occurs primarily over the tropical and sub-tropical regions.
- The lowest 8 kilometres of the atmosphere and the surface are influenced differently by factors such as stratospheric ozone depletion, atmospheric aerosols, and the El Niño phenomenon. Hence, it is physically plausible to expect that over a short time period (e.g., 20 years) there may be differences in temperature trends. In addition, spatial sampling techniques can also explain some of the differences in trends, but these differences are not fully resolved.

Snow cover and ice extent have decreased.

- Satellite data show that there are very likely to have been decreases of about 10% in the extent of snow cover since the late 1960s, and ground-based observations show that there is very likely to have been a reduction of about two weeks in the annual duration of lake and river ice cover in the mid- and high latitudes of the Northern Hemisphere, over the 20th century.
- There has been a widespread retreat of mountain glaciers in non-polar regions during the 20th century.

- Northern Hemisphere spring and summer sea-ice extent has decreased by about 10 to 15% since the 1950s. It is likely that there has been about a 40% decline in Arctic sea-ice thickness during late summer to early autumn in recent decades and a considerably slower decline in winter sea-ice thickness.

Global average sea level has risen and ocean heat content has increased.

- Tide gauge data show that global average sea level rose between 0.1 and 0.2 metres during the 20th century.
- Global ocean heat content has increased since the late 1950s, the period for which adequate observations of sub-surface ocean temperatures have been available.

Changes have also occurred in other important aspects of climate.

- It is very likely that precipitation has increased by 0.5 to 1% per decade in the 20th century over most mid- and high latitudes of the Northern Hemisphere continents, and it is likely that rainfall has increased by 0.2 to 0.3% per decade over the tropical (10°N to 10°S) land areas. Increases in the tropics are not evident over the past few decades. It is also likely that rainfall has decreased over much of the Northern Hemisphere sub-tropical (10°N to 30°N) land areas during the 20th century by about 0.3% per decade. In contrast to the Northern Hemisphere, no comparable systematic changes have been detected in broad latitudinal averages over the Southern Hemisphere. There are insufficient data to establish trends in precipitation over the oceans.
- In the mid- and high latitudes of the Northern Hemisphere over the latter half of the 20th century, it is likely that there has been a 2 to 4% increase in the frequency of heavy precipitation events. Increases in heavy precipitation events can arise from a number of causes, e.g., changes in atmospheric moisture, thunderstorm activity and large-scale storm activity.
- It is likely that there has been a 2% increase in cloud cover over mid- to high latitude land areas during the 20th century. In most areas the trends relate well to the observed decrease in daily temperature range.
- Since 1950 it is very likely that there has been a reduction in the frequency of extreme low temperatures, with a smaller increase in the frequency of extreme high temperatures.
- Warm episodes of the El Niño–Southern Oscillation (ENSO) phenomenon (which consistently affects regional variations of precipitation and temperature over much of the tropics, sub-tropics and some mid-latitude areas) have been more frequent, persistent and intense since the mid-1970s, compared with the previous 100 years.

- Over the 20th century (1900 to 1995), there were relatively small increases in global land areas experiencing severe drought or severe wetness. In many regions, these changes are dominated by inter-decadal and multi-decadal climate variability, such as the shift in ENSO towards more warm events.
- In some regions, such as parts of Asia and Africa, the frequency and intensity of droughts have been observed to increase in recent decades.

Some important aspects of climate appear not to have changed.

- A few areas of the globe have not warmed in recent decades, mainly over some parts of the Southern Hemisphere oceans and parts of Antarctica.
- No significant trends of Antarctic sea-ice extent are apparent since 1978, the period of reliable satellite measurements.
- Changes globally in tropical and extra-tropical storm intensity and frequency are dominated by inter-decadal to multi-decadal variations, with no significant trends evident over the 20th century. Conflicting analyses make it difficult to draw definitive conclusions about changes in storm activity, especially in the extra-tropics.
- No systematic changes in the frequency of tornadoes, thunder days, or hail events are evident in the limited areas analysed.

Emissions of greenhouse gases and aerosols due to human activities continue to alter the atmosphere in ways that are expected to affect the climate.

Changes in climate occur as a result of both internal variability within the climate system and external factors (both natural and anthropogenic). The influence of external factors on climate can be broadly compared using the concept of radiative forcing. A positive radiative forcing, such as that produced by increasing concentrations of greenhouse gases, tends to warm the surface. A negative radiative forcing, which can arise from an increase in some types of aerosols (microscopic airborne particles) tends to cool the surface. Natural factors, such as changes in solar output or explosive volcanic activity, can also cause radiative forcing. Characterisation of these climate forcing agents and their changes over time is required to understand past climate changes in the context of natural variations and to project what climate changes could lie ahead.

Concentrations of atmospheric greenhouse gases and their radiative forcing have continued to increase as a result of human activities.

- The atmospheric concentration of carbon dioxide (CO_2) has increased by 31% since 1750. The present CO_2 concentration has not been exceeded during the past 420,000 years and likely not during the past 20 million years. The current rate of increase is unprecedented during at least the past 20,000 years.

- About three-quarters of the anthropogenic emissions of CO_2 to the atmosphere during the past 20 years is due to fossil fuel burning. The rest is predominantly due to land-use change, especially deforestation.
- Currently the ocean and the land together are taking up about half of the anthropogenic CO_2 emissions. On land, the uptake of anthropogenic CO_2 very likely exceeded the release of CO_2 by deforestation during the 1990s.
- The rate of increase of atmospheric CO_2 concentration has been about 1.5 ppm (0.4%) per year over the past two decades. During the 1990s the year to year increase varied from 0.9 ppm (0.2%) to 2.8 ppm (0.8%). A large part of this variability is due to the effect of climate variability (e.g., El Niño events) on CO_2 uptake and release by land and oceans.
- The atmospheric concentration of methane (CH_4) has increased by 1060 ppb (151%) since 1750 and continues to increase. The present CH4 concentration has not been exceeded during the past 420,000 years. The annual growth in CH_4 concentration slowed and became more variable in the 1990s, compared with the 1980s. Slightly more than half of current CH_4 emissions are anthropogenic (e.g., use of fossil fuels, cattle, rice agriculture and landfills). In addition, carbon monoxide (CO) emissions have recently been identified as a cause of increasing CH_4 concentration.
- The atmospheric concentration of nitrous oxide (N_2O) has increased by 46 ppb (17%) since 1750 and continues to increase. The present N_2O concentration has not been exceeded during at least the past thousand years. About a third of current N_2O emissions are anthropogenic (e.g., agricultural soils, cattle feed lots and chemical industry).
- Since 1995, the atmospheric concentrations of many of those halocarbon gases that are both ozone-depleting and greenhouse gases (e.g., $CFCl_3$ and CF_2Cl_2), are either increasing more slowly or decreasing, both in response to reduced emissions under the regulations of the Montreal Protocol and its Amendments. Their substitute compounds (e.g., CHF_2Cl and CF_3CH_2F) and some other synthetic compounds (e.g., perfluorocarbons (PFCs) and sulphur hexafluoride (SF_6)) are also greenhouse gases, and their concentrations are currently increasing....

Confidence in the ability of models to project future climate has increased.

Complex physically-based climate models are required to provide detailed estimates of feedbacks and of regional features. Such models cannot yet simulate all aspects of climate (e.g., they still cannot account fully for the observed trend in the surface-troposphere temperature difference since 1979) and there are particular uncertainties associated with clouds and their interaction with radiation and aerosols. Nevertheless, confidence in the ability of these models to provide useful projections of future climate has improved due to their demonstrated performance on a range of space and time-scales.

- Understanding of climate processes and their incorporation in climate models have improved, including water vapour, sea-ice dynamics, and ocean heat transport.
- Some recent models produce satisfactory simulations of current climate without the need for non-physical adjustments of heat and water fluxes at the ocean-atmosphere interface used in earlier models.
- Simulations that include estimates of natural and anthropogenic forcing reproduce the observed large-scale changes in surface temperature over the 20th century. However, contributions from some additional processes and forcings may not have been included in the models. Nevertheless, the large-scale consistency between models and observations can be used to provide an independent check on projected warming rates over the next few decades under a given emissions scenario.
- Some aspects of model simulations of ENSO, monsoons and the North Atlantic Oscillation, as well as selected periods of past climate, have improved.

There is new and stronger evidence that most of the warming observed over the last 50 years is attributable to human activities.

The SAR concluded: "The balance of evidence suggests a discernible human influence on global climate". That report also noted that the anthropogenic signal was still emerging from the background of natural climate variability. Since the SAR, progress has been made in reducing uncertainty, particularly with respect to distinguishing and quantifying the magnitude of responses to different external influences. Although many of the sources of uncertainty identified in the SAR still remain to some degree, new evidence and improved understanding support an updated conclusion.

- There is a longer and more closely scrutinised temperature record and new model estimates of variability. The warming over the past 100 years is very unlikely to be due to internal variability alone, as estimated by current models. Reconstructions of climate data for the past 1,000 years also indicate that this warming was unusual and is unlikely to be entirely natural in origin.
- There are new estimates of the climate response to natural and anthropogenic forcing, and new detection techniques have been applied. Detection and attribution studies consistently find evidence for an anthropogenic signal in the climate record of the last 35 to 50 years.
- Simulations of the response to natural forcings alone (i.e., the response to variability in solar irradiance and volcanic eruptions) do not explain the warming in the second half of the 20th century. However, they indicate that natural forcings may have contributed to the observed warming in the first half of the 20th century.

- The warming over the last 50 years due to anthropogenic greenhouse gases can be identified despite uncertainties in forcing due to anthropogenic sulphate aerosol and natural factors (volcanoes and solar irradiance). The anthropogenic sulphate aerosol forcing, while uncertain, is negative over this period and therefore cannot explain the warming. Changes in natural forcing during most of this period are also estimated to be negative and are unlikely to explain the warming.
- Detection and attribution studies comparing model simulated changes with the observed record can now take into account uncertainty in the magnitude of modelled response to external forcing, in particular that due to uncertainty in climate sensitivity.
- Most of these studies find that, over the last 50 years, the estimated rate and magnitude of warming due to increasing concentrations of greenhouse gases alone are comparable with, or larger than, the observed warming. Furthermore, most model estimates that take into account both greenhouse gases and sulphate aerosols are consistent with observations over this period.
- The best agreement between model simulations and observations over the last 140 years has been found when all the above anthropogenic and natural forcing factors are combined. These results show that the forcings included are sufficient to explain the observed changes, but do not exclude the possibility that other forcings may also have contributed.

In the light of new evidence and taking into account the remaining uncertainties, most of the observed warming over the last 50 years is likely to have been due to the increase in greenhouse gas concentrations.

Furthermore, it is very likely that the 20th century warming has contributed significantly to the observed sea level rise, through thermal expansion of sea water and widespread loss of land ice. Within present uncertainties, observations and models are both consistent with a lack of significant acceleration of sea level rise during the 20th century.

Human influences will continue to change atmospheric composition throughout the 21st century.

Models have been used to make projections of atmospheric concentrations of greenhouse gases and aerosols, and hence of future climate, based upon emissions scenarios from the IPCC Special Report on Emission Scenarios (SRES). These scenarios were developed to update the IS92 series, which were used in the SAR and are shown for comparison here in some cases.

Greenhouse Gases

- Emissions of CO_2 due to fossil fuel burning are virtually certain to be the dominant influence on the trends in atmospheric CO_2 concentration during the 21st century.

- As the CO_2 concentration of the atmosphere increases, ocean and land will take up a decreasing fraction of anthropogenic CO_2 emissions. The net effect of land and ocean climate feedbacks as indicated by models is to further increase projected atmospheric CO_2 concentrations, by reducing both the ocean and land uptake of CO_2.

- By 2100, carbon cycle models project atmospheric CO_2 concentrations of 540 to 970 ppm for the illustrative SRES scenarios (90 to 250% above the concentration of 280 ppm in the year 1750). These projections include the land and ocean climate feedbacks. Uncertainties, especially about the magnitude of the climate feedback from the terrestrial biosphere, cause a variation of about -10 to $+30\%$ around each scenario. The total range is 490 to 1260 ppm (75 to 350% above the 1750 concentration).

- Changing land use could influence atmospheric CO_2 concentration. Hypothetically, if all of the carbon released by historical land-use changes could be restored to the terrestrial biosphere over the course of the century (e.g., by reforestation), CO_2 concentration would be reduced by 40 to 70 ppm.

- Model calculations of the concentrations of the non-CO_2 greenhouse gases by 2100 vary considerably across the SRES illustrative scenarios, with CH_4 changing by -190 to $+1,970$ ppb (present concentration 1,760 ppb), N_2O changing by $+38$ to $+144$ ppb (present concentration 316 ppb), total tropospheric O_3 changing by -12 to $+62\%$, and a wide range of changes in concentrations of HFCs, PFCs and SF_6, all relative to the year 2000. In some scenarios, total tropospheric O_3 would become as important a radiative forcing agent as CH_4 and, over much of the Northern Hemisphere, would threaten the attainment of current air quality targets.

- Reductions in greenhouse gas emissions and the gases that control their concentration would be necessary to stabilise radiative forcing. For example, for the most important anthropogenic greenhouse gas, carbon cycle models indicate that stabilisation of atmospheric CO_2 concentrations at 450, 650 or 1,000 ppm would require global anthropogenic CO_2 emissions to drop below 1990 levels, within a few decades, about a century, or about two centuries, respectively, and continue to decrease steadily thereafter. Eventually CO_2 emissions would need to decline to a very small fraction of current emissions.

Aerosols

The SRES scenarios include the possibility of either increases or decreases in anthropogenic aerosols (e.g., sulphate aerosols, biomass aerosols, black and organic carbon aerosols) depending on the extent of fossil fuel use and policies to abate polluting emissions. In addition, natural aerosols (e.g., sea salt, dust and emissions leading to the production of sulphate and carbon aerosols) are projected to increase as a result of changes in climate.

Radiative Forcing Over the 21st Century

For the SRES illustrative scenarios, relative to the year 2000, the global mean radiative forcing due to greenhouse gases continues to increase through the 21st century, with the fraction due to CO_2 projected to increase from slightly more than half to about three quarters. The change in the direct plus indirect aerosol radiative forcing is projected to be smaller in magnitude than that of CO_2.

Global average temperature and sea level are projected to rise under all IPCC SRES scenarios.

In order to make projections of future climate, models incorporate past, as well as future emissions of greenhouse gases and aerosols. Hence, they include estimates of warming to date and the commitment to future warming from past emissions.

Temperature

- The globally averaged surface temperature is projected to increase by 1.4 to 5.8°C over the period 1990 to 2100. These results are for the full range of 35 SRES scenarios, based on a number of climate models.
- Temperature increases are projected to be greater than those in the SAR, which were about 1.0 to 3.5°C based on the six IS92 scenarios. The higher projected temperatures and the wider range are due primarily to the lower projected sulphur dioxide emissions in the SRES scenarios relative to the IS92 scenarios.
- The projected rate of warming is much larger than the observed changes during the 20th century and is very likely to be without precedent during at least the last 10,000 years, based on palaeoclimate data.
- By 2100, the range in the surface temperature response across the group of climate models run with a given scenario is comparable to the range obtained from a single model run with the different SRES scenarios.
- On timescales of a few decades, the current observed rate of warming can be used to constrain the projected response to a given emissions scenario despite uncertainty in climate sensitivity. This approach suggests that anthropogenic warming is likely to lie in the range of 0.1 to 0.2°C per decade over the next few decades under the IS92a scenario....
- Based on recent global model simulations, it is very likely that nearly all land areas will warm more rapidly than the global average, particularly those at northern high latitudes in the cold season. Most notable of these is the warming in the northern regions of North America, and northern and central Asia, which exceeds global mean warming in each model by more than 40%. In contrast, the warming is less than the global mean change in south and southeast Asia in summer and in southern South America in winter.

- Recent trends for surface temperature to become more El Niño–like in the tropical Pacific, with the eastern tropical Pacific warming more than the western tropical Pacific, with a corresponding eastward shift of precipitation, are projected to continue in many models.

Precipitation

Based on global model simulations and for a wide range of scenarios, global average water vapour concentration and precipitation are projected to increase during the 21st century. By the second half of the 21st century, it is likely that precipitation will have increased over northern mid- to high latitudes and Antarctica in winter. At low latitudes there are both regional increases and decreases over land areas. Larger year to year variations in precipitation are very likely over most areas where an increase in mean precipitation is projected....

El Niño

- Confidence in projections of changes in future frequency, amplitude, and spatial pattern of El Niño events in the tropical Pacific is tempered by some shortcomings in how well El Niño is simulated in complex models. Current projections show little change or a small increase in amplitude for El Niño events over the next 100 years.
- Even with little or no change in El Niño amplitude, global warming is likely to lead to greater extremes of drying and heavy rainfall and increase the risk of droughts and floods that occur with El Niño events in many different regions.

Monsoons

It is likely that warming associated with increasing greenhouse gas concentrations will cause an increase of Asian summer monsoon precipitation variability. Changes in monsoon mean duration and strength depend on the details of the emission scenario. The confidence in such projections is also limited by how well the climate models simulate the detailed seasonal evolution of the monsoons.

Thermohaline Circulation

Most models show weakening of the ocean thermohaline circulation which leads to a reduction of the heat transport into high latitudes of the Northern Hemisphere. However, even in models where the thermohaline circulation weakens, there is still a warming over Europe due to increased greenhouse gases. The current projections using climate models do not exhibit a complete shutdown of the thermohaline circulation by 2100. Beyond 2100, the thermohaline circulation could completely, and possibly irreversibly, shut-down in either hemisphere if the change in radiative forcing is large enough and applied long enough.

Snow and Ice

- Northern Hemisphere snow cover and sea-ice extent are projected to decrease further.
- Glaciers and ice caps are projected to continue their widespread retreat during the 21st century.
- The Antarctic ice sheet is likely to gain mass because of greater precipitation, while the Greenland ice sheet is likely to lose mass because the increase in runoff will exceed the precipitation increase.
- Concerns have been expressed about the stability of the West Antarctic ice sheet because it is grounded below sea level. However, loss of grounded ice leading to substantial sea level rise from this source is now widely agreed to be very unlikely during the 21st century, although its dynamics are still inadequately understood, especially for projections on longer time-scales.

Sea Level

Global mean sea level is projected to rise by 0.09 to 0.88 metres between 1990 and 2100, for the full range of SRES scenarios. This is due primarily to thermal expansion and loss of mass from glaciers and ice caps. The range of sea level rise presented in the SAR was 0.13 to 0.94 metres based on the IS92 scenarios. Despite the higher temperature change projections in this assessment, the sea level projections are slightly lower, primarily due to the use of improved models, which give a smaller contribution from glaciers and ice sheets.

Anthropogenic climate change will persist for many centuries.

- Emissions of long-lived greenhouse gases (i.e., CO_2, N_2O, PFCs, SF_6) have a lasting effect on atmospheric composition, radiative forcing and climate. For example, several centuries after CO_2 emissions occur, about a quarter of the increase in CO_2 concentration caused by these emissions is still present in the atmosphere.
- After greenhouse gas concentrations have stabilised, global average surface temperatures would rise at a rate of only a few tenths of a degree per century rather than several degrees per century as projected for the 21st century without stabilisation. The lower the level at which concentrations are stabilised, the smaller the total temperature change.
- Global mean surface temperature increases and rising sea level from thermal expansion of the ocean are projected to continue for hundreds of years after stabilisation of greenhouse gas concentrations (even at present levels), owing to the long timescales on which the deep ocean adjusts to climate change.

- Ice sheets will continue to react to climate warming and contribute to sea level rise for thousands of years after climate has been stabilised. Climate models indicate that the local warming over Greenland is likely to be one to three times the global average. Ice sheet models project that a local warming of larger than 3°C, if sustained for millennia, would lead to virtually a complete melting of the Greenland ice sheet with a resulting sea level rise of about 7 metres. A local warming of 5.5°C, if sustained for 1,000 years, would be likely to result in a contribution from Greenland of about 3 metres to sea level rise.
- Current ice dynamic models suggest that the West Antarctic ice sheet could contribute up to 3 metres to sea level rise over the next 1,000 years, but such results are strongly dependent on model assumptions regarding climate change scenarios, ice dynamics and other factors.

Further action is required to address remaining gaps in information and understanding.

Further research is required to improve the ability to detect, attribute and understand climate change, to reduce uncertainties and to project future climate changes. In particular, there is a need for additional systematic and sustained observations, modelling and process studies. A serious concern is the decline of observational networks. The following are high priority areas for action.

- Systematic observations and reconstructions:

    - Reverse the decline of observational networks in many parts of the world.
    - Sustain and expand the observational foundation for climate studies by providing accurate, long-term, consistent data including implementation of a strategy for integrated global observations.
    - Enhance the development of reconstructions of past climate periods.
    - Improve the observations of the spatial distribution of greenhouse gases and aerosols.

- Modelling and process studies:

    - Improve understanding of the mechanisms and factors leading to changes in radiative forcing.
    - Understand and characterise the important unresolved processes and feedbacks, both physical and biogeochemical, in the climate system.
    - Improve methods to quantify uncertainties of climate projections and scenarios, including long-term ensemble simulations using complex models.

- Improve the integrated hierarchy of global and regional climate models with a focus on the simulation of climate variability, regional climate changes and extreme events.
- Link more effectively models of the physical climate and the biogeochemical system, and in turn improve coupling with descriptions of human activities.
- Cutting across these foci are crucial needs associated with strengthening international co-operation and co-ordination in order to better utilise scientific, computational and observational resources. This should also promote the free exchange of data among scientists. A special need is to increase the observational and research capacities in many regions, particularly in developing countries. Finally, as is the goal of this assessment, there is a continuing imperative to communicate research advances in terms that are relevant to decision making.

NO

Kevin A. Shapiro

Too Darn Hot?

Natives of Hawaii, inured by more than a thousand years of island life to the vagaries of the weather and the seas, have a somewhat elliptical saying: "the mists are those that know of a storm upon the water." It can be taken to mean that those nearest to something are the first to become aware of what is happening to it. Using similar reasoning, perhaps, many environmentalists today regard the small islands that dot the Pacific as a sort of planetary weathervane, outcrops of flora and fauna that are sensitive indicators of large-scale shifts in the ecological balance of the earth. If these islands are already beginning to buckle under the stresses imposed on the planet by human activity, it is a sign that we must act quickly lest catastrophe result.

An alarming presentation of this argument can be found in *Rising Waters: Global Warming and the Fate of the Pacific Islands,* an hour-long documentary that aired on PBS [in] April [2001] on Earth Day. *Rising Waters* paints a picture of island nations on the veritable brink of ruin: homes destroyed in the wake of storms or threatened by eroding shorelines, churchyards and crop-fields inundated by the rising sea, and shoals of once-vivid coral bleached by overheated waters. On camera, fishermen complain of poor hauls; a Samoan environmentalist laments the looming disappearance of his cultural heritage; Teburoro Tito, the president of tiny Kiribati, speaks glumly of the possibility that the entire populace of his cluster of atolls will have to be relocated.

What is causing this potentially immense disruption? *Rising Waters* mentions several factors, including seasonal weather fluctuations and overdevelopment, but ultimately it places most of the blame on a long chain of processes at the end of which is: global warming. The nature of this menace is well known and has been widely discussed. Increases in the industrial emission of gases like carbon dioxide (CO_2), it is said, have caused the atmosphere to absorb infrared radiation that would otherwise be reflected back into outer space. The resulting "greenhouse effect" lifts the average temperature of the earth's surface. Among the many consequences are rising sea levels caused by the melting of the polar ice caps and increases in the frequency and intensity of storm activity.

Though *Rising Waters* offers the disclaimer that the earth's climate is a complex and somewhat unpredictable system—"we don't know how it behaves completely," says Fred MacKenzie, a professor of oceanography at the University of Hawaii—its overarching message is that unregulated CO_2 emissions have already begun to heat the planet to dangerous levels. To forestall further warming, we must cut those emissions globally by as much as 80 percent over the next several decades. Alas, as *Rising Waters* notes with a hint of impending doom, the prospects for such a cut are not auspicious.

◦◦◦

On this last point, the documentary is certainly correct. Talks in the Hague on implementing the 1997 Kyoto Protocol, an international agreement aimed at reducing the CO_2 emissions of industrial nations to pre-1990 levels by the year 2012, collapsed in December, in the last month of Bill Clinton's presidency. By mid-March, the Bush administration had announced it would not seek to regulate the CO_2 emissions of power plants, provoking an outcry from environmentalists and angering European leaders who maintain (in the words of Dutch prime minister Wim Kok) that the United States is acting "irresponsibly." Two weeks later, President Bush declared that it made "no sense" for the United States to pursue implementation of the Kyoto Protocol. European governments, positively livid, dispatched an emergency delegation to Washington, but to no avail; they now plan to assemble an international coalition aimed at "shaming" the United States into reconsidering its stance. Another round of talks on Kyoto will be held in Bonn [in] July [2001], and the conflict over global warming is certain to deepen in the months and years ahead.

Against this backdrop, *Rising Waters* can only serve to underscore the now almost incessant warnings about the disaster that awaits us if we fail to change our profligate energy habits. Global warming has already been blamed for ecological hazards ranging in scale from disruptions in the migration patterns of butterflies and declining amphibian populations to extreme weather events, droughts, and food shortages in farflung portions of the globe. And the dangers that lie ahead are said to be far worse, if not horrific: famine brought on by widespread agricultural failure, an increase in epidemics of infectious disease, even mass extinctions of animal and human populations.

If anything remotely resembling this scenario is likely, it is not hard to see why so many Europeans, and with them many Americans, are apoplectic over President Bush's determination to scrap the Kyoto deal, the fruit of years of intense multinational discussions among lawmakers, economists, scientists, and environmentalists. Senator Joseph Lieberman has even promised a congressional investigation of the President's environmental decisions, declaring that they "ignore the public interest and defy common sense."

◦◦◦

Is Lieberman right? There are indeed many things about the global-warming debate that "ignore the public interest and defy common sense." But the decision to abandon the Kyoto Protocol is not one of them.

In a sense, the decision was hardly even newsworthy. The agreement has been effectively dead—at least as far as the United States is concerned—since shortly after it was negotiated in 1997. For no sooner did Clinton's negotiators return from Japan than the Senate voted 95-0 to oppose ratification of any treaty that would impose significant burdens on our national economy and that lacked "specific scheduled commitments" for emissions reductions in what are now known as "developing" countries. As Kyoto has never been amended to address these concerns, it is perplexing that any policymaker could have continued to regard the accord as viable.

Indeed, far more inscrutable than President Bush's final rejection of Kyoto is the vast amount of rhetorical and diplomatic effort that has been and continues to be expended on the agreement's behalf. Even apart from the unanimous vote in the Senate, there are serious questions about whether the provisions of the treaty could ever be implemented and enforced, and therefore about whether it really represents a workable mechanism for managing climate change.

From its very inception, as the analyst David G. Victor shows in a new monograph, the Kyoto Protocol was a product of diplomatic wishful thinking. For one thing, the limits it called for on greenhouse gas emissions were draconian. Thus, by 2012 the United States would have been required to reduce CO_2 emissions to 7 percent below 1990 levels—a modest-sounding target until one considers that by the end of 1999, emissions were already 12 percent above 1990 levels and were continuing to rise. Compliance with Kyoto would therefore have required a likely cut of as much as 30 percent by the time the treaty took effect in 2008. Not only would this cost hundreds of billions of dollars in GDP [gross domestic product] but, because most greenhouse gases are released in the course of burning fossil fuels for energy, cutbacks on such a scale would deal a major blow to significant sectors of the U.S. economy— particularly electricity generation, which is already struggling mightily to keep pace with demand.

The agreement was also exceedingly inequitable. Russia, for example, would have been required only to freeze its emissions at 1990 levels; but because the Russian economy has contracted sharply since the collapse of the Soviet Union, its emissions are already far below target, and are unlikely to recover by 2008. Though it remains a significant industrial polluter, Moscow would thus be required to do absolutely nothing. South Korea and Mexico, now formally considered "developed" countries (as defined by membership in the Organization for Economic Cooperation and Development), have for their part also not agreed to curtail emissions.

At the same time, Kyoto sets no targets at all for the developing nations, though these countries will account for half the world's greenhouse gases by 2020. The two largest such nations, India and China, have refused outright to accept any limits on their emissions output.

In short, the Kyoto Protocol demands that the United States hobble its economy with drastic cuts in energy production, while Russia, India, China, and other nations enjoy the freedom to grow untrammeled. To deal with this gross imbalance, a number of observers have proposed amending the agreement. One proposal involves altering the way emissions are accounted for—for example, by permitting industrialized countries to earn "credits" if they maintain or create carbon sinks, i.e., forest and soil zones that absorb CO_2. Another alternative would be to allow trading, whereby industrialized countries could buy the right to emit carbon dioxide from those nations whose emissions are below targeted levels.

Both of these ideas have their attractions for the United States, but they also entail immense practical and political difficulties. On the positive side, the U.S. might offset its Kyoto obligations by counting carbon sinks that resulted from intentional changes in land-use policy. If, in addition, it were permissible to count those resulting from unintentional changes (like the spontaneous reforestation of abandoned agricultural lands), we might no longer be a net emitter. But an amendment of this sort would almost certainly prove unacceptable to Europe and Japan, which, unlike the U.S., have limited capacity to plant new forests. A more fundamental problem is that the Kyoto Protocol provides no standard definitions, methods, or data for quantifying the absorption of CO_2 by trees and soils, making it easy for nations to cheat by claiming credit for carbon sinks that are short-lived or even nonexistent.

Emissions trading is beset with its own difficulties. The present terms of the Kyoto Protocol would seem to award countries with low baselines—like Russia—a windfall in fictitious credits, the sale of which would result in no reduction in global emissions whatsoever. David Victor has correctly spelled out the political implications of any such arrangement: "No Western legislature will ratify a deal that merely enriches Russia and Ukraine while doing nothing to control emissions and slow global warming."

If the most widely discussed ways of amending the Kyoto agreement are infeasible, what then? Policy analysts like Victor continue to hold out hope that the Bush administration will develop a coherent approach to global warming—perhaps a modified trading system combined with international taxes on CO_2 emissions and supplemented by investments in new technology. As for the Bush administration, the President himself has spoken of global warming as a "serious problem," and the U.S. will be participating in the upcoming talks in Bonn with the hope of finding a workable alternative to Kyoto.

The operative assumption here, of course, is that man-made climate change is a real phenomenon, and that averting catastrophe requires doing something about it, and soon. As this assumption has increasingly come to be taken for granted, disputing it has become commensurately perilous, especially for politicians. According to a 1997 poll taken for the World Wildlife Federation, two-thirds of American voters regard global warming as a "serious threat" and support an international agreement to cut greenhouse-gas emissions, even if this comes at some economic cost. A full three-quarters endorse the view that "the only scientists who do not believe global warming is happening are paid by big oil, coal, and gas companies to find the results that will protect business interests." Only 15 percent accept the statement that "scientists disagree among themselves" about the extent of the coming danger.

Clearly, climate change is no longer an issue up for grabs. Even if the public could be persuaded that the Kyoto Protocol would be disastrous for the U.S. economy and is the result of junk diplomacy, it would be far harder for a politician to make the case that the research behind Kyoto is junk science, too. But much of it is.

Let us return for a moment to those Pacific islands. It is undeniable that they have been buffeted by a series of severe storms in the past decade, accompanied by unusually intense episodes of the El Niño-Southern Oscillation (ENSO) phenomenon, a periodic fluctuation in sea temperature in the tropical Pacific that has been observed since the last century. What is not clear is whether these have anything to do with global warming.

Storm activity in the Pacific varies from year to year; 1998 saw an above-average incidence of tropical storms, while 1997 was comparatively quiet. The cause of this variation remains unknown. The ENSO phenomenon is not well understood, and it is not predicted by any model of climate change. A United Nations body called the Intergovernmental Panel on Climate Change (IPCC) has rightly observed that while many small island states fear that "global warming will lead to changes in the character and pattern of tropical cyclones (i.e., hurricanes and typhoons)," this fear is not confirmed by the most recent research. Rather, "model projections suggest no clear trend, so it is not possible to state whether the frequency, intensity, or distribution of tropical storms and cyclones will change."

And what of rising waters? In 1980, climatologists predicted that global warming would melt the polar icecaps, causing sea levels to rise more than 25 feet over the course of the next century. Such an event would undoubtedly be disastrous not only for the Pacific islands but also for densely populated coastal regions in all parts of the world.

Fortunately for those of us in Boston, Miami, New York, and Los Angeles, the deluge failed even to begin to materialize. According to the latest data, the polar icecaps do not appear to be melting at all. The 2001 IPCC report discerns "no significant trends" in the extent of Antarctic sea-ice since 1978, when reliable satellite measurements began to be taken; nor, at the other pole, is there evidence from satellite records that the air above the Arctic has warmed substantially.

With the polar caps essentially intact, it does not come as a surprise that sea levels have risen only a paltry 2 millimeters per year in the mid-1990's— roughly the same rate observed over the past 100 years. Even the gloomiest doomsayers have been compelled to jettison the dire forecasts put forward in 1980. Under the *worst*-case scenario now envisioned by the IPCC, the oceans should rise no more than a foot over the next century, not nearly enough to pose a major threat. And this forecast is in turn based on the assumption that sea levels will increase by approximately 5 millimeters per year, *give or take* 3 millimeters—in other words, the rate of rise may not change at all.

As for the climate itself, despite the alarmed rhetoric from so many quarters, we do not know for certain that it is even changing in significant ways. It is an established fact that the earth's climate has warmed slightly over the past century. Average temperatures near the surface have risen since 1900 and are now probably higher than they have been at any time in the past 600 to 1,000 years. But that statement more or less exhausts the scientific consensus. On every other important question—what the major causes of global warming are, what its effects will be, whether we should try to prevent it and, if so, how —there is considerable uncertainty.

••◉••

Most of what we "know" about the earth's future is derived from enormously sophisticated computer models that utilize millions of parameters to simulate the earth's climate. These models are still far from reliable. The editors of *Nature,* arguably the world's most prestigious scientific journal, pointed out on March 15 that "the accuracy of any model depends significantly on the quality of the underlying raw data." But the quality of the data being used for climate prediction, they go on to state, is "patchy." For example, it is not at all easy to measure the amount of sunlight absorbed by the atmosphere or reflected by its surface back into space—and yet this one key parameter alone might (or might not) account for six times the amount of energy that would be added to the climate system by the doubling of atmospheric CO_2. Similar uncertainties attend other crucial variables like the impact of differing degrees of cloud cover and water vapor.

Given the room for error, it should come as no surprise that climate-prediction models have racked up an exceedingly poor track record over the years. According to those models, the average global temperature should have increased by at least 1 degree centigrade since the beginning of the 20th century, when industrial emissions of greenhouse gases first began to rise. But the best available measurements indicate that the average global temperature has increased by only 0.5 degrees in 100 years, and much of that increase occurred before 1940—too early in the century, in other words, to have been caused by a growth in CO_2 levels.

Contrary to the simulations, moreover, the marginal uptick in surface temperatures in the years since 1970 has not been accompanied by warming of the lower atmosphere (as we know from satellite data). A pair of recent papers in the journal *Science* attempts to account for this discrepancy by locating the missing heat in the oceans, a "discovery" trumpeted by the media as yet another blow to those who remain skeptical of global warming. But this was not a discovery at all, and was not based on any finding that whatever warming may have occurred has been caused by human activity. Rather, it was merely the product of "improved" models, which have their own "improved" assumptions and their own set of poorly understood parameters.

In the face of such scientific shell-games, and in the face of the huge costs the United States has been asked to incur to combat a problem that may or may not exist, President Bush was certainly right to pull the plug on the Kyoto Protocol. But whether he will be able to stand firm against the torrent of criticism that has been unleashed against him remains an open question. According to the Natural Resources Defense Council (NRDC), the Bush administration's decision to abandon Kyoto "will have massively destructive consequences for the earth and its people." Although the IPCC has specifically rejected any direct linkage between today's local environmental perturbations and global warming, the Sierra Club is instructing its members that the apocalypse is upon us *now*, in the form of "heat waves, droughts, coastal flooding, and malaria outbreaks."

There are more narrowly partisan interests at play as well. "Democrats See Gold in Environment," ran the headline of a recent *New York Times* story describing how Bush's environmental decisions have galvanized activists in the Democratic party. Indeed, reports the *Times,* some party officials are positively "gleeful" at the political opportunities now opening up. One such official is evidently Senator Lieberman. Assuming Al Gore's mantle as the party's leading spokesman on matters environmental, Leiberman has called the decision to abandon Kyoto "flabbergasting," and is now invoking the specter of "sea levels [that] could swell up to 35 feet, potentially submerging millions of homes and coastal property."

That this is the same Joseph Lieberman who in 1997 joined 94 other Senators in voting to denounce the Kyoto Protocol suggests that when it comes to global warming we are indeed facing a rising tide—of hysteria, mixed with sheer political cynicism. As against these twin forces, it may seem hopelessly naïve to suggest that we would do better to focus on phasing out those greenhouse gases that can be eliminated at relatively low cost, like sulfur hexafluoride and perfluorocarbons, while adopting a wait-and-see attitude toward CO_2, secure in the knowledge that advances in technology and in the accuracy of prediction will allow us to address climate change more effectively and more cheaply in the future. Naïve it may be, but at present there is no basis in scientific evidence for more drastic action. All that is required is a politician tough enough and brave enough to say so.

POSTSCRIPT

Do Human Activities Threaten to Change the Global Climate?

The United Nations Conference on Environment and Development in Rio de Janeiro, Brazil, took place in 1992. High on the agenda was the problem of global warming, but despite widespread concern and calls for reductions in carbon dioxide releases, the United States refused to consider rigid deadlines or set quotas. The uncertainties seemed too great, and some thought the economic costs of cutting back on carbon dioxide might be greater than the costs of letting the climate warm.

However, James Kasting of Pennsylvania State University and James Walker of the University of Michigan warn that if one looks a little further into the future than the next century, the prospects look much more frightening. They predict that by 2100 the amount of carbon dioxide in the atmosphere will reach double its preindustrial level. By the 2200s it could be 7.6 times the preindustrial level. With draconian restrictions, however, it could be held to 4 times the preindustrial level. Global warming may therefore turn out to be much worse for the next century than anyone is predicting, although it is difficult to be at all sure of such predictions. See David Schneider, "The Rising Seas," *Scientific American* (March 1997) and Thomas R. Karl, Neville Nicholls, and Jonathan Gregory, "The Coming Climate," *Scientific American* (May 1997).

The nations that signed the UN Framework Convention on Climate Change in Rio de Janeiro in 1992 met again in Kyoto, Japan, in December 1997 to set carbon emissions limits for the industrial nations. The United States agreed to reduce its annual greenhouse gas emissions 7 percent below the 1990 level between 2008 and 2012. In November 1998 they met in Buenos Aires, Argentina, to work out practical details (see Christopher Flavin, "Last Tango in Buenos Aires," *World Watch* [November/December 1998]). Unfortunately, developing countries, where carbon emissions are growing most rapidly, face few restrictions, and political opposition in developed nations—especially in the United States—remains strong. Ross Gelbspan, in "Rx for a Planetary Fever," *American Prospect* (May 8, 2000), blames much of that opposition on "big oil and big coal [which] have relentlessly obstructed the best-faith efforts of government negotiators."

The opposition remains visible in 2001, despite the new IPCC report. Critics stress uncertainties in the data and the potential economic impacts of attempting to reduce carbon dioxide emissions. See Richard A. Kerr, "Rising Global Temperature, Rising Uncertainty," *Science* (April 13, 2001). Some feel that climate change may well be less severe than expected and also beneficial overall to agriculture and human well-being. See Patrick J. Michaels and Robert

C. Balling, Jr., *The Satanic Gases: Clearing the Air About Global Warming* (Cato Institute, 2000).

There is also opposition based on the view that the methods of reducing greenhouse gas emissions called for in the Kyoto treaty are, at root, unworkable. See Frank N. Laird, "Just Say No to Greenhouse Gas Emissions Targets," *Issues in Science and Technology* (Winter 2000–2001). However, researchers have proposed a number of innovative ways to keep from adding carbon dioxide to the atmosphere. See Howard Herzog, Baldue Eliasson, and Olav Kaarstad, "Capturing Greenhouse Gases," *Scientific American* (February 2000).

In June 2002 the U.S. Environmental Protection Agency (EPA) issued its *U.S. Climate Action Report—2002* (available at http://www.epa.gov/globalwarming/publications/car/index.html) to the United Nations. In it, the EPA admits for the first time that global warming is real and that human activities are most likely to blame. President George W. Bush immediately dismissed the report as "put out by the bureaucracy" and said he still opposes the Kyoto Protocol.

Silicon Valley Toxics Coalition

The Silicon Valley Toxics Coalition (SVTC) was formed in 1982 to engage in research, advocacy, and organizing associated with environmental and human problems caused by the rapid growth of the high-tech electronics industry; to advance environmental sustainability and clean production in the industry; and to improve health, promote justice, and ensure democratic decision making for affected communities and workers in the United States and the world.

http://www.svtc.org

Yucca Mtn. Standards

At this Web site, the U.S. Environmental Protection Agency provides information on the proposed Yucca Mountain permanent nuclear waste repository.

http://www.epa.gov/radiation/yucca/

Why Not Yucca Mountain?

On this page, the Nevada Nuclear Waste Task Force, a nonprofit organization fighting nuclear waste in Nevada and providing information about Yucca Mountain, details several reasons why the U.S. Department of Energy's plan to ship high-level nuclear waste to Yucca Mountain should not be carried out.

http://www.nvantinuclear.org

Superfund

At this Web site, the U.S. Environmental Protection Agency provides a great deal of information on the Superfund program, including material on environmental justice.

http://www.epa.gov/superfund/

Waste Prevention

This site is a valuable source of information—including publications—about dealing with municipal wastes. It is sponsored by the Local Government Commission, a nonprofit organization composed of elected officials, city and county staff, and other interested individuals who are all committed to developing and implementing local solutions to problems of state and national significance.

http://www.lgc.org/waste/index.html

PART 3

Disposing of Wastes

*T**he simple fact that people cannot live without generating waste delights archeologists, but in the modern age people generate such enormous quantities of waste that disposal has become a serious problem. The solutions include sanitary landfills (visible as sizable hills near many cities), incinerators, and recycling. Unfortunately, many wastes—sewage sludge, toxic chemicals, and radioactive nuclear waste (municipal, industrial, and military)—are too hazardous for such approaches. In some cases—such as agricultural and mining wastes—the quantities beggar the imagination. Yet they must still be disposed of as safely as possible, even when no one wants a disposal site anywhere nearby.*

This section deals with three prominent controversies concerning the disposal of hazardous, municipal, and nuclear waste.

- Hazardous Waste: Should the "Polluter Pays" Provision of Superfund Be Weakened?

- Municipal Waste: Is Recycling an Environmentally and Economically Sound Waste Management Strategy?

- Nuclear Waste: Should the United States Continue to Focus Plans for Permanent Nuclear Waste Disposal Exclusively at Yucca Mountain?

ISSUE 13

Hazardous Waste: Should the "Polluter Pays" Provision of Superfund Be Weakened?

YES: Bernard J. Reilly, from "Stop Superfund Waste," *Issues in Science and Technology* (Spring 1993)

NO: Ted Williams, from "The Sabotage of Superfund," *Audubon* (July/August 1993)

ISSUE SUMMARY

YES: DuPont corporate counsel Bernard J. Reilly argues that in defining standards and assigning costs related to waste cleanup, "Congress should focus the program on reducing real risk, not on seeking unattainable purity."

NO: *Audubon* contributing editor Ted Williams contends that insurers and polluters are lobbying to change the financial liability provisions of Superfund, and he warns against turning it into a public welfare program.

The potentially disastrous consequences of improper hazardous waste disposal burst upon the consciousness of the American public in the late 1970s. The problem was dramatized by the evacuation of dozens of residents of Niagara Falls, New York, whose health was being threatened by chemicals leaking from the abandoned Love Canal, which was used for many years as an industrial waste dump. Awakened to the dangers posed by chemical dumping, numerous communities bordering on industrial manufacturing areas across the country began to discover and report local sites where chemicals had been disposed of in open lagoons or were leaking from disintegrating steel drums. Such esoteric chemical names as dioxins and PCBs have become part of the common lexicon, and numerous local citizens' groups have been mobilized to prevent human exposure to these and other toxins.

The expansion of the industrial use of synthetic chemicals following World War II resulted in the need to dispose of vast quantities of wastes laden with organic and inorganic chemical toxins. For the most part, industry adopted a casual attitude toward this problem and, in the absence of regulatory restraint, chose the least expensive means available. Little attention was paid to the ultimate fate of chemicals that could seep into surface water or groundwater. Scientists have estimated that less than 10 percent of the waste was disposed of in an environmentally sound manner.

The magnitude of the problem is truly mind-boggling: Over 275 million tons of hazardous waste is produced in the United States each year; as many as 10,000 dump sites may pose a serious threat to public health, according to the federal Office of Technology Assessment; and other government estimates indicate that more than 350,000 waste sites may ultimately require corrective action at a cost that could easily exceed $500 billion.

Congressional response to the hazardous waste threat is embodied in two complex legislative initiatives. The Resource Conservation and Recovery Act (RCRA) of 1976 mandated action by the Environmental Protection Agency (EPA) to create "cradle to grave" oversight of newly generated waste, and the Comprehensive Environmental Response, Compensation, and Liability Act of 1980 (CERCLA), commonly called "Superfund," gave the EPA broad authority to clean up existing hazardous waste sites. The implementation of this legislation has been severely criticized by environmental organizations, citizens' groups, and members of Congress who have accused the EPA of foot-dragging and a variety of politically motivated improprieties. Less than 20 percent of the original $1.6 billion Superfund allocation was actually spent on waste cleanup.

Amendments designed to close RCRA loopholes were enacted in 1984, and the Superfund Amendments and Reauthorization Act (SARA) added $8.6 billion to a strengthened cleanup effort in 1986 and an additional $5.1 billion in 1990. While acknowledging some improvement, both environmental and industrial policy analysts remain very critical about the way that both RCRA and Superfund/SARA are being implemented. Efforts to reauthorize and modify both of these hazardous waste laws have been stalled in Congress since the early 1990s.

Although the following selections were written in 1993, they still accurately reflect the current controversy concerning the Superfund Act. In the first selection, Bernard J. Reilly argues that the legislation has turned into an "unjustifiable waste of the nation's resources at the expense of other critical society needs." He calls for major changes to "focus the program on practical risk reduction." One specific change he advocates is in the "polluter pays" provision of the law, which he argues holds companies liable for more than their fair share of the costs. In the second selection, Ted Williams acknowledges that the Superfund program has been very costly and has cleaned up little waste. But he blames this on sabotage of the law by the Bush and Reagan administrations. He specifically warns against recommendations to abandon the policy of holding polluters strictly liable for the damage they have caused, which he fears would "turn it into a public works program" whereby "citizens will pay twice, once with their environment and once with their tax money."

Bernard J. Reilly

 **YES**

Stop Superfund Waste

President Clinton's economic plan is a clear attempt to reorder federal spending priorities by putting more money into "investments" that will spur economic growth and increase national wealth, while cutting unproductive activities. One important way he could further his agenda would be to push for reform of one of today's most misguided efforts: the Superfund hazardous waste cleanup program. The President has already paid lip service to this goal, telling business leaders in a February 11 speech at the White House that, "We all know it doesn't work—the Superfund has been a disaster."

Superfund, created by the Comprehensive Environmental Response, Compensation, and Liability Act of 1980 (CERCLA) in the wake of the emergency at the Love Canal landfill in Niagara Falls, New York, was designed as a $1.6-billion program to contain the damage from and eventually clean up a limited number of the nation's most dangerous abandoned toxic waste sites. But in short order it has evolved into an open-ended and costly crusade to return potentially thousands of sites to a near-pristine condition. The result is a large and unjustifiable waste of our nation's resources at the expense of other critical societal needs.

No one questions that the nation has a major responsibility to deal with hazardous waste sites that pose a serious risk to public health and the environment. It is the manner and means by which the federal government has pursued this task, however, that are wasteful. Superfund legislation has given the U.S. Environmental Protection Agency powerful incentives and great clout to seek the most comprehensive, "permanent" cleanup remedies possible—without regard to cost or even the degree to which public health is at risk. Although the EPA does not always choose the most expensive remedial solution, there is strong evidence that, in many cases, waste sites can be cleaned up or sufficiently contained or isolated for a fraction of the cost, while still protecting the public and the environment. Further, EPA's selection of "priority" cleanup sites has been haphazard at best. Indeed, it has no system in place for determining which of those sites—or the many potential sites it has not yet characterized—pose the greatest dangers.

A 180-degree turn in policy is needed. When the Superfund program comes up for reauthorization next year, Congress should direct the EPA to abandon its pursuit of idealistic cleanup solutions and focus the program on practical risk reduction, targeting those sites that pose the greatest health risks and tying the level and cost of cleanup to the degree of actual risk. Only by making such a fundamental change can the nation maximize the benefits of its increasingly huge investment in the remediation of hazardous waste sites.

Costs Are Escalating

Estimates for cleaning up, under current practice, the more than 1,200 sites on the EPA's "national priority list" (NPL) range from $32 billion by EPA (based on a $27 million per-site cost) to $60 billion by researchers at the University of Tennessee (based on a $50 million per-site cleanup cost). These estimates are likely to be well below the ultimate cost, since EPA can add an unlimited number of sites to the list. The agency plans to add about 100 sites a year, bringing the total by the year 2000 to more than 2,100. But more than 30,000 inactive waste sites are being considered for cleanup and the universe of potential sites has been estimated at about 75,000. Most experts believe that far fewer—from 2,000 to 10,000—will eventually be cleaned up. The University of Tennessee researchers make a best guess of 3,000 sites, which would put the cost at $150 billion (in 1990 dollars) over 30 years, not including legal fees.

This $150 billion might be acceptable if the U.S. economy were buoyant and limitless funds existed for other needs. It most certainly would be justified if many sites posed unacceptable dangers to the public. But neither of these situations exists.

Skewed Priorities

A key flaw in Superfund is that most of its effort and money are directed to a relatively small number of "priority" sites, while thousands of others are ignored and, in most cases, not even sampled or studied. For this reason, it is doubtful that the NPL includes all the worst sites.

"Deadly" chemical landfills buried under residential neighborhoods have hardly been typical of the sites EPA has placed on the NPL. Indeed, EPA's efforts to create a system for ranking hazards have not been geared to actually finding the riskiest sites but to satisfying the letter of the CERCLA law. In the first ranking scheme, sites were evaluated for various threats and a score of 28.5 (on a scale of 100) was determined to be sufficient for an NPL listing. However, the listings were not necessarily based on an actual determination of the degree to which they posed threats to public health or the environment. Rather, the sites were included because Congress had determined that 400 sites must be on the NPL, and a score of 28.5 resulted in 413 listings.

Several years ago, the ranking scheme was made much more elaborate, with threats from contaminants in the air, water, and soil weighed differently. The same maximum score of 100 and listing score of 28.5 were used. Why? EPA said that it was "not because of any determination that the cutoff represented a threshold of unacceptable risk presented by the sites" but because the 28.5 score was "a convenient management tool." So much for the rigors of a system designed to cull the Love Canals from town dumps.

A 1991 report by a committee of the National Research Council (NRC) strongly faulted EPA's methods of selecting sites and setting priorities. The report said that EPA has no comprehensive inventory of waste sites, no program for discovering new sites, insufficient data for determining safe exposure levels, and an inadequate system for identifying sites that require immediate action to protect public health.

In a perfect world, every "dirty" site would be cleaned, regardless of the degree of risk it presented. In practice, this is impossible, so we should be spending more to prioritize in order to focus our limited resources on real risks.

Extreme Remedies

However it is accomplished, once a site makes the NPL, money is no object in the remediation process. This was not necessarily the case under the original 1980 Superfund law. CERCLA left some ambiguity about how extensive the cleanups had to be—whether only reasonable risks needed to be eliminated or whether the site had [to] be returned to a preindustrial condition. When it enacted the Superfund Amendments and Reauthorization Act (SARA) in 1986, however, Congress, motivated by a deep distrust of the Reagan-era EPA, took a hard-line stance. SARA, which increased funding for the program to $8.5 billion and ordered action to begin at ever more sites, directed EPA to give preference to cleanup remedies that "to the maximum extent practicable" lead to "permanent solutions." The emphasis on permanence was further reinforced by a requirement that cleanups must comply with any "applicable or relevant and appropriate requirement" (ARAR) in any other state or federal law relating to protection of public health and the environment.

SARA was deeply flawed. For one thing, it effectively forced EPA to continue remedial action even after all realistic risks at a site had been eliminated. One example is the Swope Superfund site, a former solvent reclamation facility in Pennsauken, New Jersey. Although all major sources of contamination had been removed from the site, EPA ordered the installation of a $5-million vapor extraction system to remove more contaminants. The purpose was to protect groundwater in case any private wells were sunk in the future. But EPA neglected to consider the fact that private wells had been banned in the area.

SARA's requirements also serve to exclude the use of other far less costly remedies that would give the public the same or at least acceptable protection from harm. For example, at the Bridgeport Rental and Oil Services Superfund site in Logan Township, New Jersey, EPA ordered the construction of an onsite, $100-million incinerator after PCBs were found in several sludge samples. In making its decision, EPA used the ARAR requirement to retroactively apply the federal Toxic Substances Control Act (TSCA), which requires incineration of currently generated wastes if samples indicate that PCBs in the soil exceed 500 parts per million.

The absurdity of the plan became apparent when EPA decided to create an on-site landfill to dispose of the heavy metal residues from the incineration. Given that a landfill was to be created anyway and that PCBs at the site were so scarce that EPA had to import them for trial burns of the incinerator, the agency could have opted to contain the sludge on site in the first place—using existing proven technologies—while more than adequately protecting the public at an estimated one-fifth of the cost of incineration.

A similar tale is unfolding at another Superfund site in Carlstadt, New Jersey, which is contaminated with solvents, PCBs, and heavy metals. A trench has been cut around the site to an underlying impervious soil layer and then filled with clay to prevent any migration of the contaminants. The site has also been pumped dry to protect groundwater and capped to keep out rain. Remediation work has cost about $7 million, and DuPont as well as other responsible parties have pledged to maintain these containment systems for as long as necessary. However, despite the absence of any current or reasonably foreseeable public exposures, EPA may decide to require incineration of the top 10 feet of soil at an estimated cost of several hundred million dollars. This would be a foolish waste of money.

EPA must also consider that extreme, costly remediation solutions often are not without costs of their own. Incineration, for instance, cannot destroy metals. Does the public really benefit when lead is released into the air as a byproduct? By the same token, when contaminated soil is ordered excavated and carted elsewhere, one neighborhood gets a "permanent" solution, whereas another gets a landfill with toxic residues.

Risks Exaggerated

Superfund legislation is not the only force driving EPA to seek "permanent" solutions. EPA decides on a remedy only after assessing the risks at a site. However, EPA often uses unrealistic assumptions that exaggerate the risks and lead to excessive actions. For example, according to the Hazardous Waste Cleanup Project, an industry group in which DuPont has been involved, EPA may make estimates of exposure based on a scenario in which an individual is assumed to reside near a site for 70 years, to consume two liters of water every day during those 70 years, and to obtain all of that water from groundwater at the site. It has even made exposure estimates based on the length of time a child will play (and eat dirt) on a site in the middle of an industrial location surrounded by a security fence. Each of these scenarios is highly improbable.

Questions involving risk assessment are, of course, going to be contentious ones for some time to come. Clouding the Superfund debate is the fact that there is no scientific consensus as to the precise magnitude of the dangers posed by chemicals typically found at Superfund sites.

The existence of toxic wastes at a site does not necessarily mean that they pose a threat to nearby residents. Epidemiologic studies of waste sites have severe technical limitations, and it is difficult at best to determine whether exposure to hazardous wastes can be blamed for medical problems when a long gap exists between exposure and disease. Even at such a well-known site as Times Beach, Missouri, where the entire community was evacuated, research in recent years has shown that the potential health risks were relatively small or even nonexistent.

The most comprehensive assessment of the risks from Superfund sites came in the 1991 NRC report, which concluded that "current health burdens from hazardous-waste sites appear to be small," but added that "until better evidence is developed, prudent public policy demands that a margin of safety be provided regarding public health risks from exposures to hazardous-waste sites."

No one can argue with a margin of safety. However, that is not the focus of the current Superfund program, which, far more than any other environmental program, makes no rational attempt to link costs with benefits. EPA's own Science Advisory Board, in a 1990 report that attempted to rank the environmental problems for which the agency is responsible, concluded that old toxic waste sites appeared to be "low to medium risk." Other hazards, such as radon gas in homes and cigarette smoke, were considered to pose much larger risks.

The Liability Mess

The bulk of the Superfund tab will be picked up by industry, through taxes imposed under CERCLA, out-of-pocket cleanup costs, or settlements with insurance companies. Industry recognizes that it must assume its fair share of the financial and operating burden of the cleanup effort, and it acknowledges that Superfund has compelled it to become exceptionally vigilant not only in disposing of toxic wastes but also in minimizing their generation in the first place. But it objects to a system in which EPA seemingly has put a higher priority on pinning the blame and the bill on companies than on ensuring the protection of public health.

CERCLA dictated a "polluter-pays" philosophy to deal with what had largely been lawful disposal of wastes. CERLCA and court interpretations of it also have created an extremely broad liability scheme. Virtually any company remotely involved in a site-waste generators, haulers, site owners or operators, and even, in some cases, the companies' bankers—could be held responsible. One or a few companies could be forced to pay the entire bill, even though they were only minor participants and other parties were involved—a provision called joint-and-several liability. No limits were imposed on the amount of money that could be extracted from "guilty" parties.

One problem with this liability system is that it completely lacks cost accountability. With industry paying for most of the cleanup, the funds are not in EPA's budget and thus do not have to compete in budget battles with other cash-starved federal programs. And given the strictness of the law, why should EPA regulators subject themselves to possible congressional criticism by selecting a less-than-perfect solution, especially if money is no object? But let us not kid ourselves. Although this money may seem "free" to Congress, EPA, and the public, companies must make up the difference by raising prices, cutting investment and jobs, or taking other undesirable actions.

An even more damning problem is that the liability provisions have spawned countless legal brouhahas that are consuming a large and increasing share of Superfund resources—even as the cleanup process itself has languished. (The average length of a site cleanup is 8 years, and fewer than 100 sites have been "permanently" remediated.) In the approximately 70 percent of Superfund sites that involve multiple parties, companies must fight with the EPA, among themselves, and with their insurance companies over who dumped what, when, and how much—questions extremely difficult to answer many years after the alleged "dumping" is thought to have occurred. Some experts believe that these "transaction costs" will eventually account for more than 20 percent of all Superfund expenditures. This is a boon for lawyers but a waste for the nation.

Legal costs—as well as burdensome technical and administrative expenses—could potentially be greatly reduced if Congress would allow EPA to take a more practical approach to risk reduction. Unlike other environmental laws, such as the Clean Air Act and Clean Water Act, which have sought to deal with problems in successive stages, Superfund's emphasis on finding a one-time, complete, and permanent solution magnifies the stakes to all parties, prolonging disputes and greatly increasing the costs. If companies could count on a more realistic remediation approach, they might be more willing to compromise, which could lead to faster cleanups.

The liability mess could get completely out of hand if Congress goes along with a patently unfair proposal to exempt municipalities from liability at closed municipal landfills, which account for about 20 percent of NPL sites. Municipal governments argue that most of these landfills largely contain household wastes not covered by Superfund and thus they should not be billed for the cleanup. But in many cases this is not true. For example, at the Kramer Superfund site in Mantua, New Jersey, municipal governments contributed the greatest share of hazardous substances. Despite this, EPA is no longer even naming municipalities in cost-recovery suits. (EPA's tendency to selectively enforce the law has been increasing. At Kramer, EPA sued 25 parties even though hundreds were potentially responsible.)

Industry recognizes that many municipal governments are severely strapped for revenues. Yet companies, which provide jobs and help create the tax base needed to support municipal services, should not be milked to pay for Superfund shares properly owed by others.

One last concern with the liability provisions is that they may be having a chilling effect on new investment at sites in older urban areas—areas that sorely need such investment. The reason is that any party that buys such a property would be caught in Superfund's liability web. For example, investors seeking to build a coal-fired power plant in an area with a projected need for such a use recently approached DuPont about buying a property that had been used for manufacturing for more than 100 years and clearly contains some contaminated soil. Virgin land is not needed for a site to burn coal, and risk assessments indicated that workers could be protected with commonsense steps such as paving. But efforts to get reasonable compromises from regulators on containing the site proved fruitless, and now the investment will not be made, at least in this area.

Steps to Reform

It is time for a major redirection of the Superfund program. Congress should tell EPA to abandon its focus on idealistic cleanup remedies and emphasize practical risk reduction. Instead of continuing its haphazard site selection and unjustifiably costly cleanup remedies, EPA should first define the universe of sites that may present real health risks and then take steps to deal with the most immediate dangers, taking costs into consideration. Once a national inventory has been established, extensive site evaluations can be undertaken, with the purpose of setting priorities for cleanup. Only after these actions are taken will we be able to make non-hysterical decisions as to how much we should invest in cleaning these sites, balancing such factors as risks, costs, and other societal needs.

It is particularly crucial that remedy decisions be based on the expected future use of the land and the costs and practicality of the proposed solution. If residential development is planned near the site, the cleanup may need to be extensive. In many cases, however, especially when another industrial use is planned on or near an old waste site, use of containment technologies may be sufficient to protect against risk of exposure. In the most troublesome cases, where major remediation is necessary, costs are high, and existing technology has limitations, it makes much more sense to isolate the site until more cost-effective treatment techniques are developed or increased land values justify a large investment.

In making these decisions, it would be helpful if EPA had much better information on the benefits and costs of different levels of cleanup. Currently, less than 1 percent of EPA's Superfund budget goes for research on the scientific basis for evaluating Superfund sites. Much more should be spent. EPA also should increase its research on the environmental consequences of different types of remedial actions, such as whether incineration actually increases risk by transferring hazardous substances from the ground or water into the air.

The liability provisions of the Superfund program also need to be changed. DuPont and companies in the chemical, petroleum, and other industries favor replacing the very unfair joint-and-several liability provision (making one or a few companies liable for all the costs, even though many others, often defunct, were also responsible) with proportional liability. In other words, responsible parties would pay only in proportion to the share of the cleanup costs associated with the wastes that they contributed at a site. EPA would then be forced to either find and sue all responsible parties or pay for the remainder of the cleanup costs itself. EPA is already authorized to pay for cleanup costs in cases where parties cannot be found or cannot afford to pay—shares which are often sizable. But in practice it has sought to recoup all cleanup costs under the joint-and-several provision. Proportional liability would inject more fairness into the process, and since the polluter-pays principle would be retained, it would continue to encourage responsible parties to pressure EPA to pursue the most cost-effective cleanup remedy. Most important, proportional liability would impose much-needed financial discipline on EPA, since it would be forced to pay for more of the cleanups out of its own budget. For the first time, EPA would have to consider whether the benefits were worth the costs.

Proportional liability would not, of course, solve the problem of how to divide up responsibility in the first place. One possible way out of this morass is to formalize in the law an alternative dispute resolution process in which any or all potentially responsible parties could participate. It would be chaired by neutral parties satisfactory to all. Its findings on shares could be appealed to the courts, but any party that concurred with the decision would be authorized to pay its share and exit the process. This solution would help cut site contention, reward cooperative parties, and leave messy litigation to those unwilling to pay their fair shares. It would also diminish Superfund's luster as a federally mandated entitlements program for lawyers.

More extensive reform of the liability provisions has been proposed by the insurance industry, which wants to eliminate all liability at sites in which more than one party is involved and in which waste disposal occurred prior to enactment of either CERCLA in 1980 or SARA in 1986. Site cleanup would then be paid out of the Superfund budget, financed by increased taxes on industry, including insurers. Although this proposal would eliminate contentious fights over specific site responsibility, substantially cut transaction costs, and possibly speed up site cleanups, it would be unacceptable to DuPont and other parties at Superfund sites if the new taxes were unfairly levied on the same companies already paying disproportionately large shares of the current Superfund cleanups.

Finally, the liability scheme must be changed so that prospective owners of older urban sites are not deterred from making new investments in them. New owners should certainly not be held responsible for contamination that they did not cause. One approach would be for current owners to demonstrate, before sale, that their sites, while not pristine, are adequately contained and do not pose unacceptable risks to the public. The new owner would be expected to maintain or monitor whatever containment system was developed. If EPA later did a more extensive site evaluation and determined that greater threats existed, the new owner would not have to pay. In addition, current owners should be able to make new investments in their property if they demonstrate that the sites are adequately contained.

⋅◦⊙◦⋅

The limits of our national wealth have not been so obvious since the 1930s. More than ever, we must make choices among competing, compelling demands for scarce resources. We recognize that a dollar spent on defense cannot be spent on health care. We must also recognize that a dollar spent on hazardous waste cleanup is similarly unavailable. As with other federal programs, Superfund spending must be balanced and managed. This can be done if we refocus our Superfund investment on real risks, give EPA a stake in doing its job cost-effectively, and bring more fairness into the process.

Ted Williams  **NO**

The Sabotage of Superfund

The setting was perfect: Cold rain and gull-filled mist blowing in from Buzzards Bay. Litter clinging to the bare ribs of dead brush like shards of rotten umbrella silk. Derelict, graffiti-streaked trailers stuffed to overflowing with bald truck tires. Ratty mattresses and broken easy chairs strewn about the cratered parking lot. Glass from the abandoned mill crunching under my boots and snatches of Eliot's *The Waste Land* resounding in my brain as I trudged along Wet Weather Sewage Discharge Outfall No. 022: "Sweet Thames run softly, till I end my song..."

Until this day, March 28, 1993, I had avoided Superfund sites. So this was my first visit to the waterfront of New Bedford—an impoverished, predominantly Hispanic seaport in southern Massachusetts, now as famous for the polychlorinated biphenyls (PCBs) on the bottom of its harbor as for its whaling history. PCBs, widely used in the manufacture of electrical components until banned in 1978, do hideous things to creatures that come in contact with them —such as causing their cells to proliferate wildly and warping their embryos. I wasn't about to touch anything without my rubber ice-fishing gloves.

Where the sewage dribbled into the dark Acushnet River I jumped down onto gray silt and, breathing through my teeth, scooped up five handfuls of muck. The Environmental Protection Agency's guideline for protecting marine life from chronic toxic effects of PCBs is 30 parts per trillion. I cannot accurately report the PCB content of my amateur sample (taken illegally, the EPA later informed me), but the greasy globules that floated up through the surface scum were likely very rich. Had I been able to get out into the river and upstream to the old Aerovox Inc. discharge pipe, I could have found concentrations of at least 200,000 parts per *million,* or 20 percent, among the highest ever recorded. That means that with a similar test dredging I'd have retrieved one handful of pure PCBs, along with a tangle of wriggling sludge worms, about the only creatures that can live in such habitat.

I was rinsing my gloves and boots in a rain-filled pothole when *The Waste Land's* Fisher King materialized through the gloom—a wispy, gray-haired figure in a red plaid jacket, toting a stout spinning rod. He had parked next to a sign that read in Spanish, Portuguese, and English: "Warning. Hazardous Waste. No wading, fishing, shellfishing. Per order of U.S. EPA." His name, which he printed with his forefinger on the wet trunk of his car, was Robin Rivera; he knew only enough English to make me understand that he and his family eat the fish he catches here.

Not until 1978 did the nation get angry about the indiscriminate disposal of poisonous chemicals. In that year people who lived near Hooker Chemical Company's Love Canal dump in Niagara Falls, New York, were distressed to smell vile chemical odors in their basements and observe a malevolent secretion bubbling out of the ground at a local school yard. County health officials and Hooker reps tried vainly to contain the alarm, but their assurances that all was well sounded as wrong as the uncontained leachate looked and smelled.

Eventually the citizens took their case to the young commissioner of the state Department of Environmental Conservation—Peter A. A. Berle, now president of the National Audubon Society. Although Berle had no authority to act on public health issues, he sent his people out to test houses on the strength of his environmental mandate. The benzene levels they found were, in his words, "right off the chart." Eventually 600 homes were abandoned and 2,500 residents relocated.

In response to the Love Canal horror show Congress enacted the Comprehensive Environmental Response, Compensation, and Liability Act of 1980, better known as Superfund. Amended in 1986, the law uses taxes on crude oil and 42 commercial chemicals to maintain a fund with which the EPA may, as it likes to say, "remediate" hazardous-waste sites. If perpetrators can be found and are still in business, the EPA may require one or all to clean up the entire site. This essential principle of Superfund is called joint and several liability.

For an idea of the pace at which cleanup proceeds, consider that the EPA and its contractors have been studying and planning what to do about New Bedford's harbor ever since it was declared a "National Priority" Superfund site 11 years ago. Nationwide, the EPA has spent $7.5 billion on its Superfund program, with pitiful results. In some cases remediation has created more problems than it has solved by stirring up contaminants that had been dormant. In other cases vast sums have been squandered at sites that posed little threat to the public, while deadly brews seethed nearby. Superfund contractors have consistently ripped off the EPA, billing it for everything from office parties to work they were supposed to do and didn't.

At this writing only 163 of 1,204 sites have been remediated, and in many cases polluters have been granted what the EPA calls the "containment" option —a feline approach to toxic-waste management in which they just cover their messes and walk away. The average cleanup has cost about $25 million and taken 7 to 10 years to complete.

No one remotely connected with Superfund is happy about the way it has functioned. Polluters identified by the EPA have been madly rummaging through dumps, trying to identify other polluters by their trash and so spread liability. In the process small towns, businesses, and individuals that contributed legally and insignificantly to landfills have been intimidated and assessed for cleanup costs in a fashion utterly inconsistent with the intent of Congress. Environmentalists are at the throats of insurance companies who want to do away with the polluter-pay tenet. The insurance companies are warring in court with industries to whom they have rashly sold pollution-liability policies. People who live atop and beside toxic waste claim—often correctly —that they have been ignored and lied to by the EPA, and as a result, they sometimes oppose well-advised cleanup plans.

Hearings for Superfund's 1994 reauthorization are already under way. "We all know it doesn't work," says President Bill Clinton. "Superfund has been a disaster."

_◦◆◦_

Even as I wished the Fisher King good luck, I found myself greeting the first of 94 demonstrators from the New Bedford–area citizen's group Hands Across the River. We stood in the rain, listening to fiery speeches amplified by bullhorn about the EPA's plan to dredge the five acres that contain roughly 45 percent of the PCBs extant in the 28-square-mile site, then cleanse the spoil by fire in portable incinerators set up on the downstream side of Sewage Outfall 022.

"You and I will be breathing their mistakes," bellowed rally leader Richard Wickenden. "They made their decision to incinerate behind closed doors with a total disregard for the local citizenry."

He spoke the truth. The New Bedford City Council had found out about the plan not from the EPA but from Hands Across the River, which had found out about it from federal documents at the library. "State-of-the-art incinerators" have a long history of malfunction, even when run by the most conscientious contractors. This one will be operated by Roy F. Weston, a large environmental-consulting firm based in West Chester, Pennsylvania, which in 1990 agreed to pay $750,000 to settle charges that it had defrauded the EPA by backdating data and submitting a bill for work it never did. Finally, people downwind—mostly people like Robin Rivera—are likely to be breathing mistakes, along with all manner of toxic PICs (products of incomplete combustion) that won't be monitored or even identified.

In attempting to clean up one point of pollution the EPA and its contractors will be creating others, asserted New Bedford City Councilman George Rogers. They'll be unleashing PCBs and heavy metals on moving seawater, hauling them onto the bank, then casting them to the four winds during dewatering and combustion. "They're doing this because New Bedford is a poor community; we don't have clout. They wouldn't do it in Miami Beach." He, too, spoke the truth. A study released last September by _The National Law Journal_ reveals that the EPA is lenient in penalizing polluters of minority and low-income communities and that cleanups in such areas are slower and less thorough. Toxic racism, activists call it.

Equally veracious were the allegations of David Hammond, president and founder of Hands Across the River, that polluters love incineration because their liability goes up the stack along with the toxic PICs and that the EPA has undermined Superfund's effectiveness and its own credibility by ordering remediation studies from companies that make their money remediating. In particular, Hammond is upset that Weston was hired to do the New Bedford Remedial Action Master Plan, then wound up with the $19.4 million incineration contract.

Both the city council and Hands Across the River hasten to point out that they are not against remediation. But instead of incineration they favor the "Eco Logic" process—a relatively contained heat treatment developed in Rockwood, Ontario, which combines hydrogen with PCBs to form methane and hydrogen chloride and which has been getting rave reviews in the press. "Stunning New Method Zaps Toxic Chemicals Efficiently," shouted a headline in *The Toronto Star* on January 30, 1993.

After the New Bedford speechmaking the congregation marched back and forth over the Acushnet River bridge, waving placards, obstructing traffic (much of which honked in sympathy), and chanting, "No way, EPA," and "Hey, Carol [Browner], if you please, don't you burn those PCBs. Not New Bedford, not the nation. We don't want incineration."

The citizens could scarcely have done a better, more honest job of drawing public attention to the perils of dredging and incinerating PCBs. But this doesn't mean that the EPA ought not to press ahead with its plan for New Bedford. When PCB concentrations are this high, the perils of doing something else or nothing at all are probably greater. One day the Eco Logic process may indeed be a "stunning method" of remediation. Now, despite the effusions of *The Toronto Star,* it's largely an experimental technology and therefore fraught with risk. Meanwhile, the PCBs are spreading out into the Atlantic with every tide and every storm. Humans and marine ecosystems—including half the North American population of endangered roseate terns, which nests on a single island in Buzzards Bay—don't have another decade to wait while the EPA collects data and shuffles papers.

·◄◉►·

In other contract deals the EPA has paid the New England office of Roy F. Weston, which it has criticized for poor performance, $635,000 to administer fieldwork that cost $340,000. But Weston looks like a model contractor when compared with some of the others.

Take, for example, consulting-engineering colossus CH2M Hill, which has worked on 275 major sites, including Love Canal, and which holds $1.4 billion worth of Superfund contracts. An inquiry by the House Subcommittee on Oversight and Investigations revealed that as part of alleged Superfund work, CH2M Hill billed the EPA $4,100 for tickets to basketball, baseball, and football games; $167,900 for employee parties and picnics, including the cost of reindeer suits, magicians, and a rent-a-clown; $15,000 for an office bash at a place called His Lordship; "thousands of dollars' worth" of chocolates stamped with the company logo; $63,000 for general advertising; $10,000 for a catered lobbying cruise on the Potomac; and $100 for a Christmas-party dance instructor. "I am all for rocking around the Christmas tree," commented Congressman Thomas J. Bliley Jr. (D-VA) at the hearing, "but does it have to be at the taxpayers' expense?"

Apparently yes, according to the testimony of CH2M Hill's president, Lyle Hassebroek. "No matter what differences of opinion exist on the manner in which we allocate costs," he explained, "CH2M Hill's charges to the government are fair to the taxpayer."

By no means is CH2M Hill aberrant. Last summer EPA investigators found that 23 companies hired for hazardous-waste cleanup in 1988 and 1989 spent 28 percent of their $265 million budget on wasteful administrative costs. Such inefficiency is cited by polluters and their insurance companies as a reason to "overhaul" Superfund—i.e., turn it into a public-works program whereby Uncle Sam would bail them out by picking up toxic litter (provided the offense preceded some stipulated date—1987, according to one proposal) and citizens will pay twice, once with their environment and once with their tax money.

Major polluters further foment discontent with Superfund by attempting to squeeze alleged shares of cleanup costs from everyone who might ever have sent a can of shoe polish to a landfill. The EPA and the courts don't want a nickel a day for 1,000 years and so avoid going after mom-and-pop polluters. But Mom and Pop don't know this, and technically they are liable. The real motive, charge environmental leaders, is not so much to collect money as to contrive broad support for Superfund "reform."

When Ford, Chrysler, General Motors, BASF Corporation, and Sea Ray Boats were fingered by the EPA for fouling the Metamora, Michigan, landfill with arsenic, lead, vinyl chloride, and the like, they proclaimed that 382 towns, businesses, and individuals were copolluters and tried to assess them $50 million to settle alleged liability quickly, including any unforeseen costs. Even the local Girl Scout troop was assessed $100,000. "That's a lot of cookies," declared a troop spokesperson.

In another case Doreen Merlino, the 25-year-old proprietor of a two-table pizzeria in Chadwicks, New York, offered the court officer the following plea when he served her with a two-inch-thick lawsuit in October 1990: "Aren't you at least gonna buy a pizza?" He kindly complied, but she didn't feel much better. In fact, she felt terrified. Cosmetics giant Chesebrough-Pond's and Special Metals Corporation were trying to extract $3,000 from her for helping them poison the local landfill. They weren't sure just what the trashman might have collected from her during the seven months she'd been in business—maybe pesticide containers or empty cleanser cans, they opined. But Merlino tells me she's never used pesticides and that she has always rinsed out her cleanser containers. A cover letter advised her that if she settled fast, she'd only have to pay $1,500. As it turned out, she had to pay no one save her lawyer, and none of her anger is directed at Superfund. Not all the 603 defendants were so philosophical.

When the EPA hits up polluters for toxic-waste-cleanup costs, polluters, naturally enough, hit up the insurance companies from which they have purchased pollution-liability coverage. Now the insurance companies would like to hit up the public—that is, rewrite Superfund so bygones can be bygones and taxpayers can spring for cleaning up old sites.

"Superfund's mission should be protecting human health and the environment, not fund-raising," contends the American International Group, an insurance company marshaling support for what it calls a National Environmental

Trust Fund, by which Superfund money could be raised "from all economic sectors without regard to site-specific liability" via a surcharge on commercial- and industrial-insurance premiums.

Generating pity for the insurance industry requires a greater heart than beats in the breast of environmental consultant Curtis Moore. As an aide to former Republican senator Robert Stafford of Vermont, Moore was instrumental in writing both the original Superfund law and the amended 1986 version. Insurance companies, he points out, tend not to cover purposeful acts by anyone, including God; so the policies were restricted to "sudden and accidental" pollution. "You dump crude oil on the ground for fifteen years," he says, "and over a twenty-five-year period it migrates to the water table. Would you consider that sudden? Accidental? No? Well, I got news for you: The courts do. There was a string of decisions that construed the terms *sudden* and *accidental* as covering groundwater contamination. This trend started a long time ago—in the 1970s or earlier. It was clearly discernible. Any insurance lawyer with manure for brains could see it happening. Notwithstanding, the insurance industry continued to use the terms *sudden* and *accidental* in its policies.

"So here comes Superfund, and the chemical companies start casting around, trying to figure out how they can get someone else to pick up the tab. They file suits against their insurance companies and win. Well, there's only one way to fix the insurance industry's problem. You can either shift liability or, failing that, repeal Superfund."

Presuming to speak for the insurance industry, the American International Group complains that Superfund is "bogged down in a morass of legal warfare that delays cleanup and wastes enormous financial and human resources." True enough, but what it doesn't mention is that the insurance industry has been responsible for a great deal of this legal warfare. A Rand Corporation study reveals that between 1986 and 1989 insurers spent $1.3 billion on Superfund. Of this, $1 billion went to defending themselves against their policyholders or defending their policyholders against the EPA. One leading attorney for the policyholders—Eugene Anderson, of the New York City firm of Anderson, Kill, Olick & Oshinsky—has gone so far as to suggest publicly that refusing all large claims is now seen as smart business procedure by insurers: Half the policyholders get scared away, and most of the others will settle out of court for less than full coverage.

❧◉❧

Superfund has bombed, as the President, environmentalists, inhabitants of toxic neighborhoods, brewers of toxic waste, and especially the insurance industry have observed. But it is essential to remember the difference between Superfund the law and Superfund the program.

"The law was a creation of people like Bob Stafford, Ed Muskie, Jennings Randolph, John Chafee, Jim Florio," remarks Moore. "The program was the creation of Ronald Reagan; the people who were put in charge of implementing the law six weeks after it was enacted were people who six weeks earlier had been lobbying against it. They set out with the intent of making it unworkable, and they succeeded."

There is nothing in the statute that directs the EPA's contractors to dress up like reindeer or distribute customized confections at government expense. They engage in such excess because the EPA lacks the personnel to keep them honest. Nor is there anything in the statute that mandates stonewalling and procrastination on the part of polluters. But they have learned that endless negotiation is profitable because the EPA lacks the personnel to haul more than a few of them into court. Mr. Clinton, who proposes to trim $76 million from Superfund the program, appears not to understand this.

Certainly, the statute could stand repair. It needs to define how clean is clean, provide a better, more flexible means of selecting remedies, ensure state and local participation, create incentives for companies to take voluntary action instead of suing everyone in sight. But the fact is that Superfund the law isn't broken.

Even Superfund the program, disastrous though it has been, has produced some splendid if accidental results. "Joint and several liability," says Peter Berle, "has put the fear of God into everybody, which means they are careful in ways they never were before in what they do with their waste. I also think the cost risk of inappropriate toxic-waste disposal has been the major impetus toward waste minimization. When it gets too expensive to deal with it, then you make less."

Rick Hind, toxics director for Greenpeace, agrees. "It doesn't cost you and me anything if a big company wants to spend ten million dollars on lawyers to avoid an eleven-million-dollar cleanup," he offers. "That costs the company. Good! So it costs them twice what it should. That will teach them a lesson. When a Colombian drug cartel is in court nobody cares what their legal expenses are. Nor should we care about polluters."

Insurers and polluters—not environmentalists—are the ones driving for major surgery on Superfund the law. If they are permitted to degrade it from a dedicated fund to a public-works program whereby big government passes around public-generated revenues, the hemorrhage of federal pork will make the EPA nostalgic for the days when it used taxes on crude oil and chemicals to rent clowns for CH2M Hill. If they are permitted to do away with Superfund's liability provisions and weaken its polluter-pay principle, the United States will be poisoned on a scale unimagined even in New Bedford, Massachusetts.

It may be that Superfund is mortally wounded from a dozen years of sabotage. But it also may be that it can be salvaged and made to work. We need to try. Vendors of insurance and chemicals will shriek and sob, but the law wasn't written for them. It was written for Love Canal couples forced to watch as bulldozers razed their homes, for Robin Rivera and his family, for roseate terns, for sick and deformed children, for children yet unborn.

POSTSCRIPT

Hazardous Waste: Should the "Polluter Pays" Provision of Superfund Be Weakened?

When those responsible for contaminated sites could not pay or could not be found, Superfund cleanups were to have been funded by taxes on industry (the crude oil tax, the chemical feedstock tax, the toxic chemicals importation tax, and the corporate environmental income tax). These taxes expired in 1995, by which time the fund held a $3.8 billion surplus, and although they have not yet been reinstated, activist groups such as the Public Interest Research Group (PIRG) have issued calls for the George W. Bush administration to do so without delay. But according to the PIRG, "The Bush administration opposes reauthorization of the polluter pays taxes, supports a steep increase in the amount paid by taxpayers, and has dramatically slowed down the pace of cleanups at the nation's worst toxic waste sites" (see http://www.pirg.org/enviro/superfund/). In April 2002 Senator Robert Toricelli (D-New Jersey) was expected to add reauthorization to the energy bill then being debated in Congress. The Competitive Enterprise Institute (http://www.cei.org) objects that such taxes are an assault on consumer pocketbooks. The institute also objects to CERCLA's "joint and several liability" clause, which can make minor contributors to toxic sites liable for large cleanup costs even when they acted according to all laws and regulations in force at the time. Clearly, the debate reflected in Reilly's and Williams's essays rages on.

Among the solutions that have been urged are "take-back" and "remanufacturing" practices. Gary A. Davis et al., in "Extended Product Responsibility: A Tool for a Sustainable Economy," *Environment* (September 1997), argue that such practices are crucial to the minimization of waste and describe how they are becoming more common in Europe. After "Exporting Harm" was published and drew considerable attention from the press, some industry representatives hastened to emphasize such practices as Hewlett Packard's recycling of printer ink cartridges (see Doug Bartholomew's "Beyond the Grave," *Industry Week* [March 1, 2002], which also stresses the need to minimize waste by intelligent design). Brian K. Thorn and Philip Rogerson's article "Take It Back," published in the April 2002 issue of the Institute of Industrial Engineers journal *IIE Solutions,* stresses the importance of designing for reuse or remanufacturing. Anthony Brabazon and Samuel Idowu, in "Costing the Earth," *Financial Management* (May 2001), contend that "take-back schemes may [both] provide opportunities to build goodwill and [help] companies to use resources more efficiently."

ISSUE 14

Municipal Waste: Is Recycling an Environmentally and Economically Sound Waste Management Strategy?

YES: Richard A. Denison and John F. Ruston, from "Recycling Is Not Garbage," *Technology Review* (October 1997)

NO: Chris Hendrickson, Lester Lave, and Francis McMichael, from "Time to Dump Recycling?" *Issues in Science and Technology* (Spring 1995)

ISSUE SUMMARY

YES: Environmental Defense Fund scientist Richard A. Denison and economic analyst John F. Ruston rebut a series of myths that they say have been promoted by industrial opponents in an effort to undermine the environmentally valuable and successful recycling movement.

NO: Engineering and economics researchers Chris Hendrickson, Lester Lave, and Francis McMichael assert that ambitious recycling programs are often too costly and are of dubious environmental value.

Since prehistoric times, the predominant method of dealing with refuse has been to simply dump it in some out-of-the-way spot. Worldwide, land disposal still accommodates the overwhelming majority of domestic waste. In the United States roughly 90 percent of residential and commercial waste is disposed of in some type of landfill, ranging from a simple open pit to so-called sanitary landfills, where the waste is compacted and covered with a layer of clean soil. In a small, but increasing, percentage of cases, landfills may have clay or plastic liners to reduce leaching of toxins into groundwater.

By the last quarter of the nineteenth century, odoriferous, vermin-infested garbage dumps in increasingly congested urban areas were identified as a public health threat. Large-scale incineration of municipal waste was introduced at that time in both Europe and the United States as an alternative disposal method. By 1970 more than 300 such central garbage incinerators existed in U.S. cities, in addition to the thousands of waste incinerators that had been built into large apartment buildings.

Virtually all of these early garbage furnaces were built without devices to control air pollution. During the period of heightened consciousness about urban air quality following World War II, restrictions began to be imposed on garbage burning. By 1980 the new national and local air pollution regulations had reduced the number of large U.S. municipal waste incinerators to fewer than 80. Better designed and more efficiently operated landfills took up the slack.

During the past two decades, an increasing number of U.S. cities have been unable to find suitable, accessible locations to build new landfills. This has coincided with growing concern about the threat to both groundwater and surface water from toxic chemicals in leachate and runoff from dump sites. Legislative restrictions in many parts of the country now mandate costly design and testing criteria for landfills. In many cases, communities have been forced to shut down their local landfills (some of which had grown into small mountains) and to ship their wastes tens or even hundreds of miles to disposal sites.

The lack of long-range planning coupled with skyrocketing disposal costs created a crisis situation in municipal waste management in the 1980s. Energetic entrepreneurs seized upon this situation to promote European-developed incineration technology with improved air pollution controls as the panacea for the garbage problem. Ironically, the proliferation of these new waste incinerators in the United States coincided with increasing concern in Europe about their efficiency in containing the toxic air pollutants produced by burning modern waste. Citizen groups became aware of this concern and organized opposition to incinerator construction. The industry countered with more sophisticated air pollution controls, but these trapped the toxins in the incinerator ash, which presents a troublesome and expensive disposal problem. The result has been a rapid decrease in the number of municipalities that are choosing to rely on modern incineration to solve their waste disposal problems.

Recycling, which until recently has been dismissed as a minor waste disposal alternative, is being encouraged as a major option. The Environmental Protection Agency (EPA) and several states have established hierarchies of waste disposal technologies with the goal of using waste reduction and recycling for as much as 50 percent of the material in the waste stream. Several environmental groups are urging even greater reliance on recycling, citing studies that show that more than 90 percent of municipal waste can theoretically be put to productive use if large-scale composting is included as a component of recycling.

In the following selection, Richard A. Denison and John F. Ruston maintain that most arguments against recycling, which they believe is environmentally valuable and economically viable, are self-serving attempts by the organizations that create municipal waste problems to avoid scrutiny. In the second selection, generalizing from an analysis of the recycling program of their home city, Chris Hendrickson, Lester Lave, and Francis McMichael conclude that recycling is neither cost-effective nor environmentally advantageous.

Richard A. Denison
and John F. Ruston

 **YES**

Recycling Is Not Garbage

Ever since the inception of recycling, opponents have insisted that ordinary citizens would never take the time to sort recyclable items from their trash. But despite such dour predictions, household recycling has flourished. From 1988 to 1996, the number of municipal curbside recycling collection programs climbed from about 1,000 to 8,817, according to BioCycle magazine. Such programs now serve 51 percent of the population. Facilities for composting yard trimmings grew from about 700 to 3,260 over the same period. These efforts complement more than 9,000 recycling drop-off centers and tens of thousands of workplace collection programs. According to the EPA [Environmental Protection Agency], the nation recycled or composted 27 percent of its municipal solid waste in 1995, up from 9.6 percent in 1980.

Despite these trends, a number of think tanks, including the Competitive Enterprise Institute and the Cato Institute (both in Washington, D.C.), the Reason Foundation (in Santa Monica, Calif.), and the Waste Policy Center (in Leesburg, Va.), have jumped on the anti-recycling bandwagon. These organizations are funded in part by companies in the packaging, consumer products, and waste-management industries, who fear consumers' scrutiny of the environmental impacts of their products. The anti-recyclers maintain that government bureaucrats have imposed recycling on people against their will—conjuring up an image of Big Brother hiding behind every recycling bin. Yet several consumer researchers, such as the Rowland Company in New York, have found that recycling enjoys strong support because people believe it is good for the environment and conserves resources, not because of government edict.

Alas, the debate over recycling rages on. The most prominent example was an article that appeared . . . in the *New York Times Magazine*, titled "Recycling Is Garbage," whose author, John Tierney, relied primarily on information supplied by groups ideologically opposed to recycling. Here we address the myths he and other recycling opponents promote.

Myth: The modern recycling movement is the product of a false crisis in landfill space created by the media and environmentalists. There is no shortage of places to put our trash.

Fact: Recycling is much more than an alternative to landfills. The so-called landfill crisis of the late 1980s undoubtedly lent some impetus to the recycling movement (although in many cities around the country, recycling gained momentum as an alternative to incineration, not landfills). The issues underlying the landfill crisis, however, were more about cost than space.

Landfill space is a commodity whose price varies from time to time and from place to place. Not surprisingly, prices tend to be highest in areas where population density is high and land is expensive. In the second half of the 1980s, as environmental regulations became more stringent, large numbers of old landfills began to close, and many simply filled up, particularly in the Northeast. New landfills had to meet the tougher standards; as a result, landfill prices in these regions escalated dramatically. In parts of northern New Jersey, for example, towns that shifted their garbage disposal from local dumps to out-of-state landfills found that disposal costs shot from $15–20 per ton of garbage to more than $100 per ton in a single year. Although the number of open landfills in the United States declined dramatically—according to *BioCycle* magazine, from about 8,000 in 1988 to fewer than 3,100 in 1995—huge, regional landfills located in areas where land is cheap ultimately replaced many small, unregulated town dumps. Landfill fees declined somewhat and the predicted crisis was averted. Nonetheless, the high costs of waste disposal in the Northeast and, to a lesser extent, the West Coast, have spurred local interest in recycling: two-thirds of the nation's curbside recycling programs operate in these regions.

But landfills are only part of the picture. The more important goals of recycling are to reduce environmental damage from activities such as strip mining and clearcutting (used to extract virgin raw materials) and to conserve energy, reduce pollution, and minimize solid waste in manufacturing new products. Several recent major studies have compared the lifecycle environmental impacts of the recycled materials system (collecting and processing recyclable materials and manufacturing them into usable form) with that of the virgin materials system (extracting virgin resources, refining and manufacturing them into usable materials, and disposing of waste through landfills or incineration). Materials included in the studies are those typically collected in curbside programs (newspaper, corrugated cardboard, office paper, magazines, paper packaging, aluminum and steel cans, glass bottles, and certain types of plastic bottles). The studies were conducted by Argonne National Labs, the Department of Energy and Stanford Research Institute, the Sound Resource Management Group, Franklin Associates, Ltd., and the Tellus Institute. All of the studies found that recycling-based systems provide substantial environmental advantages over virgin materials systems: because material collected for recycling has already been refined and processed, it requires less energy, produces fewer common air and water pollutants, and generates substantially less solid waste. In all, these studies confirm what advocates of recycling have long claimed: that recycling is an environmentally beneficial alternative to the extraction and manufacture of virgin materials, not just an alternative to landfills.

Myth: Recycling is not necessary because landfilling trash is environmentally safe.

Fact: Landfills are major sources of air and water pollution, including greenhouse gas emissions.

According to "Recycling Is Garbage," municipal solid waste landfills contain small amounts of hazardous lead and mercury, but studies have found that these poisons stay trapped inside the mass of garbage even in the old unlined dumps that were built before today's stringent regulations. But this statement is simply wrong. In fact, 250 out of 1,204 toxic waste sites on the Environmental Protection Agency's Superfund National Priority List are former municipal solid waste landfills. And a lot more than just lead and mercury goes into—and comes out of—ordinary landfills. The leachate that drains from municipal landfills is remarkably similar to that draining from hazardous waste landfills in both composition and concentration of pollutants. While most modern landfills include systems that collect some or all of this leachate, these systems are absent from older facilities that are still operating. Moreover, even when landfill design prevents leachate from escaping and contaminating groundwater, the collected leachate must be treated and then discharged. This imposes a major expense and burden on already encumbered plants that also treat municipal sewage.

What's more, decomposing paper, yard waste, and other materials in landfills produce a variety of harmful gaseous emissions, including volatile organic chemicals, which add to urban smog, and methane, a greenhouse gas that contributes to global warming. Only a small minority of landfills operating today collect these gases; as of 1995, the EPA estimates, only 17 percent of trash was disposed of in landfills equipped with gas-collection systems. According to a 1996 study by the EPA, landfills give off an estimated 36 percent of all methane emissions in the United States. We estimate that methane emissions from landfills in the United States are 24 percent lower than they would be if recycling were discontinued.

Myth: Recycling is not cost effective. It should pay for itself.

Fact: We do not expect landfills or incinerators to pay for themselves, nor should we expect this of recycling. No other form of waste disposal, or even waste collection, pays for itself. Waste management is simply a cost society must bear.

Unlike the alternatives, recycling is much more than just another form of solid waste management. Nonetheless, setting aside the environmental benefits, let's approach the issue as accountants. The real question communities must face is whether adding recycling to a traditional waste-management system will increase the overall cost of the system over the long term. The answer, in large part, depends on the design and maturity of the recycling program and the rate of participation within the community.

Taking a snapshot of recycling costs at a single moment early in the life of community programs is misleading. For one thing, prices of recyclable materials fluctuate, so that an accurate estimate of revenues emerges only over time. For another, costs tend to decline as programs mature and expand. Most early curbside recycling collection programs were inherently inefficient because they duplicated existing trash-collection systems. Often two trucks and crews drove down the same streets every week to collect the same amount of material that one truck used to handle. Many U.S. cities have since made their recycling collection systems more cost-effective by changing truck designs, collection schedules, and truck routes in response to the fact that picking up recyclable refuse and yard trimmings leaves less trash for garbage trucks to collect. For example, Visalia, Calif., has developed a truck that collects refuse and recyclable materials simultaneously. And Fayetteville, Ark., added curbside recycling with no increase in residential bills by cutting back waste collection from twice weekly to once.

Several major cities—Seattle, San Jose, Austin, Cincinnati, Green Bay, and Portland, Ore.—have calculated that their per-ton recycling costs are lower than per-ton garbage collection and disposal. In part, these results may reflect the overall rate of recycling: a study of recycling costs in 60 randomly selected U.S. cities by the Ecodata consulting firm in Westport, Conn., found that in cities with comparatively high levels of recycling, per-ton recycling collection costs were much lower than in cities with low recycling rates. A similar survey of 15 North Carolina cities and counties conducted by the North Carolina Department of Environment, Health, and Natural Resources found that in municipalities with recycling rates greater than 12 percent, the per-ton cost of recycling was lower than that for trash disposal. Higher rates allow cities to use equipment more efficiently and generate greater revenues to offset collection costs. If we factor in increased sales of recyclable materials and reductions in landfill disposal costs, many of these high-recycling cities may break even or make money from recycling, especially in years when prices are high.

Seattle, for example, has achieved a 39 percent recycling/composting collection rate in its residential curbside program and a 44 percent collection rate citywide. Analysis of nine years of detailed data collected by the Seattle Solid Waste Utility shows that, after a two-year startup period, recycling services saved the city's solid waste management program $1.7 to $2.8 million per year. These savings occurred during a period of reduced market prices for recyclable materials; the city's landfill fees, meanwhile, are slightly above the national average. In 1995, when prices for recyclable materials were higher, Seattle's recycling program generated savings of approximately $7 million in a total budget of $29 million for all residential solid waste management services.

To reduce the cost of recycling programs, U.S. communities need to boost recycling rates. A study of 500 towns and cities by Skumatz Economic Research Associates in Seattle, Wash., found that the single most powerful tool in boosting recycling is to charge households for the trash they don't recycle. This step raised recycling levels by 8 to 10 percent on average. These kinds of variable-rate programs are now in place in more than 2,800 communities, compared with virtually none a decade ago.

Myth: Recycled materials are worthless; there is no viable market for them.
Fact: While the prices of recycled materials fluctuate over time like those of any other commodity, the volume of major scrap materials sold in domestic and global markets is growing steadily. Moreover, many robust manufacturing industries in the United States already rely on recycled materials. These businesses are an important part of our economy and provide the market foundation for the entire recycling process.

In paper manufacturing, for example, new mills that recycle paper to make corrugated boxes, newsprint, commercial tissue products, and folding cartons generally have lower capital and operating costs than new mills using virgin wood, because the work of separating cellulose fibers from wood has already occurred. Manufacturers of office paper may also face favorable economics when using recycling to expand their mills. Overall, since 1989, the use of recycled fiber by U.S. paper manufacturers has been growing faster than the use of virgin fiber. By 1995, 34 percent of the fiber used by U.S. papermakers was recycled, compared with 23 percent a decade earlier. During the 1990s, U.S. pulp and paper manufacturers began to build or expand more than 50 recycled paper mills, at a projected cost of more than $10 billion.

Recycling has long been the lower-cost manufacturing option for aluminum smelters; and it is essential to the scrap-fired steel "mini-mills" that are part of the rebirth of a competitive U.S. steel industry. The plastics industry, however, continues to invest in virgin petrochemical plants rather than recycling infrastructure—one of several reasons why the market for recycled plastics remains limited. Another factor not addressed by the plastics industry is that many consumer products come in different types of plastic that look alike but are more difficult to recycle when mixed together. Makers and users of plastic—unlike those of glass, aluminum, steel, and paper—have yet to work together to design for recyclability.

Myth: Recycling doesn't "save trees" because we are growing at least as many trees as we cut to make paper.
Fact: Growing trees on plantations has contributed to a severe and continuing loss of natural forests.

In the southern United States, for example, where most of the trees used to make paper are grown, the proportion of pine forest in plantations has risen from 2.5 percent in 1950 to more than 40 percent in 1990, with a concomitant loss of natural pine forest. At this rate, the acreage of pine plantations will overtake the area of natural pine forests in the South during this decade, and is projected to approach 70 percent of all pine forests in the country during the next few decades. While pine plantations are excellent for growing wood, they are far less suited than natural forests to providing animal habitat and preserving biodiversity. By extending the overall fiber supply, paper recycling can help reduce the pressure to convert remaining natural forests to tree farms.

Recycling becomes even more important when we view paper consumption and wood-fiber supply from a global perspective. Since 1982, the majority of the growth in worldwide paper production has been supported by recycled fiber, much of it from the United States. According to one projection, demand for paper in Asia, which does not have the extensive wood resources of North America or northern Europe, will grow from 60 million tons in 1990 to 107 million tons in 2000. To forestall intense pressures on forests in areas such as Indonesia and Malaysia, industry analysts say that recycling will have to increase, a prediction that concurs with U.S. Forest Service projections.

> *Myth: Consumers needn't be concerned about recycling when they make purchasing decisions, since stringent U.S. regulations ensure that products' prices incorporate the costs of the environmental harms they may cause. Buying the lowest-priced products, rather than recycling, is the best way to reduce environmental impacts.*
> *Fact: Even the most regulated industries generate a range of environmental damages, or "externalities," that are not reflected in market prices.*

When a coastal wetland in the Carolinas is converted to a pine plantation, estuarine fish hatcheries and water quality may decline but the market price of wood will not reflect this hidden cost. Similarly, a can of motor oil does not cost more to a buyer who plans to dispose of it by pouring it into the gutter, potentially contaminating groundwater or surface water, than to a buyer who plans to dispose of it properly. And there is simply no way to assign a meaningful economic value to rare animal or plant species, such as those endangered by clearcutting or strip mining to extract virgin resources. While many products made from recycled materials are competitive in price and function with virgin products, buying the cheapest products available does not provide an environmental substitute for waste reduction and recycling.

> *Myth: Recycling imposes a time-consuming burden on the American public.*
> *Fact: Convenient, well-designed recycling programs allow Americans to take simple actions in their daily lives to reduce the environmental impact of the products they consume.*

In a bizarre example of research, the author of "Recycling Is Garbage" asked a college student in New York City to measure the time he spent separating materials for recycling during one week. The total came to eight minutes. The author calculated that participation in recycling cost the student $2,000 per ton of recyclable trash by factoring in janitors' wages and the rent for a square foot of kitchen space, as if dropping the newspapers on the way out the door could be equated with going to work as a janitor, or as if New Yorkers had the means to turn small, unused increments of apartment floor space into tradable commodities.

Using this logic, the author might have taken the next step of calculating the economic cost to society when the college student makes his bed and does his dishes every day. The only difference between recycling and other routine housework, like taking out the trash, is that one makes your immediate environment cleaner while the other does the same for the broader environment.

Sorting trash does take some extra effort, although most people find it less of a hassle than sorting mail, according to one consumer survey. More important, it provides a simple, inexpensive way for people to reduce the environmental impact of the products they consume.

If we are serious about lowering the costs of recycling, the best approach is to study carefully how different communities improve efficiency and increase participation rates—not to engage in debating-club arguments with little relevance to the real-world problems these communities face. By boosting the efficiency of municipal recycling, establishing clear price incentives where we can, and capitalizing on the full range of environmental and industrial benefits of recycling, we can bring recycling much closer to its full potential.

NO

Chris Hendrickson, Lester Lave, and Francis McMichael

Time to Dump Recycling?

After decades of lobbying by environmentalists and extensive experience with voluntary programs, municipal solid waste recycling has recently received widespread official acceptance. The U.S. Environmental Protection Agency (EPA) has set a national goal that 25 percent of municipal solid waste (MSW) be recycled. Forty-one states plus the District of Columbia have set recycling goals that range up to 70 percent. Twenty-nine states require municipalities or counties to enact recycling ordinances or develop recycling programs. Before celebrating this achievement, however, we need to take a hard look at the price of victory and the value of the spoils.

No one seeing the overflowing trash containers in front of each house on collection day can deny that MSW is a serious concern. Valuable resources are apparently being squandered with potentially serious environmental consequences. The popular media have carried numerous warnings that landfills are close to capacity, and we expect to find vehement local opposition to the siting of any new landfills. At first glance, recycling seems to be the perfect antidote, and it does have widespread public support.

Because it seemed to be the right thing to do, we have tolerated numerous glitches in establishing recycling programs. The supply of recycled material has grown much faster than the capacity for converting them to useful products. Prices for materials have fluctuated wildly, making planning difficult. It takes time to develop efficient collection and processing systems. But the public and policymakers have been willing to be patient as the kinks in the system are worked out. The self-evident wisdom of recycling reassured everyone that all these problems could be solved.

But as these difficulties are being resolved, we are developing a much clearer picture of the economics of recycling. Beneath the debates about markets and infrastructure lurk two fundamental questions: Is it cost effective? Does it actually preserve resources and benefit the environment? What "obviously" makes sense sometimes does not stand up to careful scrutiny.

Understanding the Problem

The U.S. gross domestic product (GDP) of $6 trillion entails a lot of "getting and spending." From short-lived items such as food and newspapers to clothing, computers, cars, household furnishings, and the buildings we live and work in, everything eventually becomes municipal solid waste. The average U.S. citizen produces 1,600 pounds of solid waste a year.

For most of our history, waste was carted to an open site outside of town and dumped there. When the public became unhappy with the smell, the appearance, and the threat to public health of these traditional dumps, EPA ruled that waste would have to be placed in engineered landfills. These sophisticated capital-intensive facilities must have liners to keep the leachate from spreading, collection and treatment systems for leachate, and covers to keep away pests and to inhibit blowing dust and debris. EPA's regulations resulted in the closure of most dumps and the elimination of the most serious environmental problems caused by MSW. Still, most people objected if their neighborhood was picked as the site of a landfill. Some analysts erred in interpreting the closure of dumps and siting difficulties as signs that the country was running short of landfills. Although a few cities, notably New York and Philadelphia, are indeed having trouble finding nearby landfills, there is no national shortage of landfills. Thus, lack of space for disposing of waste is not a rationale for recycling.

But even without a pressing need to find a new way to manage MSW, many people would promote recycling as an economically and environmentally superior strategy. Recycling is portrayed as a public-spirited activity that will generate income and conserve valuable resources. These claims need to be examined critically.

The Pittsburgh Story

To get a detailed picture of how current recycling programs work, we focus on Pittsburgh—an example of an older Northeastern city where one would expect waste disposal to be an expensive problem. In response to a state mandate, Pittsburgh introduced MSW recycling in selected districts in 1990 and gradually increased coverage of the municipality and the number of products accepted for recycling.

After studying numerous alternatives, Pittsburgh implemented a system by which recyclable trash was commingled in distinctive blue bags, separately collected at curbside, and delivered to a privately operated municipal recovery facility (MRF) for separation and eventual marketing to recyclers. The contract for operating the facility is awarded on the basis of competitive bidding. Recyclable trash is collected weekly by municipal employees using standard MSW trucks and equipment owned by the city. In addition, special leaf collections are made in the fall for composting purposes.

In 1991, the last year for which complete data are available, Pittsburgh collected 167,000 tons of curbside MSW. This represents roughly two-thirds of the city's total MSW; the other third included retail, industrial, office, and park wastes. Curbside pickup of glass, plastic, and metal produced 5,100 tons

(3.1 percent of curbside MSW) for recycling. In 1993, newsprint collection was added, and the total curbside pickup of recyclable material was 6,700 tons of newsprint and 5,300 tons of glass, plastic, and metal.

When Pittsburgh started its recycling program in 1989, it sought bids from MRF operators. The best bid was an offer to pay the city $2.18 per ton of glass, metal, and plastic delivered to the MRF facility and to charge the city $8.39 per ton to take the material if newsprint was included. The tipping fee at the landfill at the time was $24 per ton. Either option was therefore less expensive than landfilling if—and as we will see, this is a very big "if"—one does not take collection costs into account.

In the second round of bidding in 1992, the city was committed to recycling newsprint, so it solicited only bids that included newsprint. The best bid was a cost to the city of $31.60 per ton. Meanwhile, the fee for landfilling had fallen to $16.15 per ton. The city therefore had to pay almost twice as much per ton to get rid of its recyclable MSW—again, without accounting for collection costs.

The increased tipping fees for recyclable materials reflects recognition of the sorting costs associated with the Pittsburgh blue bags and the difficulties of marketing MSW recyclables. A study by Waste Management, Inc. found that the price of a typical set of recyclable MSW materials had fallen from $107 per ton (in 1992 dollars) in 1988 to $44 per ton in 1992. Prices have continued to fluctuate widely since then. Although they are high at the moment, there is no guarantee about the future.

Collector's Item

The price instability of recycled material has darkened the economic prospects for recycling and received extensive public attention. But an even more troubling problem—the cost of collecting recyclable material—has been largely overlooked. Pittsburgh's experience is particularly eye-opening. The city uses the same employees and type of equipment as it uses for regular MSW, but the trucks on the recycling collection routes use a crew of two instead of three. Using the city's own accounting figures and dividing the costs between recycling and regular collection in proportion to employee hours worked and time of truck use, we calculated total collection costs. In 1991, it cost Pittsburgh $94 per ton to collect regular MSW and $470 per ton to collect recyclable MSW. With tipping fees for recyclables now higher than those for regular MSW, the total cost of disposing of recyclable MSW is more than four times the cost for regular MSW.

Several factors account for the very large difference. First is the lower density of recyclables; a full truck will hold fewer tons of recyclables. A second reason is that the amount of material picked up at each house is much smaller (recyclable material is less than 10 percent of the total MSW in Pittsburgh) so that the truck has to travel farther and make more stops to collect each ton of recyclable MSW. Because the purpose of recycling is to preserve resources and protect the environment, it should be noted that collecting recyclable MSW results in a significant increase in fuel use and combustion emissions.

Care must be taken in generalizing from Pittsburgh, where the narrow streets and hilly terrain make collection difficult, to other cities. The cost of collecting recyclable MSW is not that high in most cities. Waste Management, Inc., reports an average collection and sorting cost of $175 per ton for recycled material, based on its experience with 5.2 million households in more than 600 communities. However, the cost of collecting regular MSW is also significantly lower elsewhere, so that the difference in the costs of collecting recyclable and regular MSW is very large everywhere. Data available for other municipalities suggests that Pittsburgh's experience is not atypical. For example, San Jose reports costs of $28 per ton to landfill versus $147 per ton to recycle.

Although the cost estimates cited above are a very rough estimate of actual costs, the difference between landfilling and recycling is so large that we are convinced that more finely tuned financial data would not have any significant effect on the bottom-line conclusion that most recycling is too expensive. City officials are apparently beginning to reach the same conclusion. After some years of experience, Pennsylvania's cities have begun to scale back their recycling program as a result of the unforeseen additional costs.

Disappointing Alternatives

Because collection accounts for such a large share of the cost of recycling, we need to look at alternatives to Pittsburgh's system of separate curbside pickup. One option would be to improve the efficiency of the current pickup system. In Pittsburgh, collection routes are determined by tradition, with little attention paid to minimizing cost. However, research by graduate students at Carnegie Mellon found that savings from improved routing and other improvements in the current collections system would be small. Although any reduction in cost would be desirable, the savings are available to regular as well as recyclable MSW collection so there should be no change in the relative costs.

A related strategy would be to decrease the frequency of collection. For example, under pressure from the city council to reduce costs, Pittsburgh adopted in mid-1994 a biweekly schedule for collecting recyclable MSW. By increasing the amount of recyclables at each residence, the density of collection has increased somewhat, but it still does not approach the density of regular MSW. Also, residences now have to store recyclable materials longer, which could weaken their willingness to participate. This might explain why Pittsburgh's 1994 collection of recyclables was 25 percent less by weight than it had been in 1993.

A second alternative for cost savings is to use a private firm for collection. This might result in marginal savings but could hardly be expected to make a significant difference. A third possibility is to use the same truck for collecting MSW and recyclables. The efficiency of this system depends on how much additional time is lost in collecting the recyclables and then dropping them off at the MRF on the way to emptying the MSW at a landfill. A few cities have adopted this approach, but no reliable economic evaluation has been done. For

Pittsburgh, we estimate that combined collection would actually increase costs by 10 percent.

Fourth, collection of recycled MSW might be abandoned in favor of disturbed dropoff stations. Households would reap the benefits of lower taxes at the expense of dropping off their recyclables. The efficiency of this system depends on the amount of recyclables to be dropped off and the number of additional miles driven. To obtain a rough estimate, assume that each household drives three extra miles (30 percent of an average shopping trip) every two weeks. The household generates 150 pounds of MSW every two weeks, of which 8 percent (12 pounds) is recyclable. Thus, the $0.90 additional driving cost amounts to $0.075 per pound or $150 per ton. Costs of dropoff center implementation and maintenance should also be added. In Wellesley, Massachusetts, the operating cost of a dropoff center is reported as $16 per ton of recycled material in 1988–1989 or roughly $18 in 1992 dollars. Thus, an estimate of the total direct cost of recycling in dropoff stations is $168 per ton. This does not include the value of volunteer labor such as sorting recyclable material and driving to the dropoff center. At $5 per hour, the labor cost is more than the vehicle costs, with a total of about $400 per ton. Having more dropoff centers would lower driving costs but add a neighborhood nuisance and increase center costs. Another consideration is that the total volume of recycled material might be much smaller because people would not want to do the extra work. Smaller volume would make it more difficult to establish a market for the recycled material and to benefit from economies of scale in processing the material.

Because the value of recycled material varies so much, efficiency might be increased by limiting collection to the most valuable materials. For example, assuming that typical MRF processing costs $150 per ton, that collecting recyclable MSW costs $75 per ton more than collecting regular MSW, and that tipping fees are about $35 per ton, the recyclable material would have to sell for at least $190 per ton to be worth separating from MSW. Only aluminum, which was selling for about $750 per ton in 1993, qualifies on this criteria. At that time, plastic was $100–$130 per ton, steel and bimetal from cans was $80 per ton, clear glass was $50 per ton, and newsprint was $30 per ton. By limiting collection to aluminum and other metal cans, plastic, and plastic containers, one could lower the separation costs at the MRF, but the unit collection cost would increase so much that it would probably dwarf the savings at the MRF. In addition, collecting only high-value materials contradicts the EPA goal of recycling 25 percent of all MSW.

A major problem with recycling is the low demand for recycled materials. For example, Germany instituted a packaging recycling program that collected essentially all used packaging, but now Europe is swamped with inexpensive (and subsidized) recycled plastic. One possible policy prescription for reducing the imposed costs of recycling is to stimulate the demand for recycled materials. For example, the federal government has changed its procurement policy to insure that 20 percent of paper purchases are of recycled pulp. In some cases, there is needless discrimination against recycled materials. However, at our estimated cost of $190 per ton for additional collection and separation costs, not many materials would be worth recycling even if demand for them surged.

Finally, we could move to a completely different arrangement such as the "take-back" system being tried in Germany in which the manufacturer is responsible for getting packaging material back from the consumer and recycling it. Germany is even considering legislation that would require manufacturers to take back and recycle their own products. In this system, firms would be required to arrange "reverse logistics" systems for collecting and eventually recycling their discarded products. For example, newspaper delivery services would have to collect used newspapers. The United States already has take-back regulation for a few particularly hazardous products such as the lead acid batteries used in automobiles. Although this approach creates strong incentives for manufacturers to reduce waste, the costs are likely to be much higher than those of the present system, because it will almost certainly require numerous collection systems.

What About the Environment?

Our analysis convinces us that recycling is substantially more expensive than landfilling MSW. But the primary motivation for recycling laws is not to save money; it is to save the environment. As it happens, saving the environment is not so different from saving money in this case. The greater costs stem from additional trucks, fuel, and sorting facilities. Every truck mile adds carcinogenic diesel particles, carbon monoxide, organic compounds, oxides of nitrogen, and rubber particles to the environment, just as building and maintaining each truck does. Collection in urban areas also increases traffic congestion and noise. Constructing, heating, and lighting for an MRF similarly use energy and other scarce resources. The variety of activities associated with the two- to four-fold increase in costs associated with recycling is almost certain to result in a net increase in resource use and environmental discharges.

For Pittsburgh and similar cities, the social cost of MSW recycling is far greater than the cost of placing the waste in landfills. No minor modifications in collection programs or prices of recycled materials are likely to change this conclusion. Approaches such as dropoff stations that attempt to hide the cost by removing it from the city ledger are likely to have the highest social cost.

Although many people object to landfill disposal, modern landfills are designed and operated to have minimal discharges to the environment. Current regulations are sufficient to minimize the environmental impacts of landfills for several decades. Nevertheless, landfills are unlikely to be the optimum long-term solutions.

The fundamental problem remains: A society in which each individual produces 1,600 pounds of MSW a year is consuming too much of our natural resources and is diminishing environmental quality. Today's MSW recycling systems are analogous to the "end-of-the-pipe" emission controls enacted 25 years ago. Air and water discharge standards were designed to stop pollution. They do so, but at a cost of about $150 billion per year. Recycling MSW lowers the amount going into landfills but at too high a cost.

EPA and some progressive companies have stressed "pollution prevention" and "green design" as the only real solution to pollution problems.

Just cleaning up Superfund waste sites has proven extraordinarily expensive. Less expensive but still inefficient is the cost of preventing environmental discharges through better management of hazardous waste. The ideal solution is to redesign production processes so that no hazardous waste is created in the first place and no money is needed for discharge control and remediation.

For MSW, this approach would mean designing consumer products to reduce waste and to facilitate recycling. The potential hazards associated with toxic materials in landfills could be reduced by eliminating the toxic components in many products. For example, stop adding cadmium to plastics to give them a shiny appearance and stop using lead pigments in paints and ink. Another example is choosing packaging to minimize the volume of waste. Finally, products can be designed so that at disposal time the high-value recyclable materials can be easily removed.

Producers and consumers don't have good information to help them make choices among materials. And even when they have the information, they are not sufficiently motivated to use it. Most consumers know that they shouldn't dump used motor oil down the drain and shouldn't put old smoke detectors or half-empty pesticide containers in their trash. If they were charged the social cost of these practices, they would find more environmentally satisfactory ways of handling these unwanted products. In some cases it may be cost effective for manufacturers to include prepaid shipping vouchers to encourage consumers to return highly toxic components such as radioactive materials in smoke detectors before disposing of a product.

The best way to inform consumers and producers and to motivate them to act in socially desirable ways is to establish a pricing mechanism for materials and products that reflects their full social cost, including resource depletion and environmental damage. Full-cost pricing of raw materials would lead producers to make more socially desirable choices of materials and lead them to designs that are easier to reuse or recycle. A major problem with the current system is that product wastes in MSW arrive at the MRF having been manufactured with little or no thought for making them easy to recycle. Full-cost pricing would change the choice of materials and design so that the MRF was an integrated part of a product's design.

Unfortunately, more research is required to determine the full cost of materials, and after that is done, it will be necessary to develop a means of implementing the concept. Neither task will be easy, but the alternative is to neglect environmental problems or to attempt to regulate every decision.

Even under the best of conditions, improved design and recycling will not eliminate the need for disposal. The waste stream will be smaller and less hazardous, but the total volume will still be daunting. We will have to come back to comparing the merits of landfills, recycling, and incineration. Changes in the waste stream will force us to examine each option with fresh eyes. At present, this might mean reserving recycling for metals, using the plastic and wood product portions of MSW as fuel for energy-producing incineration, and landfilling the rest.

MSW is a systems problem. Any one-dimensional solution, be it mandated recycling, incineration, or something else, is likely to do more harm than good. An assumed preference for recycling flies in the face of economic reality unless mechanisms can be found to greatly lower the costs of collection and sorting. The long-term answer to managing MSW is likely to include green design, materials choice, component reuse, and incineration, as well as recycling. Finding a way to use full-cost pricing so that decisions are decentralized and quickly adaptable will be the key to achieving thoughtful use of resources and improvements in environmental quality.

POSTSCRIPT

Municipal Waste: Is Recycling an Environmentally and Economically Sound Waste Management Strategy?

Denison and Ruston organize their selection around a series of claims denouncing the value of recycling that were included in "Recycling Is Garbage," the cover story by John Tierney in the June 30, 1996, issue of *The New York Times Magazine*. That story stimulated a torrent of heated responses from environmental organizations and activists who were outraged by its one-sided presentation and numerous examples of what they perceived to be misinformation. Hendrickson et al. focus on the area around Pittsburgh, Pennsylvania, where the greater availability of unpopulated areas suitable for landfills makes that option considerably more attractive and cheaper than in many other, more densely populated parts of the world. They argue that the pollution resulting from the collection and processing of recyclable materials exceeds the environmental benefits associated with recycling. Others who have examined this question maintain that the much lower pollution associated with the manufacture of paper, glass, aluminum, and steel from recycled feedstocks rather than from virgin raw materials makes the recycling of these waste-stream components highly beneficial from an environmental perspective.

"Recycling is the best-known pro-environment action that you can take without major investments in time, equipment, or social organizing," writes Fred Friedman in "Creating Markets for Recycling," *Dollars & Sense* (July 1999). However, some environmentalists distrust recycling as procorporate hype. But the demand from urban residents for recycling programs is strong, according to Kivi Leroux Miller in "Is Recycling Disposable?" *American City & County* (May 2002), and where there are markets for recycled materials, recycling works well. In "Extended Product Responsibility: A Tool for a Sustainable Economy," *Environment* (September 1997), Gary A. Davis et al. urge businesses to adopt the titular principle, which includes such practices as take-back programs and designing products for easy disassembly for reuse of parts and separation of materials. Anthony Brabazon and Samuel Idowu, in "Costing the Earth," *Financial Management* (May 2001), contend that "take-back schemes may [both] provide opportunities to build goodwill and [help] companies to use resources more efficiently." According to the Environmental Protection Agency, the United States currently recycles about 28 percent of municipal waste, double the rate of 15 years ago. And according to "Data Points: Waste for Money," *Scientific American* (July 2002), in the United States some 56,000 recycling businesses employ 1.1 million people.

ISSUE 15

Nuclear Waste: Should the United States Continue to Focus Plans for Permanent Nuclear Waste Disposal Exclusively at Yucca Mountain?

YES: Spencer Abraham, from *Recommendation by the Secretary of Energy Regarding the Suitability of the Yucca Mountain Site for a Repository Under the Nuclear Waste Policy Act of 1982* (February 2002)

NO: Jon Christensen, from "Nuclear Roulette," *Mother Jones* (September/October 2001)

ISSUE SUMMARY

YES: Secretary of Energy Spencer Abraham argues that the Yucca Mountain, Nevada, nuclear waste disposal site is suitable technically and scientifically and that its development serves the U.S. national interest in numerous ways.

NO: Science writer Jon Christensen argues that it is impossible to forecast with confidence that nuclear waste entombed in Yucca Mountain will not threaten the environment over the next 10,000 (or more) years.

Nuclear waste is generated when uranium and plutonium atoms are split to make energy in nuclear power plants, when uranium and plutonium are purified to make nuclear weapons, and when radioactive isotopes that are useful in medical diagnosis and treatment are made and used. These wastes are radioactive, meaning that as they break down they emit radiation of several kinds. Those that break down fastest are most radioactive and are said to have a short half-life (the time needed for half the material to break down). Uranium-238, the most common isotope of uranium, has a half-life of 4.5 billion years and is not very radioactive at all. Plutonium-239 (which is used in bombs) has one of 24,000 years and is radioactive enough to be quite hazardous to humans.

According to the U.S. Department of Energy, high-level waste includes spent reactor fuel (the current amount of which is 52,000 tons) and waste from weapons production (91 million gallons). Transuranic waste includes clothing, equipment, and other materials contaminated with plutonium and other radioactive materials (11.3 million cubic feet, some of which has been buried in the Waste Isolation Pilot Plant salt cavern in New Mexico). Low- and mixed-level waste includes waste from hospitals and research labs, remnants of decommissioned nuclear plants, and air filters (472 million cubic feet). The high-level waste is the most hazardous and poses the most severe disposal problems. In general, experts say, such materials must be kept away from people and other living things—with no possibility of contaminating air, water (including ground water), or soil—for 10 half-lives.

Since the beginning of the nuclear age in the 1940s, nuclear waste has been accumulating. A sense of urgency about finding a place to put the waste where it would not threaten humans or ecosystems for a quarter-million years or more has also developed. The 1982 Nuclear Waste Policy Act called for locating candidate disposal sites for high-level wastes and choosing one by 1998. Since the people of the states containing candidate sites were unhappy about it, and since many of the investigated sites were less than ideal for various reasons, the schedule proved impossible to meet. In 1987 Congress attempted to settle the matter by designating Yucca Mountain, Nevada, as the one site to be intensively studied and developed. It was scheduled to be opened for use in 2010. Risk assessment expert D. Warner North, in "Unresolved Problems of Radioactive Waste: Motivation for a New Paradigm," *Physics Today* (June 1997), asserted that the technical and political problems related to nuclear waste disposal remained formidable and that a new approach was needed. Luther J. Carter and Thomas H. Pigford, in "Getting Yucca Mountain Right," *The Bulletin of the Atomic Scientists* (March/April 1998), wrote that those formidable problems could be defeated, given technical and congressional attention, and that the Yucca Mountain strategy was both sensible and realistic. However, problems have continued to plague the project; see Chuck McCutcheon, "High-Level Acrimony in Nuclear Storage Standoff," *Congressional Quarterly Weekly Report* (September 25, 1999) and Sean Paige, "The Fight at the End of the Tunnel," *Insight on the News* (November 15, 1999).

In February 2002 Secretary of Energy Spencer Abraham recommended to the president that the United States go ahead with development of the Yucca Mountain site. His report, which is excerpted in the following selection, makes the points that a disposal site is necessary, that Yucca Mountain has been thoroughly studied, and that moving ahead with the site best serves "our energy future, our national security, our economy, our environment, and safety." Abraham further argues that objections to the site are not serious enough to stop the project. In the second selection, Jon Christensen argues that far too much confidence in Yucca Mountain's long-term safety is based on probabilistic computer models that are too uncertain to trust.

·

 **YES**

Recommendation by the Secretary of Energy Regarding the Suitability of the Yucca Mountain Site for a Repository Under the Nuclear Waste Policy Act of 1982

Introduction

For more than half a century, since nuclear science helped us win World War II and ring in the Atomic Age, scientists have known that the Nation would need a secure, permanent facility in which to dispose of radioactive wastes. Twenty years ago, when Congress adopted the Nuclear Waste Policy Act of 1982 (NWPA or "the ACT"), it recognized the overwhelming consensus in the scientific community that the best option for such a facility would be a deep underground repository. Fifteen years ago, Congress directed the Secretary of Energy to investigate and recommend to the President whether such a repository could be located safely at Yucca Mountain, Nevada. Since then, our country has spent billions of dollars and millions of hours of research endeavoring to answer this question. I have carefully reviewed the product of this study. In my judgment, it constitutes sound science and shows that a safe repository can be sited there. I also believe that compelling national interests counsel in favor of proceeding with this project. Accordingly, consistent with my responsibilities under the NWPA, today I am recommending that Yucca Mountain be developed as the site for an underground repository for spent fuel and other radioactive wastes.

The first consideration in my decision was whether the Yucca Mountain site will safeguard the health and safety of the people, in Nevada and across the country, and will be effective in containing at a minimum risk the material it is designed to hold. Substantial evidence shows that it will. Yucca Mountain is far and away the most thoroughly researched site of its kind in the world. It is a geologically stable site, in a closed groundwater basin, isolated on thousands of acres of Federal land, and farther from any metropolitan area than the great

From Spencer Abraham, *Recommendation by the Secretary of Energy Regarding the Suitability of the Yucca Mountain Site for a Repository Under the Nuclear Waste Policy Act of 1982* (February 2002). Notes omitted.

majority of less secure, temporary nuclear waste storage sites that exist in the country today.

This point bears emphasis. We are not confronting a hypothetical problem. We have a staggering amount of radioactive waste in this country—nearly 100,000,000 gallons of high-level nuclear waste and more than 40,000 metric tons of spent nuclear fuel with more created every day. Our choice is not between, on the one hand, a disposal site with costs and risks held to a minimum, and, on the other, a magic disposal system with no costs or risks at all. Instead, the real choice is between a single secure site, deep under the ground at Yucca Mountain, or making do with what we have now or some variant of it—131 aging surface sites, scattered across 39 states. Every one of those sites was built on the assumption that it would be temporary. As time goes by, every one is closer to the limit of its safe life span. And every one is at least a potential security risk—safe for today, but a question mark in decades to come.

The Yucca Mountain facility is important to achieving a number of our national goals. It will promote our energy security, our national security, and safety in our homeland. It will help strengthen our economy and help us clean up the environment.

The benefits of nuclear power are with us every day. Twenty percent of our country's electricity comes from nuclear energy. To put it another way, the "average" home operates on nuclear-generated electricity for almost five hours a day. A government with a complacent, kick-the-can-down-the-road nuclear waste disposal policy will sooner or later have to ask its citizens which five hours of electricity they would care to do without.

Regions that produce steel, automobiles, and durable goods rely in particular on nuclear power, which reduces the air pollution associated with fossil fuels—greenhouse gases, solid particulate matter, smog, and acid rain. But environmental concerns extend further. Most commercial spent fuel storage facilities are near large populations centers; in fact, more than 161 million Americans live within 75 miles of these facilities. These storage sites also tend to be near rivers, lakes, and seacoasts. Should a radioactive release occur from one of these older, less robust facilities, it could contaminate any of 20 major waterways, including the Mississippi River. Over 30 million Americans are served by these potentially at-risk water sources.

Our national security interests are likewise at stake. Forty percent of our warships, including many of the most strategic vessels in our Navy, are powered by nuclear fuel, which eventually becomes spent fuel. At the same time, the end of the Cold War has brought the welcome challenge to our Nation of disposing of surplus weapons-grade plutonium as part of the process of decommissioning our nuclear weapons. Regardless of whether this material is turned into reactor fuel or otherwise treated, an underground repository is an indispensable component in any plan for its complete disposition. An affirmative decision on Yucca Mountain is also likely to affect other nations' weapons decommissioning, since their willingness to proceed will depend on being satisfied that we are doing so. Moving forward with the repository will contribute to our global efforts to stem the proliferation of nuclear weapons in other ways, since it will encourage nations with weaker controls over their own materials to follow a similar path of permanent, underground disposal, thereby making it more difficult for these materials to fall into the wrong hands. By moving forward with Yucca Mountain, we will show leadership, set out a roadmap, and encourage other nations to follow it.

There will be those who say the problem of nuclear waste disposal generally, and Yucca Mountain in particular, needs more study. In fact, both issues have been studied for more than twice the amount of time it took to plan and complete the moon landing. My Recommendation today is consistent with the conclusion of the National Research Council of the National Academy of Sciences—a conclusion reached, not last week or last month, but 12 years ago. The Council noted "a worldwide scientific consensus that deep geological disposal, the approach being followed by the United States, is the best option for disposing of high-level radioactive waste." Likewise, a broad spectrum of experts agrees that we now have enough information, including more than 20 years of researching Yucca Mountain specifically, to support a conclusion that such a repository can be safely located there.

Nonetheless, should this site designation ultimately become effective, considerable additional study lies ahead. Before an ounce of spent fuel or radioactive waste could be sent to Yucca Mountain, indeed even before construction of the permanent facilities for emplacement of waste could begin there, the Department of Energy (DOE or "the Department") will be required to submit an application to the independent Nuclear Regulatory Commission (NRC). There, DOE would be required to make its case through a formal review process that will include public hearings and is expected to last at least three years. Only after that, if the license were granted, could construction begin. The DOE would also have to obtain an additional operating license, supported by evidence that public health and safety will be preserved, before any waste could actually be received.

In short, even if the Yucca Mountain Recommendation were accepted today, an estimated minimum of eight more years lies ahead before the site would become operational.

We have seen decades of study, and properly so for a decision of this importance, one with significant consequences for so many of our citizens. As necessary, many more years of study will be undertaken. But it is past time to stop sacrificing that which is forward-looking and prudent on the altar of a *status quo* we know ultimately will fail us. The *status quo* is not the best we can do for our energy future, our national security, our economy, our environment, and safety—and we are less safe every day as the clock runs down on dozens of older, temporary sites.

I recommend the deep underground site at Yucca Mountain, Nevada, for development as our Nation's first permanent facility for disposing of high-level nuclear waste.

Background

History of the Yucca Mountain Project and the Nuclear Waste Policy Act

The need for a secure facility in which to dispose of radioactive wastes has been known in this country at least since World War II. As early as 1957, a National Academy of Sciences report to the Atomic Energy Commission suggested burying radioactive waste in geologic formations. Beginning in the 1970s, the United States and other countries evaluated many options for the safe and permanent disposal of radioactive waste, including deep seabed disposal, remote island siting, dry cask storage, disposal in the polar ice sheets, transmutation, and rocketing waste into orbit around the sun. After analyzing these options, disposal in a mined geologic repository emerged as the preferred long-term environmental solution for the management of these wastes. Congress recognized this consensus 20 years ago when it passed the Nuclear Waste Policy Act of 1982.

In the Act, Congress created a Federal obligation to accept civilian spent nuclear fuel and dispose of it in a geologic facility. Congress also designated the agencies responsible for implementing this policy and specified their roles. The Department of Energy must characterize, site, design, build, and manage a Federal waste repository. The Environmental Protection Agency (EPA) must set the public health standards for it. The Nuclear Regulatory Commission must license its construction, operation, and closure.

The Department of Energy began studying Yucca Mountain almost a quarter century ago. Even before Congress adopted the NWPA, the Department had begun national site screening research as part of the National Waste Terminal Storage program, which included examination of Federal sites that had previously been used for defense-related activities and were already potentially contaminated. Yucca Mountain was one such location, on and adjacent to the Nevada Test Site, which was then under construction. Work began on the Yucca Mountain site in 1978. When the NWPA was passed, the Department was studying more than 25 sites around the country as potential repositories. The Act provided for the siting and development of two; Yucca Mountain was one of nine sites under consideration for the first repository program.

Following the provisions of the Act and the Department's siting Guidelines, the Department prepared draft environmental assessments for the nine sites. Final environmental assessments were prepared for five of these, including Yucca Mountain. In 1986, the Department compared and ranked the sites under construction for characterization. It did this by using a multi-attribute methodology—an accepted, formal scientific method used to help decision makers compare, on an equivalent basis, the many components that make up a complex decision. When all the components of the ranking decision were considered together, taking account of both preclosure and post-closure concerns, Yucca Mountain was the top-ranked site. The Department examined a variety of ways of combining the components of the ranking scheme; this only confirmed the conclusion that Yucca Mountain came out in first place. The EPA also looked at the performance of a repository in unsaturated tuff. The EPA noted that in its modeling in support of development of the standards, unsaturated tuff was one of the two geologic media that appeared most capable of limiting releases of radionuclides in a manner that keeps expected doses to individuals low.

In 1986, Secretary of Energy Herrington found three sites to be suitable for site characterization, and recommended the three, including Yucca Mountain, to President Reagan for detailed site characterization. The Secretary also made a preliminary finding, based on Guidelines that did not require site characterization, that the three sites were suitable for development as repositories.

The next year, Congress amended the NWPA, and selected Yucca Mountain as the single site to be characterized. It simultaneously directed the Department to cease activities at all other potential sites. Although it has been suggested that Congress's decision was made for purely political reasons, the record described above reveals that the Yucca Mountain site consistently ranked at or near the top of the sites evaluated well before Congress's action.

As previously noted, the National Research Council of the National Academy of Sciences concluded in 1990 (and reiterated [recently]) that there is "a worldwide scientific consensus that deep geological disposal, the approach being followed by the United States, is the best option for disposing of high-level radioactive waste." Today, many national and international scientific experts and nuclear waste management professionals agree with DOE that there exists sufficient information to support a national decision on designation of the Yucca Mountain site.

The Nuclear Waste Policy Act and the Responsibilities of the Department of Energy and the Secretary

Congress assigned to the Secretary of Energy the primary responsibility for implementing the national policy of developing a deep underground repository. The Secretary must determine whether to initiate the next step laid out in the NWPA—a recommendation to designate Yucca Mountain as the site for development as a permanent disposal facility. . . . Briefly, I first must determine whether Yucca Mountain is in fact technically and scientifically suitable to be a repository. A favorable suitability determination is indispensable for a positive recommendation of the site to the President. Under additional criteria I have adopted above and beyond the statutory requirements, I have also sought to determine whether, when other relevant considerations are taken into account, recommending it is in the overall national interest and, if so, whether there are countervailing arguments so strong that I should nonetheless decline to make the Recommendation.

The Act contemplates several important stages in evaluating the site before a Secretarial recommendation is in order. It directs the Secretary to develop a site characterization plan, one that will help guide test programs for the collection of data to be used in evaluating the site. It directs the Secretary to conduct such characterization studies as may be necessary to evaluate the site's suitability. And it directs the Secretary to hold hearings in the vicinity of the prospective site to inform the residents and receive their comments. It is at the completion of these stages that the Act directs the Secretary, if he finds the site suitable, to determine whether to recommend it to the President for development as a permanent repository.

If the Secretary recommends to the President that Yucca Mountain be developed, he must include with the Recommendation, and make available to the public, a comprehensive statement of the basis for his determination. If at any time the Secretary determines that Yucca Mountain is not a suitable site, he must report to Congress within six months his recommendations for further action to assure safe, permanent disposal of spent nuclear fuel and high-level radioactive waste.

Following a Recommendation by the Secretary, the President may recommend the Yucca Mountain site to Congress "if... [he] considers [it] qualified for application for a construction authorization...." If the President submits a recommendation to Congress, he must also submit a copy of the statement setting forth the basis for the Secretary's Recommendation.

A Presidential recommendation takes effect 60 days after submission unless Nevada forwards a notice of disapproval to the Congress. If Nevada submits such a notice, Congress has a limited time during which it may nevertheless give effect to the President's recommendation by passing, under expedited procedures, a joint resolution of siting approval. If the President's recommendation takes effect, the Act directs the Secretary to submit to the NRC a construction license application.

The NWPA by its terms contemplated that the entire process of siting, licensing, and constructing a repository would have been completed more than four years ago, by January 31, 1998. Accordingly, it required the Department to enter into contracts to begin accepting waste for disposal by that date.

Decision

The Recommendation

After over 20 years of research and billions of dollars of carefully planned and reviewed scientific field work, the Department has found that a repository at Yucca Mountain brings together the location, natural barriers, and design elements most likely to protect the health and safety of the public, including those Americans living in the immediate vicinity, now and long into the future. It is therefore suitable, within the meaning of the NWPA, for development as a permanent nuclear waste and spent fuel repository.

After reviewing the extensive, indeed unprecedented, analysis the Department has undertaken, and in discharging the responsibilities made incumbent on the Secretary under the Act, I am recommending to the President that Yucca Mountain be developed as the Nation's first permanent, deep underground repository for high-level radioactive waste. A decision to develop Yucca Mountain will be a critical step forward in addressing our Nation's energy future, our national defense, our safety at home, and protection for our economy and environment.

What This Recommendation Means, and What It Does Not Mean

Even after so many years of research, this Recommendation is a preliminary step. It does no more than start the formal safety evaluation process. Before a license is granted, much less before repository construction or waste emplacement may begin, many steps and many years still lie ahead. The DOE must submit an application for a construction license; defend it through formal review, including public hearings; and receive authorization from the NRC, which has the statutory responsibility to ensure that any repository built at Yucca Mountain meets stringent tests of health and safety. The NRC licensing process is expected to take a minimum of three years. Opposing viewpoints will have every opportunity to be heard. If the NRC grants this first license, it will only authorize initial construction. The DOE would have then have to seek and obtain a second operating license from the NRC before any wastes could be received. The process altogether is expected to take a minimum of eight years.

The DOE would also be subject to NRC oversight as a condition of the operating license. Construction, licensing, and operation of the repository would also be subject to ongoing Congressional oversight.

At some future point, the repository is expected to close. EPA and NRC regulations require monitoring after the DOE receives a license amendment authorizing the closure, which would be from 50 to about 300 years after waste emplacement begins, or possibly longer. The repository would also be designed, however, to be able to adapt to methods future generations might develop to manage high-level radioactive waste. Thus, even after completion of waste emplacement, the waste could be retrieved to take advantage of its economic value or usefulness to as yet undeveloped technologies.

Permanently closing the repository would require sealing all shafts, ramps, exploratory boreholes, and other underground openings connected to the surface. Such sealing would discourage human intrusion and prevent water from entering through these openings. DOE's site stewardship would include maintaining control of the area, monitoring and testing, and implementing security measures against vandalism and theft. In addition, a network of permanent monuments and markers would be erected around the site to alert future generations to the presence and nature of the buried waste. Detailed public records held in multiple places would identify the location and layout of the repository and the nature and potential hazard of the waste it contains. The Federal Government would maintain control of the site for the indefinite future. Active security systems would prevent deliberate or inadvertent human intrusion and any other human activity that could adversely affect the performance of the repository....

Nuclear Science and the National Interest

Our country depends in many ways on the benefits of nuclear science: in the generation of twenty percent of the Nation's electricity; in the operation of many of the Navy's most strategic vessels; in the maintenance of the Nation's nuclear weapons arsenal; and in numerous research and development projects, both medical and scientific. All these activities produce radioactive wastes that have been accumulating since the mid-1940s. They are currently scattered among 131 sites in 39 states, residing in temporary surface storage facilities and awaiting final disposal. In exchange for the many benefits of nuclear power, we assume the cost of managing its byproducts in a responsible, safe, and secure fashion. And there is a near-universal consensus that a deep geologic facility is the only scientifically credible, long-term solution to a problem that will only grow more difficult the longer it is ignored.

Energy Security

Roughly 20 percent of our country's electricity is generated from nuclear power. This means that, on average, each home, farm, factory, and business in America runs on nuclear fuel for a little less than five hours a day.

A balanced energy policy—one that makes use of multiple sources of energy, rather than becoming dependent entirely on generating electricity from a single source, such as natural gas—is important to economic growth. Our vulnerability to shortages and price spikes rises in direct proportion to our failure to maintain diverse sources of power. To assure that we will continue to have reliable and affordable sources of energy, we need to preserve our access to nuclear power.

Yet the Federal government's failure to meet its obligation to dispose of spent nuclear fuel under the NWPA—as it has been supposed to do starting in 1998—is placing our access to this source of energy in jeopardy. Nuclear power plants have been storing their spent fuel on site, but many are running out of space to do so. Unless a better solution is found, a growing number of these plants will not be able to find additional storage space and will be forced to shut down prematurely. Nor are we likely to see any new plants built.

Already we are facing a growing imbalance between our projected energy needs and our projected supplies. The loss of existing electric generating capacity that we will experience if nuclear plants start going off-line would significantly exacerbate this problem, leading to price spikes and increased electricity rates as relatively cheap power is taken off the market. A permanent repository for spent nuclear fuel is essential to our continuing to count on nuclear energy to help us meet our energy demands.

National Security

Powering the Navy Nuclear Fleet

A strong Navy is a vital part of national security. Many of the most strategically important vessels in our fleet, including submarines and aircraft carriers, are nuclear powered. They have played a major role in every significant military action in which the United States has been involved for some 40 years, including our current operations in Afghanistan. They are also essential to our nuclear deterrent. In short, our nuclear-powered Navy is indispensable to our status as a world power.

For the nuclear Navy to function, nuclear ships must be refueled periodically and the spent fuel removed. The spent fuel must go someplace. Currently, as part of a consent decree entered into between the State of Idaho and the Federal Government, this material goes to temporary surface storage facilities at the Idaho National Environmental and Engineering Laboratory. But this cannot continue indefinitely, and indeed the agreement specifies that the spent fuel must be removed. Failure to establish a permanent disposition pathway is not only irresponsible, but could also create serious future uncertainties potentially affecting the continued capability of our Naval operations.

Allowing the Nation to Decommission Its Surplus Nuclear Weapons and Support Nuclear Non-Proliferation Efforts

A decision now on the Yucca Mountain repository is also important in several ways to our efforts to prevent the proliferation of nuclear weapons. First, the end of the Cold War has brought the welcome challenge to our country of disposing of surplus weapons-grade plutonium as part of the process of decommissioning weapons we no longer need. Current plans call for turning the plutonium into "mixed-oxide" or "MOX" fuel. But creating MOX fuel as well as burning the fuel in a nuclear reactor will generate spent nuclear fuel, and other byproducts which themselves will require somewhere to go. A geological repository is critical to completing disposal of these materials. Such complete disposal is important if we are to expect other nations to decommission their own weapons, which they are unlikely to do unless persuaded that we are truly decommissioning our own.

A respository is important to non-proliferation for other reasons as well. Unauthorized removal of nuclear materials from a repository will be difficult even in the absence of strong institutional controls. Therefore, in countries that lack such controls, and even in our own, a safe repository is essential in preventing these materials from falling into the hands of rogue nations. By permanently disposing of nuclear weapons materials in a facility of this kind, the United States would encourage other nations to do the same.

Protecting the Environment

An underground repository at Yucca Mountain is important to our efforts to protect our environment and achieve sustainable growth in two ways. First, it will allow us to dispose of the radioactive waste that has been building up in our country for over fifty years in a safe and environmentally sound manner. Second, it will facilitate continued use and potential expansion of nuclear power, one of the few sources of electricity currently available to us that emits no carbon dioxide or other greenhouse gases.

As to the first point: While the Federal government has long promised that it would assume responsibility for nuclear waste, it has yet to start implementing an environmentally sound approach for disposing of this material. It is past time for us to do so. The production of nuclear weapons at the end of the Second World War and for many years thereafter has resulted in a legacy of high-level radioactive waste and spent fuel, currently located in Tennessee, Colorado, South Carolina, New Mexico, New York, Washington, and Idaho. Among these wastes, approximately 100,000,000 gallons of high-level liquid waste are stored in, and in some instances have leaked from, temporary holding tanks. In addition to this high-level radioactive waste, about 2,100 metric tons of solid, unreprocessed fuel from a plutonium-production reactor are stored at the Hanford Nuclear Reservation, with another 400 metric tons stored at other DOE sites.

In addition, under the NWPA, the Federal government is also responsible for disposing of spent commercial fuel, a program that was to have begun in 1998, four years ago. More than 161 million Americans, well more than half the population, reside within 75 miles of a major nuclear facility—and, thus, within 75 miles of that facility's aging and temporary capacity for storing this material. Moreover, because nuclear reactors require abundant water for cooling, on-site storage tends to be located near rivers, lakes, and seacoasts. Ten *closed* facilities, such as Big Rock Point, on the banks of Lake Michigan, also house spent fuel and incur significant annual costs without providing any ongoing benefit. Over the long-term, without active management and monitoring, degrading surface storage facilities may pose a risk to any of 20 major U.S. lakes and waterways, including the Mississippi River. Millions of Americans are served by municipal water systems with intakes along these waterways. In recent letters, Governors Bob Taft of Ohio and John Engler of Michigan raised concerns about the advisability of long-term storage of spent fuel in temporary systems so close to major bodies of water. The scientific consensus is that disposal of this material in a deep underground repository is not merely the safe answer and the right answer for protecting our environment but the *only* answer that has any degree of realism.

In addition, nuclear power is one of only a few sources of power available to us now in a potentially plentiful and economical manner that could drastically reduce air pollution and greenhouse gas emissions caused by the generation of electricity. It produces no controlled air pollutants, such as sulfur and particulates, or greenhouse gases. Therefore, it can help keep our air clean, avoid generation of ground-level ozone, and prevent acid rain. A repository of Yucca Mountain is indispensable to the maintenance and potential expansion of the use of this environmentally efficient source of energy....

Summary

In short, there are important reasons to move forward with a repository at Yucca Mountain. Doing so will advance our energy security by helping us to maintain diverse sources of energy supply. It will advance our national security by helping to provide operational certainty to our nuclear Navy and by facilitating the decommissioning of nuclear weapons and the secure disposition of nuclear materials. It will help us clean up our environment by allowing us to close the nuclear fuel cycle and giving us greater access to a form of energy that does not emit greenhouse gases. And it will help us in our efforts to secure ourselves against terrorist threats by allowing us to remove nuclear materials from scattered above-ground locations to a single, secure underground facility. Given the site's scientific and technical suitability, I find that compelling national interests counsel in favor of taking the next step toward siting a repository at Yucca Mountain.

NO

Jon Christensen

Nuclear Roulette

We're on our way to Yucca Mountain. And there are some things you should know before we get there. That is, before the Bush administration and Congress decide once and for all to entomb the nuclear age's most deadly legacy in the Nevada desert about 100 miles northwest of Las Vegas. The most important thing to remember is this: It's not about-Yucca Mountain. And yet it is.

The other thing to keep in mind is that there is not just one Yucca Mountain. There are three. There is Yucca Mountain the place, a heap of ash and rubble that was blasted from a volcano some 12 million years ago and cemented together and eroded over eons into the shape of a wave breaking westward across a desert sea. There is little love lost for this Yucca Mountain, even among Nevadans like me who cultivate a taste for such unworldly landscapes. If I didn't have to come to Yucca Mountain, I wouldn't. And neither would you.

But we do. Because there is another Yucca Mountain. And this Yucca Mountain is the political answer to the question of what to do with spent fuel from 118 commercial nuclear reactors, 10 nuclear-weapons plants, and 37 research reactors around the country. This Yucca Mountain offers salvation for a nuclear industry poised for a comeback—and for politicians from states that don't want the highly radioactive waste stockpiled within their borders. Nearly 20 years ago, the federal government signed a contract promising that it would take charge of the spent fuel, which is now stored in dry casks and cooling pools at the plant sites. The political Yucca Mountain is the reason we are here.

And finally, there is Yucca Mountain the computer model. This Yucca Mountain is the most difficult to see, let alone understand. It is the virtual product of a program called a Monte Carlo simulation that calculates how much risk the real mountain's specific flaws—water percolating through the rock, groundwater flowing beneath, potential earthquakes and volcanic eruptions—will pose over the thousands of years that the waste will remain dangerously radioactive. In its ethereal way, this ghost of Yucca Mountain embodies both the technocratic hubris and the gambler's faith in the odds that have brought us to the brink of a decision whose consequences, as acknowledged by everyone involved, we cannot foresee.

By the end of [2001], the Department of Energy is scheduled to issue its final recommendation on turning Yucca Mountain into the nation's first and only high-level radioactive-waste repository—a permanent graveyard for 70,000 tons of some of the most deadly and long-lasting toxins ever made. There is very little doubt about which way the recommendation will go. President Bush has called nuclear power "a major component" of his energy plan; the administration wants to extend the licenses of existing reactors and encourage the building of new ones. And as Energy Secretary Spencer Abraham told CNN's "Moneyline" in May, "If we can't find a repository for the waste, then it is very unlikely we would see new plants built."

Once the recommendation is made and the president formally endorses it, the state of Nevada will most certainly file a formal objection. That protest will send the decision to Congress, where a simple majority of both the House and the Senate will be all that is needed to override the state's pro forma veto.

Congress doesn't much care about the real Yucca Mountain. Earlier this year, an Energy Department document put it succinctly: "The technical suitability of the site is less of a concern to Congress than whether the nuclear waste problem can be solved at an affordable price in both financial and political terms." (Officials quickly disavowed the memo, blaming a contractor for the inadvertently telling wording.)

The "technical suitability" of Yucca Mountain, however, is what has proved most difficult to establish. Over 20 years of poking and prodding, this spot has become one of the most intensely studied pieces of real estate in the world, at a cost of close to $3.4 billion so far (and an estimated $50 billion more if the repository is built). Researchers have found that the mountain is crisscrossed by earthquake faults and that there are dormant volcanoes nearby. But the main concern in this arid spot turns out to be water.

Only an average six inches of rain fall on Yucca Mountain each year, barely enough to keep a sparse covering of grass and creosote bush alive. At first, the volcanic ash that makes up the mountain was thought to be so tightly compressed that what little water there is would not flow through the layers of rock. But as geologists dug into the mountain, they found that the rock is riddled with fractures. On average, they discovered a fracture every couple of inches. And they found water moving through the fractures.

They thought the water was moving slowly. But in 1996, they found chlorine 36—an isotope left by atmospheric bomb testing at the nearby Nevada Test Site in the 1950s—in water sampled at the level where waste would be stored, 800 feet underground. That meant rainwater could percolate down to the waste-storage area in just 50 years, and in another 50 years or so could reach the aquifer 1,000 feet farther down.

Originally, scientists also believed that if contamination escaped to the aquifer, most of it would cling to the rock and was unlikely to reach the nearby Amargosa Valley, now home to 1,500 people and a dairy farm that produces 41,000 gallons of milk a day. But studies have since found that plutonium from underground bomb tests hitched a ride on microscopic specks of clay suspended in groundwater and moved nearly a mile in 30 years—much faster than expected.

Add all of that up, and Yucca Mountain no longer looks like the perfect site to bury material that by law must be kept isolated from the environment for

10,000 years (though the half-lives of some of the most potent elements in the waste, such as plutonium, are much longer). In fact, the Energy Department has essentially conceded as much. It now asserts that what will protect the waste is not the mountain itself, but a special kind of canister made from a nickel-based metal called Alloy 22. The department says the metal—which has been around for a few decades but tested for just three years—will last about 12,000 years.

That is where Yucca Mountain the computer model (officially known as a "total system performance assessment") comes into the picture. The analysis uses a Monte Carlo simulation, a technique commonly employed in science and business to model the probability of various outcomes in a complex situation. Take the probability that water will drip through the cracks in the mountain and onto the waste canisters; mix that with the likelihood that the canisters will corrode; add to that the probability that water will carry the contamination to the aquifer below; and finally, factor in the chances that a family living nearby will drink that water. Incorporate the possibility of a volcanic eruption, and of another ice age making this a much wetter place, and then throw in the probability that a future prospector—let's call him the "unluckiest man in the world," as Energy Department scientists do—will decide to drill or dig at the site.

After sampling all of these variables many times, as if drawing cards for hundreds of poker hands, the Monte Carlo simulation spits out a probability curve. It estimates that radiation is unlikely to leak from the site for the next 10,000 years (if the canisters last that long; if they don't, all bets are off). By then, the model suggests, the radiation will have diminished, and contamination from the repository will be partially absorbed in the rock and diluted in the water under Yucca Mountain. So the dose to a hypothetical family in Amargosa Valley won't rise above 15 millirems—the maximum allowed by the Environmental Protection Agency—for hundreds of thousands of years.

"In some sense, it is science fiction to project out 300,000 years," Abraham Van Luik, the official in charge of the modeling, once commented while showing me around Yucca Mountain. "It gets more and more difficult to defend your assumptions as you move into the future." But, he hastened to add, "our modeling is overconservative. Absolutely no one is going to get hurt by this repository for hundreds of thousands of years."

The problem, some experts warn, is that "absolutely" is not something that can be said about a model based on probabilities. The Energy Department likes to say that the model has revealed "no showstoppers" at Yucca Mountain —no single factor that would disqualify the site. But Rodney Ewing, a nuclear-waste management expert who served on the peer review panel for the Yucca Mountain model in 1998, says the computer simulation wouldn't know a showstopper if it saw one. "The uncertainty in these analyses was so large as to make them unusable," explains Ewing, who in 1999 published a scathing article in the journal *Science*, criticizing the department's reliance on the computer model. "One should not expect greater success with such a prediction than we have in other fields," he says, "such as predicting which presidential candidate gets the electoral votes from Florida."

In fact, says Ewing, we have a lot more practice predicting electoral results than we do forecasting how a complicated combination of geology and engineering will perform over thousands of years. "If an airplane were built in this way," he says, "that is, smaller versions of the plane hadn't been test-flown, but you were assured that, good and competent engineers and scientists had modeled the plane's ability to fly, would you fly on the first airplane based on these analyses?"

If that prospect makes you nervous, just try to remember this: It's not about Yucca Mountain.

And yet it is.

POSTSCRIPT

Nuclear Waste: Should the United States Continue to Focus Plans for Permanent Nuclear Waste Disposal Exclusively at Yucca Mountain?

Abraham notes that the state of Nevada has the right to object to his recommendation. Not surprisingly, Nevada governor Kenny Guinn did exactly that on April 8, 2002. On May 8 the House of Representatives promptly voted to set aside the veto, and on July 9 the Senate voted to do the same. News reports said that this ends "years of political debate over nuclear waste disposal," but Nevada still has half a dozen lawsuits challenging the project pending.

Even those who favor using Yucca Mountain for high-level nuclear waste disposal admit that in time the site is bound to leak. The intensity of the radioactivity emitted by the waste will decline rapidly as short–half-life materials decay, and by 2300, when the site is expected to be sealed, that intensity will be less than 5 percent of the initial level. After that, however, radiation intensity will decline much more slowly. The nickel-alloy containers for the waste are expected to last at least 10,000 years, but they will not last forever. The Department of Energy's computer simulations predict that the radiation released to the environment will rise rapidly after about 100,000 years, with a peak annual dose after 400,000 years that is about double the natural background exposure. Many people are skeptical that the site can be protected for any significant fraction of such time periods. These are among the considerations that lead James Flynn et al., in "Overcoming Tunnel Vision," *Environment* (April 1997), to urge stopping work on the Yucca Mountain project and rethinking the entire nuclear waste disposal issue. On the other hand, Jonah Goldberg, in "Dead and Buried," *National Review* (April 8, 2002), contends that such considerations are irrelevant and that critics exaggerate the dangers of storing waste at Yucca Mountain.

The nuclear waste disposal problem in the United States is real, and it must be dealt with. If it is not, America may face the same kinds of problems created by the former Soviet Union, which disposed of some nuclear waste simply by dumping it into the sea. For a recent summary of the nuclear waste problem and the disposal controversy, see Michael E. Long, "Half Life: The Lethal Legacy of America's Nuclear Waste," *National Geographic* (July 2002). Gary Taubes, in "Whose Nuclear Waste?" *Technology Review* (January/February 2002), argues that a whole new approach may be necessary.

On the Internet ...

National Councils for Sustainable Development

Since the 1992 Earth Summit in Rio de Janeiro, over 70 countries have established multi-stakeholder groups known as National Councils for Sustainable Development (NCSDs) to promote and implement sustainable development. This is the Earth Council's NCSD site, which is dedicated to facilitating and coordinating the work of the NCSDs.

http://www.ncsdnetwork.org

United Nations Environment Programme

The United Nations Environment Programme "works to encourage sustainable development through sound environmental practices everywhere. Its activities cover a wide range of issues, from atmosphere and terrestrial ecosystems, the promotion of environmental science and information, to an early warning and emergency response capacity to deal with environmental disasters and emergencies."

http://www.unep.org

International Institute for Sustainable Development

The International Institute for Sustainable Development advances sustainable development policy and research by providing information and engaging in partnerships worldwide. According to this site, the institute "promotes the transition toward a sustainable future [and seeks] to demonstrate how human ingenuity can be applied to improve the well-being of the environment, economy and society."

http://www.iisd.org/default.asp

Population Council

Established in 1952, "the Population Council is an international, nonprofit institution that conducts research on three fronts: biomedical, social science, and public health. This research—and the information it produces—helps change the way people think about problems related to reproductive health and population growth." Many of the council's publications are available online.

http://www.popcouncil.org

The Heritage Foundation

The Heritage Foundation is a think tank whose mission is to formulate and promote conservative public policies on many issues, including environmental ones. Its work is based on the principles of free enterprise, limited government, individual freedom, and traditional American values.

http://www.heritage.org

PART 4

Potential Solutions

*P*eople argue over whether or not certain matters (such as global warming) are truly problems that warrant concern, but arguments also rage over what to do about recognized problems. For instance, is biodiversity best protected by limiting what humans do to the environment or by limiting the number of humans who do it? Must government regulate industry to prevent pollution, or can industry be trusted to solve the problem by itself? What is the best approach to bringing industry into line—economic incentives or threats of fines? Is it even possible to have an industrial civilization that treads lightly enough to last for generations to come?

These and other questions are not easy ones to answer. But answers are needed.

- Is Limiting Population Growth a Key Factor in Protecting the Global Environment?

- Will Pollution Rights Trading Effectively Control Environmental Problems?

- Will Voluntary Action by Industry Reduce the Need for Future Environmental Regulation?

- Is Sustainable Development Compatible With Human Welfare?

ISSUE 16

Is Limiting Population Growth a Key Factor in Protecting the Global Environment?

YES: Paul R. Ehrlich and Anne H. Ehrlich, from "The Population Explosion: Why We Should Care and What We Should Do About It," *Environmental Law* (vol. 27, no. 4, 1997)

NO: Stephen Moore, from "Body Count," *National Review* (October 25, 1999)

ISSUE SUMMARY

YES: Population biologists Paul R. Ehrlich and Anne H. Ehrlich argue that if humanity fails to reduce the impact of population in terms of both numbers and resource consumption, it faces the prospect of environmental disaster.

NO: Stephen Moore, director of the Cato Institute, argues that human numbers pose no threat to human survival or the environment but that efforts to control population do threaten human freedom and worth.

In 1798 the British economist Thomas Malthus published his *Essay on the Principle of Population*. In it, he pointed with alarm at the way the human population grew geometrically (a hockey-stick curve of increase) and at how agricultural productivity grew only arithmetically (a straight-line increase). It was obvious, he said, that the population must inevitably outstrip its food supply and experience famine. Contrary to the conventional wisdom of the time, population growth was not necessarily a good thing. Indeed, it led inexorably to catastrophe. For many years, Malthus was something of a laughingstock. The doom he forecast kept receding into the future as new lands were opened to agriculture, new agricultural technologies appeared, new ways of preserving food limited the waste of spoilage, and the developed nations underwent a "demographic transition" from high birth rates and high death rates to low birth rates and low death rates.

Demographers initially attributed the demographic transition to increasing prosperity and predicted that as prosperity increased in countries whose populations were rapidly growing, birth rates would surely fall. Later, some scholars analyzed the historical data and concluded that the transition had actually preceded prosperity. The two views have contrasting implications for public policy designed to slow population growth—economic aid or family planning aid—but neither has worked very well. In 1994 the UN Conference on Population and Development, which was held in Cairo, Egypt, concluded that better results would follow from improving women's access to education and health care.

Should we be trying to slow or reverse population growth? In the 1968 book *The Population Bomb* (Ballantine Books), Paul R. Ehrlich warned that unrestricted population growth would lead to both human and environmental disaster. But some religious leaders oppose population control because family planning is against God's will. Furthermore, minority groups and developing nations contend that they are unfairly targeted by family planning programs.

The world's human population has grown tremendously. In Malthus's time, there were about 1 billion human beings on earth. By 1950 there were a little over 2.5 billion. In 1999 the tally passed 6 billion. By 2025 it will be over 8 billion. Statistics like these, which are presented in *World Resources 2000-2001*, a biennial report of the World Resources Institute in collaboration with the United Nations Environment and Development Programmes (Oxford University Press, 2000), are positively frightening. By 2050 the UN expects the world population to be between 9 and 10 billion and to still be rising; some estimates peg the 2050 population as high as 12 billion. While global agricultural production has also increased, it has not kept up with rising demand, and—because of the loss of topsoil to erosion, the exhaustion of aquifers for irrigation water, and the high price of energy for making fertilizer (among other things)—the prospect of improvement seems exceedingly slim to many observers.

What will happen to the environment? The earth already faces global climate change, air pollution, resource depletion, loss of species, and more, and the more people there are on the planet, the greater the effect they must have. Some say that current trends suggest that population will level off and even decline before catastrophe strikes, but most observers are not that optimistic. However, there is hope. For example, the United Nations Development Programme, the United Nations Environment Programme, the World Bank, and the World Resources Institute analyzed world ecosystems and concluded that although there are many signs of trouble, once overuse is controlled, many ecosystems can recover. See *World Resources 2000-2001: People and Ecosystems: The Fraying Web of Life* (World Resources Institute, 2000).

In the following selections, Paul R. Ehrlich and Anne H. Ehrlich argue that "population growth may be the paramount force moving humanity inexorably towards disaster." They therefore maintain that it is essential to reduce the impact of population in terms of both numbers and resource consumption. Stephen Moore, on the other hand, argues that human numbers pose no threat to human survival or the environment but that efforts to control population do threaten human freedom and worth.

Paul R. Ehrlich and Anne H. Ehrlich  **YES**

The Population Explosion

Almost everyone has heard about the population explosion, but few people understand its significance. Following is a brief overview of the basic problem caused by the rapid increase in human numbers from roughly one billion people in 1800 to some six times that number less than two centuries later. Half of that growth has occurred just since 1960, and it appears that, at a minimum, two to three billion more people—and possibly several billion beyond that—will be added to the population before growth ends (assuming a disastrous die-off can be averted). The significance of numbers in the billions is often difficult to grasp; suffice it to say the world is annually adding roughly the population equivalent of present day Germany, and that perhaps thirty to fifty more 'Germanys' are likely to be added to the population before reduced birth rates can bring growth to a halt.

I. Why Is the Population Explosion Important?

If one asks this question of an acquaintance, the answer often focuses on crowding. A resident of the San Francisco Bay area might mention the perpetual traffic jams on the freeways. Or there might be some comment about starvation in Africa, or the flow of refugees into California from Mexico and the resultant strain on school and health budgets. While there is a population component in those problems, such answers do not get to the most important consequences of overpopulation. Moreover, debates about the roles of women in society, and particularly battles over whether women should be required by the government to carry fetuses to term, have given many Americans the misimpression that population problems should be viewed mainly as issues of the reproductive rights of individuals. In the extreme, this narrow focus has led the uninitiated to claim that there is no connection between the size of the human population and environmental problems. All these views miss the main point.

A. Population Impact on Life-Support Systems

The overriding reason to care about the population explosion is its contribution to the expanding scale of the human enterprise and thus to humanity's impact on the environmental systems that support civilization. The number of people (*P*), multiplied by per capita affluence (*A*) or consumption, in turn multiplied by an index of the environmental damage caused by the technologies employed to service the consumption (*T*), gives a measure of the environmental impact (*I*) of a society. This is the basic $I = P \times A \times T$ identity, often just called the "*I = PAT* equation." A useful surrogate for the $A \times T$ of the $I = PAT$ equation is per capita energy consumption (E_{pc}), Almost all of a society's most environmentally damaging activities involve the mobilization and use of energy at high levels, including the manufacture and powering of vehicles, machinery, and appliances; constructing and maintaining infrastructure; lighting and heating buildings; converting forests into paper, furniture, and homes; producing inputs for, and processing and distributing outputs from, high-yield agriculture; and so on.

The surrogate formula has some drawbacks, however. At the lowest levels of development, energy use probably underestimates environmental impact. For example, very poor people can cause serious environmental damage by cutting down trees for fuelwood. At the highest development levels, energy use probably overestimates environmental impact: a given amount of energy use in Western Europe, Japan, or the United States undoubtedly provides more benefits and does less damage than the same amount used in Poland or Russia because of much greater efficiency and stricter environmental regulation. Yet, despite these imperfections, for comparisons between nations or for intertemporal comparisons, energy use seems to be *a priori* a reasonable measure that correlates with many types of environmental damage. It certainly is the most readily available statistic with those characteristics.

Employing energy use as the standard, the scale of the human enterprise has grown about twenty-fold since 1850. During that time, per capita energy consumption has risen about five-fold globally, and the population has grown about four-fold. Roughly then, population growth can be considered to be responsible for about 45% of humanity's environmental peril: the combined risks accrued as a result of increasing worldwide environmental impacts. The risks arise from human-caused worldwide changes such as widespread habitat destruction (e.g., deforestation, desertification, urban construction), alteration of the composition and geochemical processes of the atmosphere (e.g., the addition of excess greenhouse gases, depletion of stratospheric ozone, generation of air pollution), overdrafts of groundwater, soil depletion and erosion, water pollution, disruption of the hydrologic, carbon, and nitrogen cycles (among others), and general toxification of the planet. These and many other factors combine into an unprecedented assault on the life-support systems of civilization: the global cycles and natural ecosystems that supply indispensable goods and services to humanity.

These mostly unappreciated but indispensable benefits include the maintenance of the quality of the atmosphere, regulation of the climate, provision of food from the sea, replenishment of soils, control of pests, and other vital underpinnings of agriculture, production of timber, medicines, and myriad other industrial materials, and regulation of freshwater flows (including controlling floods and droughts) and other forms of weather amelioration. Natural ecosystems also maintain a vast genetic library from which humanity has already derived all manner of things, including domesticated plants and animals, and which is essential to their continued usefulness.

The $I = PAT$ equation carries an especially important lesson for Americans. It is customary to think of poor nations as overpopulated compared to rich ones, but in terms of global environmental impact, exactly the opposite is true. It is true that most European nations and Japan have greatly slowed, halted, or even reversed their population growth, while most developing nations continue to expand their population sizes at rates of 1.5% to 3.5% per annum.

But when consumption, the $A \times T$ factor (E_{pc}), is considered, an entirely different picture emerges. Thus, around 1990, the average American used some 11,000 watts (11.1 kilowatts, kW) per person, more than ten times as much energy as the average citizen of a developing country. In actuality, the gap between the United States and developing nations is often much wider. For example, the United States in 1990 used 195 times as much commercial energy per capita as Madagascar, 20 times that of Zambia, and 13 times that of China. In other cases, it was narrower: eight times that of Malaysia and six times that of Mexico. Using commercial energy as a measure excludes the use of gathered wood, crop residues, and animal wastes as fuel by poor farmers, so the actual per capita energy consumption in very poor countries is somewhat understated. And in some developing nations such as China, Indonesia, and Malaysia, commercial energy use is growing very rapidly. Nevertheless, the overall picture is quite clear.

The United States already has the world's third largest population, 268 million people. China is number one with 1.24 billion, India number two with 970 million, and Indonesia number four with 205 million. Compared to other industrialized countries, the American population is growing at a record rate of more than one percent per year (if immigration is included). When the population figures are added to energy consumption, it is easy to see why the United States can be called the most overpopulated nation.

By assaulting earth's ecosystems, humanity is, in essence, sawing off the limb on which it perches. Population growth is clearly a major force behind the saw. The chances of successfully feeding and otherwise caring for an expanding population are being continuously diminished. That is why all human beings should care deeply about the population explosion and, because of their own disproportionate environmental impacts, why Americans should show particular concern.

II. What Should Be Done About the Population Explosion?

It is a great deal easier to explain why the population explosion should be a critical issue for all of humanity than it is to find one's way through the manifold issues of what ought to be done about it.

A. Basic Goals

The easiest answer to the question above is move as rapidly as is humanely possible toward an optimum sustainable population size. But this vague answer immediately raises a series of obvious questions: What is an optimum sustainable population size? What steps would move society in that direction? How does one establish what is humane? Science can put theoretical bounds on the answer to the first question, since there is a biophysical upper limit on the number of people that could be supported over the long term with a given set of technologies and social (including political and economic) arrangements.

But most people would probably agree that there is a considerable difference between the *largest* sustainable population and an *optimal* sustainable population. Few would find supporting the maximum number of human beings, in a situation somewhat analogous to the way battery chickens are reared, to be optimal. Many would desire varied diets, comfortable homes, opportunities for travel and solitude, uncrowded living conditions, and other amenities, all of which would reduce the number of people that could be sustained. With an approximation of current conditions, it has been estimated that the upper bound of an optimum population, one that would in some sense allow for a maximum *quality* rather than *quantity* of human life over the long term, would be in the vicinity of two billion people, about one third the current number.

That estimate is based largely on patterns of energy use. Today, humanity is using roughly 13,000,000,000,000 watts of energy. In more convenient notation, that is 13×10^{12} watts, or 13 terawatts (tW). With that much energy use, humanity has developed a quite unsustainable civilization. Indeed, it is only able to maintain some 5.8 billion people today, with a billion or so undernourished and in desperate poverty, by using up its natural capital.

The most important forms of natural capital are productive agricultural soils, fossil groundwater, and biodiversity. Soils, which normally are generated on a scale of inches per millennium, are in many places being eroded at rates of inches per decade. At least twenty-five billion tons of topsoil are lost annually, and some estimates range far higher. In many areas, groundwater that accumulated during the ice ages is overdrawn. In the southern high plains of the United States, the Ogallala aquifer is naturally recharged at a rate of about one-half inch per year, but is being pumped at a rate of four to six feet per year to irrigate crop fields. This "mining" of the aquifer produces a net withdrawal about equal to the flow of the Colorado River. Similar overdrafts of groundwater are occurring in many areas of the world, including parts of India and China. Depletion and degradation of natural resources is, of course, a story as old as civilization, what is new is the unprecedentedly colossal and planet-wide scale on which it is occurring today.

The loss of populations, species, and communities of plants, animals, and microorganisms that are working parts of our life-support systems (and thus partly responsible for the delivery of ecosystem services) is the most irreversible loss of all. Just one element of biodiversity, species diversity, is disappearing at a rate estimated to be 1000 to 10,000 times the "background" rate, which is the more or less constant extinction rate that biologists presume to occur naturally over time. Populations, another critical element, are disappearing even faster. We are witnessing the greatest biological cataclysm of the last sixty-five million years—since an apparent collision with an extraterrestial object exterminated the dinosaurs and much of the rest of Earth's flora and fauna.

In short, in a 13 tW world, humanity is unable to maintain itself on its natural "income," the sustainable flow of solar energy and cycles of elements in the biosphere. Like a profligate son, humanity is spending its inheritance of capital, in essence bragging each year that it is writing bigger and bigger checks on its "account" while paying no heed to the plummeting "balance." This behavior is supported by diverse claims that fly in the face of all of environmental science: technology can save humanity because resources are infinite, population growth can continue for another seven billion years, and there is no need to worry about the state of the environment.

All the degradation and depletion of natural resources that should be constantly renewed through solar energy and ecosystem functions, as well as the generation of various forms of pollution and the buildup of greenhouse gases in the atmosphere leading to global climate change, are largely being driven by civilization's use of 13 tW of non-solar energy, mostly from fossil fuels. Thus, it would seem that a 4.5 to 6 tW world, given substantial changes in human behavior, *might* be sustainable while providing everyone in a moderately large population with a life of reasonably high quality. But will it be possible to get there without wrecking our life-support systems? Since it seems nearly inevitable that the human population will not stop expanding until it has reached at least eight to ten billion people several decades from now, it is clear that one or two additional centuries would be required for a gradual, humane reduction to an optimum population size, assuming that some way is found to support the gigantic overshoot of carrying capacity.

Energy expert John P. Holdren of the Kennedy School at Harvard University has shown a way that a successful transition to a sustainable pattern of energy use might be achieved. But the Holdren scenario is based on what seem today to be extremely optimistic assumptions. It envisions at least a temporary increase to a 27 tW world—a frightening prospect to those of us watching the trends in a 13 tW world. In the latest version of the Holdren scenario, developing nations would develop fast enough to *increase* their per capita energy use by about 2 percent per year between 1990 and 2040, raising it from 1.0 to 2.5 kilowatts (kW). Simultaneously, the industrialized nations would strive to *reduce* their per capita use through increased efficiency, dropping their energy use per person from 7.5 to 4.1 kW, while maintaining or enhancing the quality of life.

Rich and poor nations would converge on an average per-person energy use of 3 kW during the remainder of the 21st century. Since energy use is a reasonable surrogate for estimating availability of the various physical ingredients of human well-being, the Holdren scenario represents an elimination of international inequity. In our view, that is a key requirement for creating conditions in which any significant amount of biodiversity can be maintained. Meanwhile, in the scenario, the world population peak size of nine billion people would be passed by 2100—a reasonably achievable goal—then a slow decline would begin. During the peak period, total energy use thus would be 9 billion $\times$ 3 kW, or 27 tW. . . .

Remember that this version of Holdren's scenario assumes that the total human population size can be contained at approximately nine billion. The scenario also assumes that a high standard of living can be achieved with a per capita rate of energy use of only one fourth to one third of that now seen in the United States. This assumption seems plausible, based on technologies already in existence. But it might entail redevelopment of the United States into a lifestyle and infrastructure centered around people rather than automobiles, so that virtually everyone could walk or bicycle to their workplaces or work at home using advanced communications systems.

Such a change might seem disastrous to those who think gross national product must grow no matter what, but the change could be highly beneficial for the quality of life. Indeed, depending on assumptions made about major energy sources and end uses in the future—which sources and technologies supply what fraction of energy and for what use—efficiency alone could make available the equivalent of 10 to 40 tW by the year 2050. Holdren's scenario depends on increased efficiency supplying the services of some 40 tW in today's technologies by 2100—the difference between the scenario's 27 tW and the 67.5 tW that would be required to give nine billion people a lifestyle resembling that of Europe or North America in the 1990s, powered by 1990 technologies using 7.5 kW per capita.

Even with such enormous gains in efficiency, it must be emphasized that Holdren's scenario still yields a total energy use more than twice that of 1990, a situation that would produce horrendous environmental impacts unless the mix of energy sources and technologies were substantially different from today's. Fortunately, policy analysts already widely recognize that the mix must be changed (although this has not been generally recognized by decision makers). The main thrusts behind this recognition are the clear limits to readily accessible supplies of petroleum and natural gas, and increasing public opposition to unacceptable environmental risks and tradeoffs such as oil spills in fragile coastal or polar areas or the sacrifice of prime farmland to strip-mine coal. Needless to say, such a shift will be essential if biodiversity is not to suffer catastrophic losses, leaving surviving restoration ecologists struggling to stabilize a weedy world with crippled ecosystem services. . . .

B. Control of Conception

The key to any humane management of human population size is regulation of birth rates. The objective is to avoid a death-rate solution to the population outbreak in which billions of people perish prematurely and in misery. This means that people must have both the knowledge and means to control their reproduction. Human beings have exercised some control over their reproduction for at least thousands and perhaps hundreds of thousands of years. The techniques employed have ranged from crocodile dung suppositories in ancient Egypt to infanticide from hunter-gatherer times up to 1979 in China, and have varied in both their efficacy and social acceptability just as modern techniques do. In the 20th century, the story of birth control in the now-industrialized nations has been one of gradual acceptance of modern forms of contraception, strongly associated with the movement for women's liberation and an assertion of women's rights to determine the number and timing of children they bear.

In the United States and Western Europe, cultural and religious-based biases against contraception are no longer taken seriously by the majority of people. For example, despite the Catholic Church's prohibitions against contraception and abortion, the average completed family sizes in Catholic Italy and Spain are among the smallest in the world (1.2 children in 1997), and both contraception and abortion are widely used in the United States, a country with a substantial Catholic population. In the United States, Catholic women have slightly smaller families than Protestant women. . . .

C. Family Size Preferences

Providing the means to control fertility can help reduce birth rates, since there is still a very large unmet demand for contraception. Increasing accessibility of birth control will do little good, however, unless people choose to limit their family sizes to an average level that eventually will produce zero population growth and then declining populations worldwide. Ideally, completed family sizes everywhere should be reduced to an average in the vicinity of 1.5 children for several generations. Globally, the average today is three. Total fertility rates (TFRs), in essence average family sizes, are much higher in many poor countries: Nigeria has an average of 6.2, Iraq, 5.7, Pakistan, 5.6, and Honduras, 5.2. India's TFR has fallen to 3.5. Yet some countries that have low incomes are doing considerably better: Indonesia's TFR has fallen to 2.9, Costa Rica's to 2.8, Brazil's to 2.5, Thailand's to 1.9, and China's to 1.8. Rates are generally lower in high income countries, with France at 1.7, Japan at 1.5, Germany at 1.3, and Italy at 1.2. The United States remains an outlier among the rich, with a TFR of 2.0.

Many factors appear to influence family size preferences. One is the stage of development of a nation. In the West, both birth and death rates were high in the eighteenth century. As the industrial revolution spread, accompanied by rapid urbanization, rising education rates, and improved public health conditions, first death rates and then birth rates slowly declined.

In non-industrialized nations after World War II, the deployment of very efficacious "death control" technologies, especially antibiotics and the use of diclorodiphenyltrichloroethane (DDT) against mosquitoes that transmit malaria, led to precipitous drops in death rates. Most of the factors that lead to lower birth rates were not present, however, and the result was the post-war population explosion in non-industrial countries. In some regions, fertility remained very high for several decades after death rates had fallen to levels below those that had preceded falling birth rates in more developed societies. Conversely, in other nations, fertility began to fall before much progress had been made in industrializing or raising income levels....

D. Development Economics

This apparently contradictory situation left demographers and economists scratching their heads; if "development" was supposed to be the magic bullet that reduced fertility, why did some underdeveloped societies have falling birth rates, while others, seemingly more advanced, still have high ones? Eventually, it was learned that some important key factors, besides the obvious one of lower infant and child mortalities, are involved in motivating couples to have fewer children. The education of women, and their ability to participate outside of the home, in society, appear to be especially important. Some studies show that when women are educated and economically active, not only do birth rates fall, but their societies benefit broadly and develop more rapidly. The perceived value of children to their parents is also important. When young children contribute to their family's income as farm and household help or in commercial activity, rather than requiring financial support while in school, parents are motivated to have more of them....

III. What Needs to Be Done

The cultural norms for family size in any society can be modified under appropriate circumstances. Indeed, only a handful of countries still have pre-industrial birth rates of six or more children per family. Those countries are generally characterized by extreme poverty. Typically, they lack basic amenities such as safe drinking water, adequate food, health care, and education. Thus, the prerequisites for success in ending population growth, remarkably akin to those required for successful modernization, are the following: basic health care and sanitation, education and economic opportunities for both sexes, local control over supporting resources, and fair and responsible government.

There are some societies where the basic amenities, including education for women, are provided and the birth rate still remains high. Typically, these are societies hindered by a powerful tradition of male dominance and female dependence. Often the family planning programs are weak or nonexistent. The provision of family planning information and services is important for enabling people to implement their family size decisions, backed up by safe abortion services where the culture permits.

These basic requirements have for decades been reasonably well fulfilled in most of the industrialized world, and birth rates accordingly are almost uniformly low, or in some cases well below replacement. Many other nations have recently reached or passed transitional stages of development during which these prerequisites were largely met, and they too have fertility rates at or near replacement. However, since their fertility has only recently fallen, a good deal of momentum remains in the population structure, and substantial population growth will still occur. Thus, demographers project that China, which reduced its total fertility rate (TFR) from nearly 5.0 in 1970 to about 2.3 by 1980, will grow from today's 1.2 billion to about 1.6 billion before growth stops. Enormous momentum, of course, remains in the developing nations that still have high birth rates; they may more than double their present population sizes before growth ends.

Projections for sub-Saharan Africa, for example, indicate that the region's population of about 600 million in 1997 will more than double by 2025 and continue growing for decades thereafter. Even though some African nations recently have begun to show declines in birth rates, the average family size for the region is still six. Persisting poverty, inadequate food supplies, poor health standards, and high illiteracy rates are typical of the region, which also has a strong tradition of male dominance. Reversing the downward spiral of poverty and hunger in sub-Saharan Africa will require a huge effort to modernize economies, including the agricultural sectors, reverse the degradation of the natural resource base, improve health and reduce infant and child mortality rates, educate millions of people and develop employment opportunities for them, reform corrupt and brutal governments, and promote family planning. Accomplishing all this in the face of continuing rapid population growth and the spread of AIDS and other serious diseases will be no easy task.

Family planning programs succeed best when supported by government policies that promote smaller families. Political leadership in many societies has been shown to have potent effects; enlistment of support from religious leaders and institutions also can be helpful. For example, when Mexico launched its Responsible Parenthood program in the late 1960s, public support of most of the country's bishops was a critical element in its rapid acceptance by Mexicans. In those African nations where family planning is at last being accepted by increasing numbers of women, public support from government leaders has also been an important factor....

B. Overall Strategies

... [T]o avoid the most disastrous possible outcomes of the human predicament, efforts must be made to reduce *all three* of the major factors in the $I = PAT$ identity, not merely to concentrate on population control. If the size of the human population were gradually reduced to two billion people, and all those people attempted to live like Americans of today, then that population would not be sustainable; among other things, it would be using almost twice as much energy as the present global population of nearly six billion.

Thus, in addition to trying to reduce its numbers, and long before it passes its peak population size, civilization should strive to reduce its aggregate consumption and deploy the most environmentally benign technologies possible. Reductions in consumption, however, should first be concentrated in the rich nations. Such reductions are needed not only to lower overall impacts from further population expansion, to which we are demographically committed, but to help offset increasing consumption in developing nations. Most people in poor countries have access to too few resources and too little energy; meeting their needs will inevitably require increases in their per capita consumption. More efficient, environmentally benign technologies, however, could help contain the impacts that would inevitably attend an increase in per capita consumption.

IV. Conclusion

Population growth may be the paramount force moving humanity inexorably towards disaster. Although both overconsumption and the use of needlessly environmentally damaging technologies are major contributors, population growth is one of the most difficult to end rapidly because of its built-in momentum. But overconsumption may prove even more difficult to end. Theoretically, it could be ended in a few decades, as the substantial transformations of the U.S. economy at the beginning and end of World War II demonstrated. Yet what is physically and economically feasible may be socially much more difficult to achieve. Nevertheless, if we care about the world our descendants will inherit, our only responsible choice is to try.

Body Count

At a Washington reception, the conversation turned to the merits of small families. One woman volunteered that she had just read Bill McKibben's environmental tome, *Maybe One,* on the benefits of single-child families. She claimed to have found it "ethically compelling." I chimed in: "Even one child may put too much stress on our fragile ecosystem. McKibben says 'maybe one.' I say, why not none?" The response was solemn nods of agreement, and even some guilt-ridden whispers between husbands and wives.

McKibben's acclaimed book is a tribute to the theories of British economist Thomas Malthus. Exactly 200 years ago, Malthus—the original dismal scientist—wrote that "the power of population is . . . greater than the power in the earth to produce subsistence for man." McKibben's application of this idea was to rush out and have a vasectomy. He urges his fellow greens to do the same—to make single-child families the "cultural norm" in America.

Now, with the United Nations proclaiming that this month we will surpass the demographic milestone of 6 billion people, the environmental movement and the media can be expected to ask: Do we really need so many people? A recent AP headline lamented: "Century's growth leaves Earth crowded—and noisy." Seemingly, Malthus has never had so many apostles.

In a rational world, Malthusianism would not be in a state of intellectual revival, but thorough disrepute. After all, virtually every objective trend is running in precisely the opposite direction of what the widely acclaimed Malthusians of the 1960s—from Lester Brown to Paul Ehrlich to the Club of Rome—predicted. Birth rates around the world are lower today than at any time in recorded history. Global per capita food production is much higher than ever before. The "energy crisis" is now such a distant memory that oil is virtually the cheapest liquid on earth. These facts, collectively, have wrecked the credibility of the population-bomb propagandists.

Yet the population-control movement is gaining steam. It has won the hearts and wallets of some of the most influential leaders inside and outside government today. Malthusianism has evolved into a multi-billion-dollar industry and a political juggernaut.

Today, through the U.S. Agency for International Development (AID), the State Department, and the World Bank, the federal government pumps some

350 million tax dollars a year into population-containment activities. The Clinton administration would be spending at least twice that amount if not for the efforts of two Republican congressmen, Chris Smith of New Jersey and Todd Tiahrt of Kansas, who have managed to cut off funding for the most coercive birth-reduction initiatives.

Defenders of the U.N. Population Fund (UNFPA) and other such agencies insist that these programs "protect women's reproductive freedom," "promote the health of mothers," and "reduce infant mortality." Opponents of international "family planning," particularly Catholic organizations, are tarred as anti-abortion fanatics who want to deprive poor women of safe and cheap contraception. A 1998 newspaper ad by Planned Parenthood, entitled "The Right Wing Coup in Family Planning," urged continued USAID funding by proclaiming: "The very survival of women and children is at stake in this battle." Such rhetoric is truly Orwellian, given that the entire objective of government-sponsored birth-control programs has been to invade couples' "reproductive rights" in order to limit family size. The crusaders have believed, from the very outset, that coercion is necessary in order to restrain fertility and avert global eco-collapse.

The consequences of this crusade are morally atrocious. Consider the one-child policy in China. Some 10 million to 20 million Chinese girls are demographically "missing" today because of "sex-selective abortion of female fetuses, female infant mortality (through infanticide or abandonment), and selective neglect of girls ages 1 to 4," according to a 1996 U.S. Census Bureau report. Girls account for over 90 percent of the inmates of Chinese orphanages —where children are left to die from neglect.

Last year, Congress heard testimony from Gao Xiao Duan, a former Chinese administrator of the one-couple, one-child policy. Gao testified that if a woman in rural China is discovered to be pregnant without a state-issued "birth-allowed certificate," she typically must undergo an abortion—no matter how many months pregnant she is. Gao recalled, "Once I found a woman who was nine months' pregnant but did not have a birth-allowed certificate. According to the policy, she was forced to undergo an abortion surgery. In the operating room, I saw how the aborted child's lips were sucking, how its limbs were stretching. A physician injected poison into its skull, and the child died and was thrown into the trash can."

The pro-choice movement is notably silent about this invasion of women's "reproductive rights." In 1989, Molly Yard, of the National Organization for Women, actually praised China's program as "among the most intelligent in the world." Stanford biologist Paul Ehrlich, the godfather of today's neo-Malthusian movement, once trumpeted China's population control as "remarkably vigorous and effective." He has congratulated Chinese rulers for their "grand experiment in the management of population."

Last summer, Lisa McRee of *Good Morning America* started an interview with Bill McKibben by asking, in all seriousness, "Is China's one-child policy a good idea for every country?" She might as well have asked whether every country should have gulags.

Gregg Easterbrook, writing in the Nov. 23, 1998 *New Republic*, correctly lambasted China for its "horrifying record on forced abortion and sterilization." But even the usually sensible Easterbrook offered up a limp apology for the one-child policy, writing that "China, which is almost out of arable land, had little choice but to attempt some degree of fertility constraint." Hong Kong has virtually no arable land, and 75 times the population density of mainland China, but has one of the best-fed populations in the world.

These coercive practices are spreading to other countries. Brian Clowes writes in the *Yale Journal of Ethics* that coercion has been used to promote family planning in at least 35 developing countries. Peru has started to use sterilization as a means of family planning, and doctors have to meet sterilization quotas or risk losing their jobs. The same is true in Mexico.

In disease-ridden African countries such as Nigeria and Kenya, hospitals often lack even the most rudimentary medical care, but are stocked to the rafters with boxes of contraceptives stamped "UNFPA" and "USAID." UNFPA boasts that, thanks to its shipments, more than 80 percent of the women in Haiti have access to contraceptives; this is apparently a higher priority than providing access to clean water, which is still unavailable to more than half of the Haitian population.

Population-control groups like Zero Population Growth and International Planned Parenthood have teamed up with pro-choice women in Congress—led by Carolyn Maloney of New York, Cynthia McKinney of Georgia, and Connie Morella of Maryland—to try to secure $60 million in U.S. funding for UNFPA over the next two years. Maloney pledges, "I'm going to do whatever it takes to restore funding for [UNFPA]" this year.

Support for this initiative is based on two misconceptions. The first is the excessively optimistic view that (in the words of a *Chicago Tribune* report) "one child zealotry in China is fading." The Population Research Institute's Steve Mosher, an authority on Chinese population activities, retorts, "This fantasy that things are getting better in China has been the constant refrain of the one-child apologists for at least the past twenty years." In fact, after UNFPA announced in 1997 that it was going back into China, state councillor Peng Peiyun defiantly announced, "China will not slacken our family-planning policy in the next century."

The second myth is that UNFPA has always been part of the solution, and has tried to end China's one-child policy. We are told that it is pushing Beijing toward more "female friendly" family planning. This, too, is false. UNFPA has actually given an award to China for its effectiveness in population-control activities—activities far from female-friendly. Worse, UNFPA's executive director, Nafis Sadik, is, like her predecessors, a longtime apologist for the China program and even denies that it is coercive. She is on record as saying—falsely—that "the implementation of the policy is purely voluntary. There is no such thing as a license to have a birth."

Despite UNFPA's track record, don't be surprised if Congress winds up re-funding it. The past 20 years may have demonstrated the intellectual bankruptcy of the population controllers, but their coffers have never been more flush.

American billionaires, past and present, have devoted large parts of their fortunes to population control. The modern-day population-control movement dates to 1952, when John D. Rockefeller returned from a trip to Asia convinced that the teeming masses he saw there were the single greatest threat to the earth's survival. He proceeded to divert hundreds of millions of dollars from his foundation to the goal of population stabilization. He was followed by David Packard (co-founder of Hewlett-Packard), who created a $9 billion foundation whose top priority was reducing world population. Today, these foundations are joined by organizations ranging from Zero Population Growth (ZPG) to Negative Population Growth (which advocates an optimal U.S. population size of 150 million–120 million fewer than now) to Planned Parenthood to the Sierra Club. The combined budget of these groups approaches $1 billion.

These organizations tend to be extremist. Take ZPG. Its board of directors passed a resolution declaring that "parenthood is not an inherent right but a privilege" granted by the state, and that "every American family has a right to no more than two children."

"Population growth is analogous to a plague of locusts," says Ted Turner, a major source of population-movement funding. "What we have on this earth today is a plague of people. Nature did not intend for there to be as many people as there are." Turner has also penned "The Ted Commandments," which include "a promise to have no more than two children or no more than my nation suggests." He recently reconsidered his manifesto, and now believes that the voluntary limit should be even lower—just *one* child. In Turner's utopia, there are no brothers, sisters, aunts, or uncles.

Turner's $1 billion donation to the U.N. is a pittance compared with the fortunes that Warren Buffett (net worth $36 billion) and Bill Gates (net worth roughly $100 billion) may bestow on the cause of population control. Buffett has announced repeatedly that he views overpopulation as one of the greatest crises in the world today. Earlier this year, Gates and his wife contributed an estimated $7 billion to their foundation, of which the funding of population programs is one of five major initiatives.

This is a massive misallocation of funds, for the simple reason that the overpopulation crisis is a hoax. It is true that world population has tripled over the last century. But the explanation is both simple and benign: First, life expectancy—possibly the best overall numerical measure of human well-being—has almost doubled in the last 100 years, and the years we are tacking on to life are both more active and more productive. Second, people are wealthier—they can afford better health care, better diets, and a cleaner environment. As a result, infant-mortality rates have declined nearly tenfold in this century. As the late Julian Simon often explained, population growth is a sign of mankind's greatest triumph—our gains against death.

We are told that this good news is really bad news, because human numbers are soon going to bump up against the planet's "carrying capacity." Pessimists worry that man is procreating as uncontrollably as John B. Calhoun's famous Norwegian rats, which multiply until they die off from lack of sustenance. Bill McKibben warns that "we are adding another New York City every month, a Mexico every year, and almost another India every decade."

But a closer look shows that these fears are unfounded. Fact: If every one of the 6 billion of us resided in Texas, there would be room enough for every family of four to have a house and one-eighth of an acre of land—the rest of the globe would be vacant. (True, if population growth continued, some of these people would eventually spill over into Oklahoma.)

In short, the population bomb has been defused. The birth rate in developing countries has plummeted from just over 6 children per couple in 1950 to just over 3 today. The major explanation for smaller family sizes, even in China, has been economic growth. The Reaganites were right on the mark when, in 1984, they proclaimed this truth to a distraught U.N. delegation in Mexico City. (The policy they enunciated has been memorably expressed in the phrase "capitalism is by far the best contraceptive.") The fertility rate in the developed world has fallen from 3.3 per couple in 1950 to 1.6 today. These low fertility rates presage declining populations. If, for example, Japan's birth rate is not raised at some point, in 500 years there will be only about 15 Japanese left on the planet.

Other Malthusian worries are similarly wrongheaded. Global food prices have fallen by half since 1950, even as world population has doubled. The dean of agricultural economists, D. Gale Johnson of the University of Chicago, has documented "a dramatic decline in famines" in the last 50 years. Fewer than half as many people die of famine each year now than did a century ago—despite a near-quadrupling of the population. Enough food is now grown in the world to provide every resident of the planet with almost four pounds of food a day. In each of the past three years, global food production has reached new heights.

Overeating is fast becoming the globe's primary dietary malady. "It's amazing to say, but our problem is becoming overnutrition," Ho Zhiqiuan, a Chinese nutrition expert, recently told *National Geographic*. "Today in China obesity is becoming common."

Millions are still hungry, and famines continue to occur—but these are the result of government policies or political malice, not inadequate global food production. As the International Red Cross has reported, "the loss of access to food resources [during famines] is generally the result of intentional acts" by governments.

Even if the apocalyptic types are correct and population grows to 12 billion in the 21st century, so what? Assuming that human progress and scientific advancement continue as they have, and assuming that the global march toward capitalism is not reversed, those 12 billion people will undoubtedly be richer, healthier, and better fed than the 6 billion of us alive today. After all, we 6 billion are much richer, healthier, and better fed than the 1 billion who lived in 1800 or the 2 billion alive in 1920.

The greatest threat to the planet is not too many people, but too much statism. The Communists, after all, were the greatest polluters in history. Economist Mikhail Bernstam has discovered that market-based economies are about two to three times more energy-efficient than Communist, socialist, Maoist, or "Third Way" economies. Capitalist South Korea has three times the population density of socialist North Korea, but South Koreans are well fed while 250,000 North Koreans have starved to death in the last decade.

Government-funded population programs are actually counterproductive, because they legitimize command-and-control decision-making. As the great development economist Alan Rufus Waters puts it, "Foreign aid used for population activities gives enormous resources and control apparatus to the local administrative elite and thus sustains the authoritarian attitudes corrosive to the development process."

This approach usually ends up making poor people poorer, because it distracts developing nations from their most pressing task, which is market reform. When Mao's China established central planning and communal ownership of agriculture, tens of millions of Chinese peasants starved to death. In 1980, after private ownership was established, China's agricultural output doubled in just ten years. If Chinese leaders over the past 30 years had concentrated on rapid privatization and market reform, it's quite possible that economic development would have decreased birth rates every bit as rapidly as the one-child policy.

The problem with trying to win this debate with logic and an arsenal of facts is that modern Malthusianism is not a scientific theory at all. It's a religion, in which the assertion that mankind is overbreeding is accepted as an article of faith. I recently participated in a debate before an anti-population group called Carrying Capacity Network, at which one scholar informed me that man's presence on the earth is destructive because *Homo sapiens* is the only species without a natural predator. It's hard to argue with somebody who despairs because mankind is alone at the top of the food chain.

At its core, the population-control ethic is an assault on the principle that every human life has intrinsic value. Malthusian activists tend to view human beings neither as endowed with intrinsic value, nor even as resources, but primarily as consumers of resources. No wonder that at last year's ZPG conference, the Catholic Church was routinely disparaged as "our enemy" and "the evil empire."

The movement also poses a serious threat to freedom. Decisions on whether to have children—and how many—are among are the most private of all human choices. If governments are allowed to control human reproduction, virtually no rights of the individual will remain inviolable by the state. The consequence, as we have seen in China, is the debasement of human dignity on a grand scale.

Another (true) scene from a party: A radiant pregnant woman is asked whether this is her first child. She says, no, in fact, it is her sixth. Yuppies gasp, as if she has admitted that she has leprosy. To have three kids—to be above replacement level—is regarded by many as an act of eco-terrorism.

But the good news for this pregnant woman, and the millions of others who want to have lots of kids, is that the Malthusians are simply wrong. There is no moral, economic, or environmental case for small families. Period.

If some choose to subscribe to a voluntary one-child policy, so be it. But the rest of us—Americans, Chinese, and everybody else—don't need or want Ted Turner or the United Nations to tell us how many kids to have. Congress should not be expanding "international family planning" funding, but terminating it.

Congress may want to consider a little-known footnote of history. In time, Thomas Malthus realized that his dismal population theories were wrong. He awoke to the reality that human beings are not like Norwegian rats at all. Why? Because, he said, man is "impelled" by "reason" to solve problems, and not to "bring beings into the world for whom he cannot provide the means of support." Amazingly, 200 years later, his disciples have yet to grasp this lesson.

POSTSCRIPT

Is Limiting Population Growth a Key Factor in Protecting the Global Environment?

The "I = PAT" equation discussed by the Ehrlichs is crucial to understanding human population. If we face a crisis, it is not *just* because our numbers are so huge, nor is it *just* because of the damaging technologies those numbers deploy to extract a living from the world. It is a combination of the two, as Joel E. Cohen discusses in *How Many People Can the Earth Support?* (W. W. Norton, 1996). The key to answering Cohen's titular question is human choice, and the choices are ones that must be made in the near future.

Some commentators still stress the role of population by itself; see Werner Fornos, "No Vacancy," *The Humanist* (July/August 1998). Malcolm Potts, in "The Unmet Need for Family Planning," *Scientific American* (January 2000), stresses the need for improved control of human fertility. Andrew R. B. Ferguson, in "Perceiving the Population Bomb," *World Watch* (July/August 2001), sets the maximum sustainable human population at 2.1 billion. Another important writer on the topic is Garrett Hardin, whose influential essay "The Tragedy of the Commons," *Science* (December 13, 1968) describes the consequences of using self-interest alone to guide the exploitation of publicly owned resources, such as air and water.

The Winter 1994 issue of *The Amicus Journal* carries a special section entitled "Population, Consumption and Environment," which includes Paul Harrison's "Sex and the Single Planet: Need, Greed, and Earthly Limits." In "Population Fictions: The Malthusians Are Back in Town," *Dollars & Sense* (September/October 1994), Betsy Hartmann argues that the real problem is not how many people there are but that controls over resource consumption are inadequate.

The *AAAS Atlas of Population and Environment* by Paul Harrison and Fred Pearce (University of California Press, 2002) uses six case studies to provide a useful overview of trends in global population, natural resources, land use, environmental quality, ecosystems, and biodiversity. The United Nations Environmental Programme's *Global Environmental Outlook 3* (Earthscan, 2002), produced as a "global state of the environment report" in preparation for the World Summit on Sustainable Development in Johannesburg, South Africa, states, "Population size to a great extent governs demand for natural resources and material flows. Population growth enlarges the challenge of improving living standards and providing essential social services, including housing, transport, sanitation, health, education, jobs and security. It can also make it harder to deal with poverty."

ISSUE 17

Will Pollution Rights Trading Effectively Control Environmental Problems?

YES: Charles W. Schmidt, from "The Market for Pollution," *Environmental Health Perspectives* (August 2001)

NO: Brian Tokar, from "Trading Away the Earth: Pollution Credits and the Perils of 'Free Market Environmentalism,'" *Dollars & Sense* (March/April 1996)

ISSUE SUMMARY

YES: Freelance science writer Charles W. Schmidt argues that economic incentives such as emissions rights trading offer the most useful approaches to reducing pollution.

NO: Author, college teacher, and environmental activist Brian Tokar maintains that pollution credits and other market-oriented environmental protection policies do nothing to reduce pollution while transferring the power to protect the environment from the public to large corporate polluters.

Following World War II the United States and other developed nations experienced an explosive period of industrialization accompanied by an enormous increase in the use of fossil fuel energy sources and a rapid growth in the manufacture and use of new synthetic chemicals. In response to growing public concern about the pollution and other forms of environmental deterioration resulting from this largely unregulated activity, the U.S. Congress passed the National Environmental Policy Act of 1969. This legislation included a commitment on the part of the government to take an active and aggressive role in protecting the environment. The next year the Environmental Protection Agency (EPA) was established to coordinate and oversee this effort. During the next two decades an unprecedented series of legislative acts and administrative rules were promulgated, placing numerous restrictions on industrial and commercial activities that might result in the pollution, degradation, or contamination of land, air, water, food, and the workplace.

Such forms of regulatory control have always been opposed by the affected industrial corporations and developers as well as by advocates of a free-market policy. More moderate critics of the government's regulatory program recognize that adequate environmental protection will not result from completely voluntary policies. They suggest that a new set of strategies is needed. Arguing that "top down, federal, command and control legislation" is not an appropriate or effective means of preventing ecological degradation, they propose a wide range of alternative tactics, many of which are designed to operate through the economic marketplace. The first significant congressional response to these proposals was the incorporation of tradable pollution emission rights into the 1990 Clean Air Act amendments as a means for achieving the set goals for reducing acid rain–causing sulfur dioxide emissions. More recently, the 1997 international negotiations on controlling global warming in Kyoto, Japan, resulted in a protocol that includes emissions trading as one of the key elements in the plan to limit the atmospheric buildup of greenhouse gases.

Despite past difficulties in obtaining compliance with or enforcing strict statutory pollution limits, the idea of using such market-based strategies as the trading of pollution control credits or the imposition of pollution taxes has won limited acceptance from some major mainstream environmental organizations. Many environmentalists, however, continue to oppose the idea of allowing anyone to pay to pollute, either on moral grounds or because they doubt that these tactics will actually achieve the goal of controlling pollution. Diminishment of the acid rain problem is often cited as an example of how well emission rights trading can work, but in "Dispelling the Myths of the Acid Rain Story," *Environment* (July–August 1998), Don Munton argues that other control measures, such as switching to low-sulfur fuels, deserve much more of the credit for reducing sulfur dioxide emissions.

In "A Low-Cost Way to Control Climate Change," *Issues in Science and Technology* (Spring 1998), Byron Swift argues that the "cap-and-trade" feature of the U.S. Acid Rain Program has been so successful that a similar system for implementing the Kyoto Protocol's emissions trading mandate as a cost-effective means of controlling greenhouse gases should work. In March 2001 the U.S. Senate Committee on Agriculture, Nutrition, and Forestry held a "Hearing on Biomass and Environmental Trading: Opportunities for Agriculture and Forestry," in which witnesses urged Congress to encourage trading for both its economic and its environmental benefits. Richard L. Sandor, chairman and chief executive officer of Environmental Financial Products LLC, said that "200 million tons of CO_2 could be sequestered through soils and forestry in the United States per year. At the most conservative prices of $20–$30 per ton, this could potentially generate $4–$6 billion in additional agricultural income."

In the following selections, Charles W. Schmidt describes the use of economic incentives to motivate corporations to reduce pollution, and he argues that emissions trading schemes represent "the most significant developments" in this area. Brian Tokar has a much more negative assessment of sulfur dioxide pollution credit trading. He argues that such "free-market environmentalism" tactics fail to reduce pollution while turning environmental protection into a commodity that corporate powers can manipulate for private profit.

Charles W. Schmidt

The Market for Pollution

Throughout much of its short history, environmental protection in the United States has been guided by a traditional paradigm based on strict regulatory guidelines for reducing emissions and punishments for noncompliance. Experts credit this traditional approach with improvements in air and water quality evident since the U.S. Environmental Protection Agency (EPA) was created more than 30 years ago. Tough environmental standards imposed under programs such as the Clean Water Act and the Clean Air Act filled a regulatory void and forced industries to cut their emissions or face heavy fines. Many of the greatest gains were seen with respect to point sources such as smokestacks and effluent pipes that could be easily monitored. But beyond the avoidance of penalties, industries regulated under those so-called command-and-control programs had little motivation to develop advanced pollution control technologies, which produced little economic gain.

Today, many stakeholders believe a more modern framework based on economic incentives that allow companies to profit from achieving environmental goals will build on the achievements of the past and allow for even greater improvements in environmental protection. Types of incentives vary widely, but they all share one thing in common: they attach a monetary value to the act of reducing pollution. In a January 2001 document titled *The United States Experience with Economic Incentives for Protecting the Environment,* the EPA described several types of incentives, including fees and taxes levied on pollutant releases, tax rebates for environmental technologies, and the trading of air emissions permits on the open market.

Attention is increasingly turning to the use of economic incentives in the wake of President George W. Bush's pledge to make them a foundation of his environmental policy. During the 2000 presidential campaign, Bush said that under his watch government would "set high environmental standards and provide market-based incentives to develop new technologies . . . so that Americans could meet and exceed those standards."

Business organizations have responded warmly to the administration's support for incentives. For example, the Business Roundtable, a Washington, D.C.–based nonprofit organization of "CEOs committed to improving public policy," released a statement on 17 May 2001 that "applauds President Bush

From Charles W. Schmidt, "The Market for Pollution," *Environmental Health Perspectives*, vol. 109, no. 8 (August 2001).

for incorporating the use of new technologies, as well as incentives that spur technological innovation, as the cornerstone of the administration's national energy policy."

Among the environmental community, the idea that market instruments could be used to control pollution was initially greeted with skepticism and even hostility. But over time, support has risen to a level that Joseph Goffman, a senior attorney with the public interest group Environmental Defense in Washington D.C., describes as "lukewarm to enthusiastic in many cases."

According to Goffman, economic incentives motivate companies to reduce pollution quickly and to exceed environmental standards whenever possible. This is in contrast to command-and-control approaches, which he says stifle innovation while encouraging polluters to do little more than meet minimum requirements. Under a traditional system, the EPA not only sets environmental standards, it often describes how companies should achieve them—a scenario sometimes described as "technology forcing."

Goffman suggests the downside to this approach is that the EPA usually only sets standards that can be met with current technology. This means companies have to wait for the agency to finish a technology review before either the EPA or the states revise a given standard. "With incentive programs," he says, "you don't have this kind of chicken-and-egg mentality. The agency sets a target and leaves the means of compliance up to industry. Companies want to profit from pollution control, so they invest more resources in technology development." Furthermore, Goffman adds, market forces naturally gravitate toward the least-cost option for reducing pollution, while traditional regulatory strategies lock companies into technologies that become progressively less effective, and thus less attractive, over time.

Most experts suggest it's too soon to gauge where and how incentive programs will grow under the Bush administration. This is because a host of key positions at the EPA and other agencies remain unfilled, and policy directions have yet to be fully clarified. However, Bush's commitment to market forces is undiminished, as indicated by comments from White House spokesperson Marcy Viana, who, referring to the president's position on global warming during an interview on 4 June 2001, said, "[He is] committed to reducing greenhouse gas emissions by drawing on the power of the market and the power of technology."

Emissions Trading Schemes

The most significant developments in incentive programs have occurred in the area of emissions trading, through which air pollutants are viewed as tradable commodities, each with its own regional, national, and even international markets. In an emissions trading program, companies that emit less than their assigned limits, or caps, of a pollutant can sell residual allowances on the open market or bank them for future transactions. This gives other, higher-polluting facilities a choice: either buy allowances and continue releasing the same pollutant or clean their own emissions—whichever is cheaper. The only stipulation is that regional environmental quality continue to meet mandated standards.

These so-called cap-and-trade schemes aren't new. The best-known example is the Acid Rain Program established under the Clean Air Act amendments of 1990, which allows electric utilities to trade allowance credits in sulfur dioxide (SO_2). Many experts point to this initiative, which achieved dramatic reductions in SO_2 at lower costs than expected, as an emissions trading success story. The EPA estimates that since the program was formalized in 1995, annual emissions of SO_2 have fallen by 4 million tons, while rainfall acidity in the Northeast has dropped by 25%. Dallas Burtraw, a senior fellow at Resources for the Future in Washington, D.C., says the program works well because it's simple, it sets firm environmental targets, it keeps transaction costs to a minimum, and it's transparent—meaning that information on available allowances and credit trades is freely available to the public.

The success of the Acid Rain Program has fueled the development of similar initiatives within the private sector. Undeterred by President Bush's rejection of the Kyoto Protocol, a diverse group of 34 major companies called the Chicago Climate Exchange (CCX) recently announced an emissions trading scheme for carbon dioxide and other greenhouse gases. Boasting high-profile members such as BP, Ford Motor Company, DuPont, and International Paper, this effort aims to reduce greenhouse gas emissions to 5% below 1999 levels by 2005. The CCX's role will be similar to that of an organized commodity exchange— it will establish the requisite technical infrastructure, common standards, and a computerized platform through which participants can trade in emissions reductions.

Richard Sandor, project leader at the CCX, points to the following hypothetical trade as an example of how the system will work: Two companies, a manufacturer with advanced pollution control technology and a power plant with older controls, agree to cut their combined emissions of greenhouse gases by three tons each for a total of six tons. Taking advantage of its superior technology, the manufacturer can cut its own emissions by five tons at minimal cost while the power plant can only reduce its own emissions cost-efficiently by one ton. But by purchasing the rights to the additional two tons from the manufacturer, the power plant pays for another company to reduce greenhouse gases on its behalf. In this win–win situation, the manufacturer takes in revenues for reducing pollution while the power plant avoids higher costs by passing off its emissions reductions agreement to another source.

According to Sandor, the CCX will facilitate trades among seven midwestern states that together comprise the fourth-largest trading bloc in the world. The CCX also plans to include Brazil as a member, indicating the organization hopes to achieve an international presence. Says Sandor, "We've had a fantastic response from industry. We expect to be in the design phase for 12 months and to begin trading by 2002."

The states have also gotten into the game. In Southern California, a cap-and-trade program known as the Regional Clean Air Incentives Market, or RECLAIM, is being used to control SO_2 and nitrogen oxide (NO_x) air emissions from 360 industrial facilities, including power plants, in Los Angeles and the San Bernardino Valley. A coalition known as the Ozone Transport Commission, comprising the environmental agencies from 13 northeastern and midwestern states and the federal EPA, has developed a cap-and-trade program for NO_x. And elsewhere, in Chicago, a cap-and-trade program for volatile organic compounds was established by the Illinois EPA in early 2000.

The states have, for the most part, had a measure of success with these programs. The Ozone Transport Commission announced on 10 May 2001 that NO_x emissions for 1999 and 2000 were less than half those reported in 1990, before the cap-and-trade system was implemented. California's RECLAIM system has been in operation since 1993 but is just now beginning to demonstrate results. The reason for the delay, says Sam Atwood, spokesperson for the Diamond Bar–based South Coast Air Quality Management District, which coordinates RECLAIM, is that state-mandated "allocations" (a state term that defines the emissions that can be traded under the cap) for SO_2 and NO_x have only recently been set at levels below actual emissions released by industry. For several years after the program was initiated, facilities regulated under RECLAIM were allowed to emit SO_2 and NO_x at unusually high levels to cushion the economic shock of a recession that took place during the early 1990s. "By dropping the allocation levels below real emissions, we're just starting to cross over to the point where the incentive begins to kick in," says Atwood. "This is when we expect to see voluntary improvements in technology."

The Question of Mobile Sources

In a recent and somewhat controversial trend, emissions trading schemes have begun incorporating mobile sources, such as cars and trucks. Under this approach, stationary sources such as factories can obtain emission credits from regulators by paying to have old, highly polluting vehicles taken off the road. For example, RECLAIM recently issued a rule allowing stationary sources to receive mobile source credits by replacing diesel-fueled heavy-duty vehicles with cleaner-running alternatives.

Burtraw suggests this practice provides a major opportunity for cost savings. "It can be a lot less expensive to reduce emissions from mobile sources than stationary sources," he explains. But he concedes that adding mobile sources to the mix doesn't come without its own unique set of challenges. "People are all too willing to bring in an old lemon that barely runs so they can collect $500 from a utility company," he says. In a case like this, the emissions reduction is negligible because the car isn't driveable anyway.

Goffman says programs that include mobile sources need to incorporate safeguards to prevent this kind of abuse. The challenges exist, he says, but solutions are available if the systems are well designed at the outset. The South Coast Air Quality Management District, for example, only agrees to pay credits for cars that could continue running for three years or more.

Trading Issues

Despite a generally positive response from the stakeholder community, emissions trading still raises a number of important concerns. Perhaps the greatest worry is that it might lead to "hot spots," or areas of high pollutant exposure. A company that cuts its emissions in half might help reduce average air pollution concentrations in a particular region, but this means little to those who live close to an older facility that buys credits rather than upgrading its pollution control technology.

John Walke, director of clean air projects with the National Resources Defense Council in Washington, D.C., suggests that environmental justice problems could arise if the dirtier facilities are located close to poor communities. "There are a lot of fundamental issues that need to be addressed with these systems," he says. "One is the extent to which pollution sources may be heavily localized in a particular area. It's important to consider how much pollution the neighboring communities are already saddled with."

And what about facilities located upwind of residential communities? Should they be allowed to purchase air pollution credits if downwind populations don't experience the benefit of cleaner emissions? Experts suggest the answer is no, and that hot spots can be avoided with effective planning. Suellen Keiner, director of the Center for the Economy and the Environment at the National Academy of Public Administration, a public interest group based in Washington, D.C., says potential solutions include discouraging trades across long distances and on-site review of credit uses to protect against hot spots.

Another incentive category that tends to trouble environmentalists is "open market" emissions trading, which is a scheme developed by the EPA in 1995. Unlike cap-and-trade programs, neither the overall sectors nor the individual trading sources regulated under an open market trading system are subject to a cap. Rather, any source that finds that its actual rate of emissions is below permitted levels for even a short time is eligible for credit that it can save for later or sell to another source. A chief concern is that under these schemes industry sets the standard for emissions allowances—not the regulatory agency. This is critical, given widespread agreement among stakeholders that health-protective standards should be set by the government on behalf of the public, while the means of compliance is left to the regulated community.

Burtraw says monitoring emissions under an open market system is particularly challenging. "Unlike cap-and-trade programs, which are often targeted toward large stationary sources that can be monitored at the stack, open trading is geared toward smaller sources, for example dry cleaners," he explains. "It's difficult and expensive to monitor actual emissions from these sources, so they tend to be estimated based on economic activity and the use of a given technology. On paper, open market trading seems promising, but in practice monitoring is often poor, and emissions inventories are weak."

Responding to New Jersey's announcement of an open market trading system for NO_x, approved by the EPA in July 2001, Environmental Defense called on the agency to withhold additional pending approvals in states including Michigan, New Hampshire, and Illinois. Also critical of open market trading is the Washington, D.C.–based organization Public Employees for Environmental Responsibility. This group, which says it represents anonymous EPA employees who fear the repercussions of speaking out publicly, issued a white paper in June 2000 called *Trading Thin Air* in which they claim that state and federal agencies don't have the ability to monitor these programs. According to the paper, open market trading could "cripple enforcement of the Clean Air Act against stationary sources of pollution."

Despite the uproar, many experts believe open market systems will improve over time. "I do have a healthy dose of skepticism about open market trading," says Burtraw. "It isn't based on sound policy and shouldn't be used on a wide scale. But I also see it as a way to include in trading programs a variety of smaller sources of emissions for which there do not exist emission inventories. At best, open market trading should be viewed as a transitional stepping stone to some better-developed institution that will emerge in the future."

Outlook for the Future

When applied to the nation as a whole, the EPA suggests in its April 2001 report that "the potential savings from widespread use of economic incentives . . . could be almost one-fourth of the approximately $200 billion per year currently spent on environmental pollution control in the United States." In applying these tools, the EPA recommends that regulators consider their use in the context of political acceptability, potential for stimulating technological improvements, and enforceability. A number of important questions need to be considered. How many sources are there for each pollutant? Does a unit of pollution from each source have the same health and ecologic impact regardless of where it's released? Who's being affected by the pollution, and will the program reduce these impacts?

A key point raised by Burtraw is that incentives are a tool—not a solution. "You can compare incentives to a hammer," he says. "You can use a hammer to build a house, or you can use it to pull out the nails. This is the big issue we're facing now—if we use the incentives to back away from emissions reductions, then we're using the hammer to pull out the nails. But if we use incentives to aggressively pursue emissions reduction in the most cost-effective way, then we're building a stronger house for the future."

Brian Tokar

Trading Away the Earth: Pollution Credits and the Perils of "Free Market Environmentalism"

The Republican takeover of Congress has unleashed an unprecedented assault on all forms of environmental regulation. From the Endangered Species Act to the Clean Water Act and the Superfund for toxic waste cleanup, laws that may need to be strengthened and expanded to meet the environmental challenges of the next century are instead being targeted for complete evisceration.

For some activists, this is a time to renew the grassroots focus of environmental activism, even to adopt a more aggressively anti-corporate approach that exposes the political and ideological agendas underlying the current backlash. But for many, the current impasse suggests that the movement must adapt to the dominant ideological currents of the time. Some environmentalists have thus shifted their focus toward voluntary programs, economic incentives and the mechanisms of the "free market" as means to advance the cause of environmental protection. Among the most controversial, and widespread, of these proposals are tradeable credits for the right to emit pollutants. These became enshrined in national legislation in 1990 with President George Bush's amendments to the 1970 Clean Air Act.

Even in 1990, "free market environmentalism" was not a new phenomenon. In the closing years of the 1980s, an odd alliance had developed among corporate public relations departments, conservative think tanks such as the American Enterprise Institute, Bill Clinton's Democratic Leadership Council (DLC), and mainstream environmental groups such as the Environmental Defense Fund. The market-oriented environmental policies promoted by this eclectic coalition have received little public attention, but have nonetheless significantly influenced debates over national policy.

Glossy catalogs of "environmental products," television commercials featuring environmental themes, and high profile initiatives to give corporate officials a "greener" image are the hallmarks of corporate environmentalism in the 1990s. But the new market environmentalism goes much further than these showcase efforts. It represents a wholesale effort to recast environmental protection based on a model of commercial transactions within the marketplace.

From Brian Tokar, "Trading Away the Earth: Pollution Credits and the Perils of 'Free Market Environmentalism,'" *Dollars & Sense* (March/April 1996). Copyright © 1996 by Economic Affairs Bureau, Inc. Reprinted by permission. *Dollars & Sense* is a progressive economics magazine published six times a year.

"A new environmentalism has emerged," writes economist Robert Stavins, who has been associated with both the Environmental Defense Fund and the DLC's Progressive Policy Institute, "that embraces ... market-oriented environmental protection policies."

Today, aided by the anti-regulatory climate in Congress, market schemes such as trading pollution credits are granting corporations new ways to circumvent environmental concerns, even as the same firms try to pose as champions of the environment. While tradeable credits are sometimes presented as a solution to environmental problems, in reality they do nothing to reduce pollution —at best they help businesses reduce the costs of complying with limits on toxic emissions. Ultimately, such schemes abdicate control over critical environmental decisions to the very same corporations that are responsible for the greatest environmental abuses.

How It Works, and Doesn't

A close look at the scheme for nationwide emissions trading reveals a particular cleverness; for true believers in the invisible hand of the market, it may seem positively ingenious. Here is how it works: The 1990 Clean Air Act amendments were designed to halt the spread of acid rain, which has threatened lakes, rivers and forests across the country. The amendments required a reduction in the total sulfur dioxide emissions from fossil fuel burning power plants, from 19 to just under 9 million tons per year by the year 2000. These facilities were targeted as the largest contributors to acid rain, and participation by other industries remains optional. To achieve this relatively modest goal for pollution reduction, utilities were granted transferable allowances to emit sulfur dioxide in proportion to their current emissions. For the first time, the ability of companies to buy and sell the "right" to pollute was enshrined in U.S. law.

Any facility that continued to pollute more than its allocated amount (roughly half of its 1990 rate) would then have to buy allowances from someone who is polluting less. The 110 most polluting facilities (mostly coal burners) were given five years to comply, while all the others would have until the year 2000. Emissions allowances were expected to begin selling for around $500 per ton of sulfur dioxide, and have a theoretical ceiling of $2000 per ton, which is the legal penalty for violating the new rules. Companies that could reduce emissions for less than their credits are worth would be able to sell them at a profit, while those that lag behind would have to keep buying credits at a steadily rising price. For example, before pollution trading every company had to comply with environmental regulations, even if it cost one firm twice as much as another to do so. Under the new system, a firm could instead choose to exceed the mandated levels, purchasing credits from the second firm instead of implementing costly controls. This exchange would save money, but in principle yield the same overall level of pollution as if both companies had complied equally. Thus, it is argued, market forces will assure that the most cost-effective means of reducing acid rain will be implemented first, saving the economy billions of dollars in "excess" pollution control costs.

Defenders of the Bush plan claimed that the ability to profit from pollution credits would encourage companies to invest more in new environmental technologies than before. Innovation in environmental technology, they argued, was being stifled by regulations mandating specific pollution control methods. With the added flexibility of tradeable credits, companies could postpone costly controls—through the purchase of some other company's credits—until new technologies became available. Proponents argued that, as pollution standards are tightened over time, the credits would become more valuable and their owners could reap large profits while fighting pollution.

Yet the program also included many pages of rules for extensions and substitutions. The plan eliminated requirements for backup systems on smokestack scrubbers, and then eased the rules for estimating how much pollution is emitted when monitoring systems fail. With reduced emissions now a marketable commodity, the range of possible abuses may grow considerably, as utilities will have a direct financial incentive to manipulate reporting of their emissions to improve their position in the pollution credits market.

Once the EPA actually began auctioning pollution credits in 1993, it became clear that virtually nothing was going according to their projections. The first pollution credits sold for between $122 and $310, significantly less than the agency's estimated minimum price, and by 1995, bids at the EPA's annual auction of sulfur dioxide allowances averaged around $130 per ton of emissions. As an artificial mechanism superimposed on existing regulatory structures, emissions allowances have failed to reflect the true cost of pollution controls. So, as the value of the credits has fallen, it has become increasingly attractive to buy credits rather than invest in pollution controls. And, in problem areas air quality can continue to decline, as companies in some parts of the country simply buy their way out of pollution reductions.

At least one company has tried to cash in on the confusion by assembling packages of "multi-year streams of pollution rights" specifically designed to defer or supplant purchases of new pollution control technologies. "What a scrubber really is, is a decision to buy a 30-year stream of allowances," John B. Henry of Clean Air Capital Markets told the *New York Times,* with impeccable financial logic. "If the price of allowances declines in future years," paraphrased the *Times,* "the scrubber would look like a bad buy."

Where pollution credits have been traded between companies, the results have often run counter to the program's stated intentions. One of the first highly publicized deals was a sale of credits by the Long Island Lighting Company to an unidentified company located in the Midwest, where much of the pollution that causes acid rain originates. This raised concerns that places suffering from the effects of acid rain were shifting "pollution rights" to the very region it was coming from. One of the first companies to bid for additional credits, the Illinois Power Company, canceled construction of a $350 million scrubber system in the city of Decatur, Illinois. "Our compliance plan is based almost totally on purchase of credits," an Illinois Power spokesperson told the *Wall Street Journal.* The comparison with more traditional forms of commodity trading came full circle in 1991, when the government announced that the entire system for trading and auctioning emissions allowances would be administered by the Chicago Board of Trade, long famous for its ever-frantic markets in everything from grain futures and pork bellies to foreign currencies.

Some companies have chosen not to engage in trading pollution credits, proceeding with pollution control projects, such as the installation of new scrubbers, that were planned before the credits became available. Others have switched to low-sulfur coal and increased their use of natural gas. If the 1990 Clean Air Act amendments are to be credited for any overall improvement in the air quality, it is clearly the result of these efforts and not the market in tradeable allowances.

Yet while some firms opt not to purchase the credits, others, most notably North Carolina-based Duke Power, are aggressively buying allowances. At the 1995 EPA auction, Duke Power alone bought 35% of the short-term "spot" allowances for sulfur dioxide emissions, and 60% of the long-term allowances redeemable in the years 2001 and 2002. Seven companies, including five utilities and two brokerage firms, bought 97% of the short term allowances that were auctioned in 1995, and 92% of the longer-term allowances, which are redeemable in 2001 and 2002. This gives these companies significant leverage over the future shape of the allowances market.

The remaining credits were purchased by a wide variety of people and organizations, including some who sincerely wished to take pollution allowances out of circulation. Students at several law schools raised hundreds of dollars, and a group at the Glens Falls Middle School on Long Island raised $3,171 to purchase 21 allowances, equivalent to 21 tons of sulfur dioxide emissions over the course of a year. Unfortunately, this represented less than a tenth of one percent of the allowances auctioned off in 1995.

Some of these trends were predicted at the outset. "With a tradeable permit system, technological improvement will normally result in lower control costs and falling permit prices, rather than declining emissions levels," wrote Robert Stavins and Brad Whitehead (a Cleveland-based management consultant with ties to the Rockefeller Foundation) in a 1992 policy paper published by the Progressive Policy Institute. Despite their belief that market-based environmental policies "lead automatically to the cost-effective allocation of the pollution control burden among firms," they are quite willing to concede that a tradeable permit system will not in itself reduce pollution. As the actual pollution levels still need to be set by some form of regulatory mandate, the market in tradeable allowances merely gives some companies greater leverage over how pollution standards are to be implemented.

Without admitting the underlying irrationality of a futures market in pollution, Stavins and Whitehead do acknowledge (albeit in a footnote to an Appendix) that the system can quite easily be compromised by large companies' "strategic behavior." Control of 10% of the market, they suggest, might be enough to allow firms to engage in "price-setting behavior," a goal apparently sought by companies such as Duke Power. To the rest of us, it should be clear that if pollution credits are like any other commodity that can be bought, sold and traded, then the largest "players" will have substantial control over the entire "game." Emissions trading becomes yet another way to assure that large corporate interests will remain free to threaten public health and ecological survival in their unchallenged pursuit of profit.

Trading the Future

Mainstream groups like the Environmental Defense Fund (EDF) continue to throw their full support behind the trading of emissions allowances, including the establishment of a futures market in Chicago. EDF senior economist Daniel Dudek described the trading of acid rain emissions as a "scale model" for a much more ambitious plan to trade emissions of carbon dioxide and other gases responsible for global warming. This plan was unveiled shortly after the passage of the 1990 Clean Air Act amendments, and was endorsed by then-Senator Al Gore as a way to "rationalize investments" in alternatives to carbon dioxide-producing activities.

International emissions trading gained further support via a U.N. Conference on Trade and Development study issued in 1992. The report was co-authored by Kidder and Peabody executive and Chicago Board of Trade director Richard Sandor, who told the *Wall Street Journal*, "Air and water are simply no longer the 'free goods' that economists once assumed. They must be redefined as property rights so that they can be efficiently allocated."

Radical ecologists have long decried the inherent tendency of capitalism to turn everything into a commodity; here we have a rare instance in which the system fully reveals its intentions. There is little doubt that an international market in "pollution rights" would widen existing inequalities among nations. Even within the United States, a single large investor in pollution credits would be able to control the future development of many different industries. Expanded to an international scale, the potential for unaccountable manipulation of industrial policy by a few corporations would easily compound the disruptions already caused by often reckless international traders in stocks, bonds and currencies.

However, as long as public regulation of industry remains under attack, tradeable credits and other such schemes will continue to be promoted as market-savvy alternatives. Along with an acceptance of pollution as "a by-product of modern civilization that can be regulated and reduced, but not eliminated," to quote another Progressive Policy Institute paper, self-proclaimed environmentalists will call for an end to "widespread antagonism toward corporations and a suspicion that anything supported by business was bad for the environment." Market solutions are offered as the only alternative to the "inefficient," "centralized," "command-and-control" regulations of the past, in language closely mirroring the rhetoric of Cold War anti-communism.

While specific technology-based standards can be criticized as inflexible and sometimes even archaic, critics choose to forget that in many cases, they were instituted by Congress as a safeguard against the widespread abuses of the Reagan-era EPA. During the Reagan years, "flexible" regulations opened the door to widely criticized—and often illegal—bending of the rules for the benefit of politically favored corporations, leading to the resignation of EPA administrator Anne Gorsuch Burford and a brief jail sentence for one of her more vocal legal assistants.

The anti-regulatory fervor of the present Congress is bringing a variety of other market-oriented proposals to the fore. Some are genuinely offered to further environmental protection, while others are far more cynical attempts to replace public regulations with virtual blank checks for polluters. Some have proposed a direct charge for pollution, modeled after the comprehensive pollution taxes that have proved popular in Western Europe. Writers as diverse as Supreme Court Justice Stephen Breyer, American Enterprise Institute economist Robert Hahn and environmental business guru Paul Hawken have defended pollution taxes as an ideal market-oriented approach to controlling pollution. Indeed, unlike tradeable credits, taxes might help reduce pollution beyond regulatory levels, as they encourage firms to control emissions as much as possible. With credits, there is no reduction in pollution below the threshold established in legislation. (If many companies were to opt for substantial new emissions controls, the market would soon be glutted and the allowances would rapidly become valueless.) And taxes would work best if combined with vigilant grassroots activism that makes industries accountable to the communities in which they operate. However, given the rapid dismissal of Bill Clinton's early plan for an energy tax, it is most likely that any pollution tax proposal would be immediately dismissed by Congressional ideologues as an outrageous new government intervention into the marketplace.

Air pollution is not the only environmental problem that free marketeers are proposing to solve with the invisible hand. Pro-development interests in Congress have floated various schemes to replace the Endangered Species Act with a system of voluntary incentives, conservation easements and other schemes through which landowners would be compensated by the government to protect critical habitat. While these proposals are being debated in Congress, the Clinton administration has quietly changed the rules for administering the Act in a manner that encourages voluntary compliance and offers some of the very same loopholes that anti-environmental advocates have sought. This, too, is being offered in the name of cooperation and "market environmentalism."

Debates over the management of publicly-owned lands have inspired far more outlandish "free market" schemes. "Nearly all environmental problems are rooted in society's failure to adequately define property rights for some resource," economist Randal O'Toole has written, suggesting a need for "property rights for owls and salmon" developed to "protect them from pollution." O'Toole initially gained the attention of environmentalists in the Pacific Northwest for his detailed studies of the inequities of the U.S. Forest Service's long-term subsidy programs for logging on public lands. Now he has proposed dividing the National Forest system into individual units, each governed by its users and operated on a for-profit basis, with a portion of user fees allocated for such needs as the protection of biological diversity. Environmental values, from clean water to recreation to scenic views, should simply be allocated their proper value in the marketplace, it is argued, and allowed to out-compete unsustainable resource extraction. Other market advocates have suggested far more sweeping transfers of federal lands to the states, an idea seen by many in the West as a first step toward complete privatization.

Market enthusiasts like O'Toole repeatedly overlook the fact that ecological values are far more subjective than the market value of timber and minerals removed from public lands. Efforts to quantify these values are based on various sociological methods, market analysis and psychological studies. People are asked how much they would pay to protect a resource, or how much money they would accept to live without it, and their answers are compared with the prices of everything from wilderness expeditions to vacation homes. Results vary widely depending on how questions are asked, how knowledgeable respondents are, and what assumptions are made in the analysis. Environmentalists are rightfully appalled by such efforts as a recent Resources for the Future study designed to calculate the value of human lives lost due to future toxic exposures. Outlandish absurdities like property rights for owls arouse similar skepticism.

The proliferation of such proposals—and their increasing credibility in Washington—suggest the need for a renewed debate over the relationship between ecological values and those of the free market. For many environmental economists, the processes of capitalism, with a little fine tuning, can be made to serve the needs of environmental protection. For many activists, however, there is a fundamental contradiction between the interconnected nature of ecological processes and an economic system which not only reduces everything to isolated commodities, but seeks to manipulate those commodities to further the single, immutable goal of maximizing individual gain. An ecological economy may need to more closely mirror natural processes in their stability, diversity, long time frame, and the prevalence of cooperative, symbiotic interactions over the more extreme forms of competition that thoroughly dominate today's economy. Ultimately, communities of people need to reestablish social control over economic markets and relationships, restoring an economy which, rather than being seen as the engine of social progress, is instead, in the words of economic historian Karl Polanyi, entirely "submerged in social relationships."

Whatever economic model one proposes for the long-term future, it is clear that the current phase of corporate consolidation is threatening the integrity of the earth's living ecosystems—and communities of people who depend on those ecosystems—as never before. There is little room for consideration of ecological integrity in a global economy where a few ambitious currency traders can trigger the collapse of a nation's currency, its food supply, or a centuries-old forest ecosystem before anyone can even begin to discuss the consequences. In this kind of world, replacing our society's meager attempts to restrain and regulate corporate excesses with market mechanisms can only further the degradation of the natural world and threaten the health and well-being of all the earth's inhabitants.

POSTSCRIPT

Will Pollution Rights Trading Effectively Control Environmental Problems?

Does pollution rights trading give major corporate polluters too much power to control and manipulate the market for emission credits? This is one of the key issues that continues to inspire developing countries to withhold their endorsement of the greenhouse gas emissions trading provisions of the Kyoto Protocol. The evidence that Tokar cites, which is primarily based on short-term experience with trading in sulfur dioxide pollution credits, does not appear to fully justify the broad generalizations he makes about the inherent perils in market-based regulatory plans. Recent assessments of the Acid Rain Program by the EPA and such organizations as the Environmental Defense Fund are more positive. So is the corporate world: In "Economic Man, Cleaner Planet," *The Economist* (September 29, 2001), it is asserted that economic incentives have proved very useful and that "market forces are only just beginning to make inroads into green policymaking." In March 2002 *Pipeline & Gas Journal* reported that "despite uncertainty surrounding U.S. and international environmental policies, companies in a wide range of industries—especially those in the energy field—are increasingly using emission reduction credits as a way to meet the challenges of cutting greenhouse gas emissions."

The position of those who are ideologically opposed to pollution rights is concisely stated in Michael J. Sandel's op-ed piece "It's Immoral to Buy the Right to Pollute," *The New York Times* (December 15, 1997). In "Selling Air Pollution," *Reason* (May 1996), Brian Doherty supports the concept of pollution rights trading but argues that the kind of emission cap imposed in the case of sulfur dioxide is an inappropriate constraint on what he believes should be a completely free-market program. Richard A. Kerr, in "Acid Rain Control: Success on the Cheap," *Science* (November 6, 1998), contends that emissions trading has greatly reduced acid rain and that the annual cost has been about a tenth of the $10 billion initially forecast. According to Barry D. Solomon and Russell Lee, in "Emissions Trading Systems and Environmental Justice," *Environment* (October 2000), "a significant part of the opposition to emissions trading programs is a perception that they do little to reduce environmental injustice and can even make it worse." However, Byron Swift, in "Allowance Trading and Potential Hot Spots—Good News From the Acid Rain Program," *Environment Reporter* (May 12, 2000), argues that the success of the EPA's emission trading program has not led to the creation of pollution "hot spots" as feared by some critics. On the other hand, EPA researchers recently reported that even though acid emissions were down dramatically, lakes remain affected by past emissions. See Leslie Roberts, "Acid Rain: Forgotten, Not Gone," *U.S. News & World Report* (November 1, 1999).

ISSUE 18

Will Voluntary Action by Industry Reduce the Need for Future Environmental Regulation?

YES: Raymond J. Patchak and William R. Smith, from *ISO 14000 Perspective: So Long! Command and Control . . . Hello! ISO 14000* (December 1998)

NO: Linda Greer and Christopher van Löben Sels, from "When Pollution Prevention Meets the Bottom Line," *Environmental Science and Technology* (vol. 31, no. 9, 1997)

ISSUE SUMMARY

YES: Certified hazardous materials managers Raymond J. Patchak and William R. Smith describe the voluntary ISO 14000 environmental program developed by the International Organization for Standardization. They assert that this initiative will result in increased environmental protection by permitting industry more flexibility in achieving pollution prevention than current "command and control" regulations do.

NO: Environmental Defense Fund scientist Linda Greer and project analyst Christopher van Löben Sels conclude from a case study of a Dow Chemical facility that not even projected cost savings will ensure that a corporation will adopt a voluntary pollution prevention plan.

Although the actions and lifestyles of individuals make significant contributions to environmental degradation, the major contributors to local, regional, and global pollution of air, land, and water are commercial and industrial activities. Until recently, governmental efforts to protect the environment have principally taken the form of legislated or court-mandated restrictions on such activities. Applicants for the siting of large new facilities of all types must usually satisfy federal, state, and local environmental requirements and undergo extensive reviews to assess the potential for serious adverse environmental impacts. Limits have been set on the emissions of a variety of pollutants. Toxicity testing has been prescribed for suspect industrial and agricultural chemicals. And standards have been set for the safe disposal of hazardous wastes.

Until recently, the response of industry to environmental regulation was usually strident and strictly one-dimensional—adamant opposition. Denials of the severity, or even the existence, of the negative environmental impact of industrial development were the first line of defense. If overwhelming evidence rendered such assertions implausible, then corporate spokespeople would often argue that the costs of regulatory compliance were simply too great and that the public should accept some decrease in air or water quality as the inevitable price of an increased standard of living.

As the environmental movement has matured, the arguments of all parties to the debate have become more sophisticated. Independent of their personal views, industrial leaders have responded to increasing environmental concerns among consumers by adopting strategies designed to portray their corporations as ecologically responsible—a posture that is referred to as "green" in the contemporary vernacular. Although they continue to lobby against most proposed environmental regulations, many corporate decision makers now try to implement and publicize alternative, less damaging means of production that will not lower their profit margins.

A significant manifestation of this change of attitude is the highly publicized Responsible Care program, initiated in the United States in 1988 by the Chemical Manufacturers Association (CMA). As the 10 "ethical principles" that were established by the CMA to guide its members indicate, a major thrust of that program is to persuade the industry to voluntarily change its priorities and practices in ways that would enhance environmental protection. The CMA has openly admitted that one of the prime motivating factors for the Responsible Care initiative is the desire to counter the low esteem accorded to the chemical industry by the public. Recent polls indicate that despite the CMA's reports of many positive achievements resulting from the program, it has not made much progress in the public relations arena.

A more global effort along the same lines is code 14000 of the International Organization for Standardization (ISO). This voluntary code, issued in its final form in 1996, establishes environmental management standards. A professed goal of supporters of this effort is to persuade governmental environmental agencies to accept self-regulation by industries that make a commitment to comply with ISO 14000. Thus far, more companies in Europe and Asia than in the United States have adopted the standards.

Raymond J. Patchak and William R. Smith enthusiastically support the market-driven ISO 14000 code. In the following selection, they argue that ISO 14000 is a more flexible approach to environmental protection and that it will prove to be more effective than the current "command and control" regulatory strategy. An implicit assumption of those who argue in favor of voluntary industrial environmental initiatives is that corporations will adopt more ecologically sound development strategies when it is profitable for them to do so. In the second selection, Linda Greer and Christopher van Löben Sels report on their study of a Dow Chemical facility's response to a money-saving pollution protection plan, which challenges that assumption.

Raymond J. Patchak and
William R. Smith

 **YES**

So Long! Command and Control...
Hello! ISO 14000

The regulatory system currently in use by the United States Environmental Protection Agency (EPA), as well as other U.S. regulatory agencies, can be characterized as one of command and control. Developments in the regulatory approach taken by other industrialized nations and the advent of environmental management systems have ignited a process of critical review of our command and control system. Many new regulatory programs instituted in Europe seek to capitalize on the synergistic effect that can be obtained when you get both the regulated community and regulatory agencies working together to solve environmental problems. This idea is the basis for a new international environmental standard called ISO 14000. Let's take a quick look at both of these environmental systems.

Command and Control

The current system of environmental protection in the United States is, as a whole, the most advanced in the world. Environmental regulations have achieved a great deal in turning back the effects of toxic substances on the environment. Laws such as the Toxic Substances Control Act (TSCA) and the Resource Conservation and Recovery Act (RCRA) have been good control measures for the treatment and disposal of hazardous substances and wastes, while laws such as the Clean Air Act (CAA), the Safe Drinking Water Act (SDWA), and the Clean Water Act (CWA) have been instrumental in responding to and reducing toxic substance releases into the environment. One can easily argue that these regulations have contributed significantly to decreasing the proliferation of disasters like the ones at Love Canal, Times Beach and the burning of the Cuyahoga River in Ohio.

Two shortcomings exist in the current command and control approach to environmental protection. One is that regulations are often too rigid and complicated to allow innovation and common sense to play a role in environmental protection. Second, the current regulatory scheme does not foster a cooperative relationship between regulating agencies and industry. As a result, instead of working together to find common solutions to a problem, industry and the

EPA all too often find themselves battling it out in the court system. Although these shortcomings have been reduced in recent years, the fact that the EPA is involved in some 600 lawsuits at any given time should be evidence enough to show that there is a fundamental conflict between business issues and regulatory constraints. The sad part about this situation is that valuable resources are being wasted in the courts rather than responding to the problem at hand.

Industry has made great progress in its efforts to learn how to protect the environment. They have spent hundreds of billions of dollars to decrease the release of toxic substances into the environment, while also developing technologies to reduce or eliminate hazardous waste generation. Many industry groups such as the Chemical Manufacturers Association have developed initiatives, which utilize pollution prevention programs as the cornerstone for their environmental protection efforts. These types of industry-led initiatives, coupled with advances in technology, are changing the way that many companies view and are responding to their environmental obligations.

The EPA Administrator, Carol M. Browner, has supported industry in furthering their environmental protection efforts by agreeing to look at beneficial ways to modify the current system. Ms. Browner's Common Sense Initiative (CSI) is a fundamentally different approach to environmental protection. In the development phase of CSI, the EPA worked with six pilot industries to look at the regulations that are impacting their businesses and to identify ways to change the ones that are complicated and inconsistent. The CSI program is attempting to promote creativity and encourage the development of innovative technologies by allowing industry more flexibility in meeting stronger environmental objectives. In this fashion, the goal of the CSI program is to develop a comprehensive strategy for environmental protection that will result in a cleaner environment at less cost.

What Is ISO 14000?

In the spirit of the Common Sense Initiative, ISO 14000 signifies a new generation in environmental protection. This standard is directed at establishing a link between business and environmental management for all companies no matter their size or purpose. ISO 14000 provides industry with a system to track, manage, and improve environmental performance without conflicting with the business priorities of an operation. Business considerations, flexibility, continual improvement and a simplistic approach to environmental protection are the main differences between ISO 14000 and the command and control system currently in place.

By design, ISO 14000 is a set of simplified environmental management standards that takes into account business and economic considerations while improving on already established environmental protection programs. In this context it should be pointed out that ISO 14000 standards are not intended to supersede current state and federal regulations. In fact, the ISO 14001 Specifications specifies the incorporation of these regulations as an integral part of a facility's EMS program.

In recent years there has been increasing interest and commitment to improve environmental management practices. This interest is demonstrated in collaborative international events such as the NAFTA Montreal Protocol and the mandates set during the 1992 Earth Summit in Rio de Janeiro. The birthplace of ISO 14000 can be traced back to the environmental goals established during the Earth Summit. At this Summit the United Nations convened representatives from the world's industrialized countries to discuss global environmental issues and to develop a means to meet their basic goal of "sustainable development."

At the conclusion of the Earth Summit, the International Organization for Standardization (ISO) set out to develop a group of international standards that meet the goals of the United Nations Conference. ISO established Technical Committee (TC) 207. Its mission, "to establish management tools and systems that organizations can voluntarily use for their own purposes which may, over time, improve their environmental performance levels." In this mission TC 207 created and is developing the ISO 14000 series of standards.

TC 207 is comprised of representatives from the ISO member nations. The United States's representative to TC 207 is the American National Standards Institute (ANSI). ANSI is responsible for fielding U.S. participation in the development of these international standards. They receive assistance within the U.S. through its Technical Advisory Group (TAG), which is administered by the American Society of Testing and Materials (ASTM). The TAG is further divided into Sub-TAGs, which are responsible for reviewing and commenting on progressive drafts of the individual policy documents that compose ISO 14000. Once completed the draft documents are voted on by the entire international membership of TC 207. Upon approval by the membership the documents become final standards.

ISO 14000 is the name for a family of environmental standards. This family is composed of five major components, each of which has one or more policy documents. The five major components are as follows:

- Environmental Management Systems (EMS)
- Environmental Auditing (EA)
- Environmental Performance Evaluation (EPE)
- Life-Cycle Assessment (LCA)
- Environmental Labeling (EL)

The **Environmental Management System (EMS)**, which incorporates policy documents numbered 14001 & 14004, were released as final standards in September of 1996. The EMS is the building block on which the other four components are incorporated. Through the EMS a company can identify its environmental goals and establish a program for monitoring their progress in reaching these goals.

Although individual EMS programs will differ from one organization to the next, they will all consist of seven core components. These include the following: Identification of the company's environmental policy, its objectives and targets, guidelines for identifying environmental concerns and applicable regulations, implementing procedures for controlling process and activities impacting the environment, a program for internal and external auditing of the system, clear assignment of responsibility and accountability, and a requirement for periodic review of the EMS by top level management.

The **Environmental Auditing (EA)** standards are composed of documents 14010 through 14012, and were released for publication as final standards in October 1996. These documents detail the requirements for the general principles of auditing. They include guidelines for conducting audits of EMS programs, and criteria for evaluating the qualifications of environmental auditors.

The **Environmental Performance Evaluation (EPE)** guidance is scheduled for publication in the second quarter of 1999. This standard includes document ISO 14031 as well as a technical report unofficially designated as TR 14032. EPE is an internal management process that uses indicators to provide management of an organization with reliable and verifiable information comparing the organization's past and present environmental performance with Management's environmental performance goals. ISO 14031 defines EPE as a "process to facilitate management decisions regarding an organization's environmental performance by selecting indicators, collecting and analyzing data, assessing information against environmental performance criteria, reporting and communicating, and periodic review and improvement of this process." These documents contain guidance for a process that identifies and quantifies the impact that a company has on the environment. These measurements are made against baseline levels, and the results are evaluated based on improvements from these levels.

The **Life Cycle Assessment (LCA)** guidance can be found in document numbers 14040 through 14043. Although no date has yet been set for final publication of ISO 14043, the ISO 14040 and ISO 14041 documents were released in late 1997 and 1998 respectively. These documents provide a means for determining what effects the products that are manufactured will have on today's environment, as well as that of future generations. This assessment looks at the impacts that are associated with the entire life of a product from raw material acquisition through production, use and disposal.

The **Environmental Labeling (EL)** standard is contained in documents 14020, ISO 14021, ISO 14024 and 14025. The ISO 14020 standard, published in late 1998, provides general principles of environmental labeling. The draft international standards ISO 14021 and ISO 14024 are expected for final release in early 1999. The overall goal of the environmental labeling standards is to provide manufacturers with a tool to assess and verify the accuracy of product environmental claims and to encourage the demand for those products that cause less stress on the environment. The intent of developing these EL standards is to stimulate the growth of market-driven continuous improvement of environmental performance.

In its development of ISO 14000, TC 207 was influenced by the British Standard BS 7750: "Environmental Management System" and the Global Environmental Management Initiative (GEMI). These systems utilize a consensus approach for determining their operational effectiveness. For example, GEMI is composed of 21 leading companies dedicated to fostering environmental excellence in businesses worldwide. It is intended to promote a worldwide business ethic for environmental management and sustainable development. The standard places an emphasis on individual companies and the incorporation of environmental goals into the company's overall corporate goals. These standards also set up a third party review process, to establish whether a company is working towards its goals.

How Will ISO 14000 Benefit an Organization?

ISO 14000 is not a rigid system of regulations, it is a flexible standard designed to fit any size and type of operation. Through this type of system a company can gain many benefits. These benefits can come in the form of improvements in public relations, improvements in management effectiveness, decreases in non-compliance fines, and improvements in marketing and customer relations.

The implementation of an effective ISO 14001 program can provide assurance to consumers that a company is committed to being not just environmentally friendly, but environmentally protective as well. In the same fashion, by maintaining conformance with the standard a company can make a much stronger case for their commitment to protecting the environment. ISO 14001 also makes regulatory compliance a more integral part of the business operation, and thus the likelihood of a violation can be minimized along with the stiff fines and penalties that normally accompany them.

Part of the ISO 14001 standard requires that a program for continuous improvement be developed and implemented throughout the company. The elements of a well orchestrated continuous improvement program can save even a small organization thousands of dollars per year in compliance and pollution control costs. These savings can come about as a part of improved compliance and in the form of reductions in fines and penalties, waste disposal costs, energy consumption and raw materials costs. For instance the implementation of an effective waste minimization program will not only decrease disposal costs, but can also reduce regulatory burden on the operation.

Other regulatory benefits related to implementing the ISO 14001 standard include reducing the number of audits by regulators. The EPA has already proposed, as part of their environmental auditing policy statement, that companies with internal self auditing programs, like ISO 14001, could be subject to less regulatory scrutiny and compliance audits.

Recent history has shown us that a company's market segment can also benefit by establishing a program like ISO 14000. This is exactly what happened when the International Quality Management Practices standard (ISO 9001) was released. As companies established and certified their ISO 9001/2 programs, they became strong believers in the importance of the program and in turn many required their major suppliers to become certified. This move is already underway with regard to ISO 14000, as many companies and government agencies in the U.S. and Europe are already posturing in this direction. Specific details about these companies are included in the next section.

One of the other business incentives for adopting and implementing ISO 14000 is that continuous improvement is crucial for maintaining prosperity and growth. As mentioned earlier in the article, ISO 14001 was developed in part from the United Kingdom BS 7750 Standard. Those companies that have already implemented the BS 7750 standard have reported improvements in the following areas:

- Productivity
- Waste reduction
- Paperwork declines, and
- Public relations

Industries' Acceptance of ISO 14000

As more and more companies have become familiar with the many benefits of ISO 14000 its acceptance has increased steadily. Companies within the U.S. are already lagging behind their European counterparts in taking the first step towards ISO 14000 certification. Many companies in Europe have established EMS programs and are certified to the draft ISO 14000 standards. The enthusiasm experienced in Europe has started to overflow into the U.S. This fact becomes apparent when you look at the SGS-Thompson Microelectronics facility in Rancho Bernardo, California. SGS-Thompson is the first U.S. manufacturer to have an EMS program certified; this occurred on January 3, 1996. SGS-Thompson, which is a corporation based in France, already has 5 facilities certified to the draft standards. Further, company representatives vow that all 16 SGS-Thompson facilities located in U.S., Europe, Southeast Asia, and North Africa will be certified by the end of 1997. As of December 1998 there have been over 5,000 facilities registered to the ISO 14001 standard worldwide.

According to a company official at SGS-Thompson, many of their major suppliers and contractors may find themselves forced to implement ISO 14000 standards by 1999 to maintain contractual preference. The official explains, "We assign an overall score to our suppliers based on many elements, including quality and environmental performance.... While an EMS program alone does not seem that important, [because of this scoring system] it may be the deciding factor when we decide who gets equipment and manufacturing supplier contracts."

Businesses are not the only ones that foresee the positive changes that can come about as a part of ISO 14000 implementation. Agencies with the United States government are also boarding the band wagon. Two of these agencies include the Department of Energy (DOE) and the Environmental Protection Agency (EPA). The DOE is encouraging all major contractors to implement an EMS program. Sources at the DOE have stated that specific contractors will be required to have ISO 14001 certification. The mood at the EPA can perhaps be best exemplified by the testimony of one of their representatives, who stated that "the old 'command and control' method of environmental regulation, so important in the first 25 years of the environmental movement, will occupy less time and fewer resources as companies start to use ISO 14000 and the next generation of environmental protection tools."

The EPA, which is studying how the ISO 14000 standards will influence and affect their operations, have already come out in support of them. This support has been heard at all levels including Carol M. Browner. Recently this support was testified to during the March 6, 1996 hearing in front of the Senate Environmental Resources & Energy Committee. During this hearing, James M. Seif, Secretary for the Pennsylvania Department of Environmental Protection, testified that ISO 14000 "represents the next generation of tools needed to more effectively achieve our environmental protection goals." Further, Mr. Seif said, "These new tools approach environmental protection in an entirely new way using performance-based environmental objectives, positive incentives to comply, external validation, a flexible approach to implementation and systems which constantly look for new opportunities to prevent pollution and reduce environmental compliance cost."

In the international arena, ISO 14000 is being seriously considered for adoption under NAFTA [North American Free Trade Agreement] and various GATT [General Agreement on Tariffs and Trade] trade agreements, to prevent the development of artificial trade barriers from country-specific environmental requirements. The standard is also viewed in Europe as meeting major components of the European Union's ECO-Management and Auditing Scheme (EMAS) regulations. And much like the proliferation of ISO 9000, it is expected that multi-national companies will require their suppliers to be ISO 14000 certified in order to do business in Europe.

In the U.S. a number of organizations that directly represent environmental and/or technical based members have already come out in support of ISO 14000. A partial list of these organizations includes:

- The American Society for Testing and Materials (ASTM),
- The American Society for Quality (ASQ),
- The Registration and Accreditation Board (RAB),
- The National Registry of Environmental Professionals (NREP),
- The American Forestry and Paper Association,
- The Academy of Certified Hazardous Materials Managers (ACHMM), and
- The American National Standards Institute (ANSI).

ISO 14000: A Business Decision

ISO 14000 is designed as a market driven approach to environmental protection. This system has the potential to be many times more effective in achieving significant environmental improvements than the current command and control approach.

Business success can be attributed partly to growth, strategic planning, and by maintaining a competitive edge in the market place. Organizations that fall in line behind the competition in market place developments often do not achieve a competitive advantage like the leaders in the market. ISO 14000 is seen by many organizations not as a new environmental compliance burden, but as a market place trend, which can provide for a more productive operation. Only those organizations which look at ISO 14000 as a business decision will be able to utilize it to the fullest extent. These organizations will be able to capitalize on the long term cost savings while at the same time broaden their market and improve their customer relations.

**Linda Greer and
Christopher van Löben Sels**

 **NO**

When Pollution Prevention
Meets the Bottom Line

What if a manufacturer learned that there were untapped opportunities to reduce waste and emissions within a plant that would also significantly cut costs? Conventional wisdom is that the company would seize on such opportunities and implement them.

But the reality is those opportunities are not always taken. A case study completed in 1996 at a Dow Chemical facility showed that certain pollution prevention strategies would save the company more than $1 million a year, approximately 10–20% of the existing environmental expenditures at the plant. Process changes would have eliminated 500,000 pounds (lb) of waste and allowed the company to shut down a hazardous waste incinerator. Surprisingly, these benefits were not enough of an incentive to outweigh other corporate priorities and the potential loss of future business that might have accompanied the incinerator's shutdown.

These findings came out of a collaborative study. In 1993, the pollution prevention pilot program (4P) was begun by the Natural Resources Defense Council (NRDC), an environmental advocacy group, and Dow Chemical, Monsanto, Amoco, and Rayonier Paper. Study participants, who were all interested in pollution prevention in a real-life industrial setting, wanted to know the reason for the lack of widespread reliance on promising pollution prevention techniques. Was it because there was not much to be gained environmentally or economically by using this environmental management technique? Was it because there were government regulations acting at cross purposes, incorporating barriers to implementation? If these factors did not explain the problem, what did?

Pollution prevention is conceptually quite different from pollution control, which relies on capturing emissions generated in processing before their release into the environment. Pollution prevention seeks opportunities to minimize reliance on toxic chemicals, increase efficiency, and decrease waste and emissions. Instead of focusing on changes required for environmental and health reasons, pollution prevention planners also identify opportunities to save money, making the process a potential "win-win" for industry and environmentalists.

This approach has failed, however, to take hold in the business world and at EPA and most state agencies. In fact, total waste production reported to the Toxics Release Inventory (TRI) in 1995 is up 6% from 1991, even though industry's TRI emissions have decreased. Some believe that, ironically, EPA'S regulations are responsible for this trend, because its highly prescriptive end-of-the-pipe nature discourages companies from implementing more holistic, innovative ideas at their plants. Others believe the more important obstacles to pollution prevention lie within the companies themselves. They suggest that the companies do not prioritize waste reduction initiatives in their business operations.

Texas Chemical Plant Selected

The study organizers picked a Dow Chemical Company facility in La Porte, Tex., as the primary site to evaluate. Located near the Houston ship channel, this relatively small, well-run chemical manufacturing operation produces methylene diamine diisocyanate (MDI), the major ingredient of foamed and thermoplastic polyurethane. Polyurethane foams appear in a variety of rigid foam products, from automobile parts to insulation in water heaters and picnic coolers. Polyurethane thermoplastic resins are used in tool handles and other clear plastics.

Dow sells most of the MDI from the plant as raw material to companies that combine it with various polyols to create foam and plastic; the rest is combined with polyols on site for some smaller volume Dow product lines. The La Porte facility's gross annual revenues are more than $350 million per year, and its estimated annual environmental expenditures are $5 million to $10 million.

Dow has several voluntary environmental improvement goals: reduction of dioxin emissions; decreased reliance on incinerators throughout the company; and, by 2005, a 50% decrease in the amount of waste generated prior to treatment. The La Porte facility's environmental staff are more interested in pollution prevention than most people in industry, making them good study participants. Because the La Porte facility's manufacturing operations are not especially unusual, study organizers thought that the results of this case study would be broadly applicable.

Dow La Porte's basic manufacturing process first combines formaldehyde and aniline to form methylenedianiline (MDA). The carbon atom from formaldehyde forms the methylene bridge between the two aniline molecules. MDA is then purified, placed in solution in monochlorobenzene (MCB), and reacted with phosgene (produced on site) to form monomeric and polymeric methylenebis(phenylisocyanate) (MDI and PMDI, respectively). In the final process step, MDI and PMDI are purified and then sold.

In 1993, La Porte's TRI releases for this process totaled 506,457 lb, well below the 1989 level of 1,137,300 lb (Table 1). Most of these releases were to air, followed by water. Because these emissions put the Dow La Porte facility in the top 4% of TRI facilities for total releases and transfers, its industrial operations were significant locally and nationally.

Table 1

Toxic Releases and Transfers, in Thousands of Pounds, From a Dow Chemical Facility

| | | | | Year | | | |
Category	1987	1988	1989	1990	1991	1992	1993
MCB air releases	712.0	980.0	1036.0	876.0	520.0	462.0	406.0
MDI transfers	620.0	426.0	630.4	181.0	230.0	182.6	227.0
Water releases	89.7	4.8	4.0	3.1	3.2	0.7	0.1
Other air releases	167.9	135.1	97.3	106.7	224.7	184.7	100.3
Other transfers	137.0	226.3	122.0	72.8	125.5	147.8	227.0
TOTALS	**1726.6**	**1772.2**	**1889.7**	**1239.6**	**1103.4**	**977.8**	**960.4**

Toxic releases from Dow Chemical's facility in La Porte, Tex., have declined steadily since 1989. Much of the decline has come from cutting monochlorobenzene (MCB) releases.

Source: Dow Chemical

At the time of the study, MCB was the leading chemical released annually from La Porte. In 1993, 406,000 lb of MCB (80% of the facility's total releases) were emitted (Table 1). In addition to being a toxic chemical, MCB is a volatile organic compound, emissions of which affect the region's ability to meet its ozone attainment levels. (After this study was completed, MCB emissions were substantially reduced; most are now captured and vented to the hazardous waste incinerator.)

La Porte treated about 1.5 million lb of TRI waste at the site in 1993, nearly three times as much as it released to the environment. Almost all of this treatment occurred in an on-site hazardous waste incinerator, covered by a Resource Conservation and Recovery Act (RCRA) permit. Phosgene and methanol provided the largest quantities of wastes burned, followed by 170,000 lb of MCB (see Table 2). (The amount of MCB burned today is greater than 170,000 lb, because additional quantities are now being captured.)

Assessment of Pollution Prevention

La Porte already had a pollution prevention plan, as required by the state of Texas, and Dow was planning to capture most of the remaining MCB air emissions and incinerate them on site. However, this action was on hold pending a decision to upgrade the plant's section that produced these continuing emissions. All the other chemicals in Dow's existing pollution prevention plan were ozone depleters, required to be phased down under the Montreal Protocol.

Conventional wisdom suggests that good opportunities to reduce waste and emissions have already been identified by large, environmentally sensitive companies. In fact, when this study began, plant personnel said they believed no other "low-hanging fruit" remained at the plant; that is, no other opportunities to reduce wastes and emissions remained that could be readily implemented to the financial or environmental benefit of the facility.

Pollution prevention literature, however, suggests that conventional wisdom might be wrong, and that various barriers within companies (1–3) or in government regulations (4) keep many important opportunities from being identified or implemented. To find out whether this was the case at La Porte, the study team first examined its pollution issues, reviewed existing pollution prevention plans, and assessed further opportunities for prevention. Once the "fact pattern" was established, we identified various barriers to expanded use of pollution prevention and sought agreement on recommendations to further its use.

In the first phase of the project, to understand its environmental impact, the coverage of existing regulations, and the plant's view of opportunities and barriers to additional environmental improvement, we submitted written questions to the facility and obtained an extensive, documented response. We then toured the plant to clarify the written responses to our questions. From this work, we characterized the status of the plant before the project's pollution prevention assessment.

In the second phase of this 18-month project, a third-party pollution prevention assessor, Bill Bilkovich of Environmental Quality Consultants, Tallahassee, Fla., went to the plant to seek pollution prevention opportunities. Bilkovich spent about 300 hours at the plant, talking with the staff and investigating opportunities. Our group met many times with the consultant.

Table 2

TRI Wastes (in Pounds) Incinerated at Dow Chemical Facility in 1993

Ammonia	15,000
Aniline	6800
Chlorine	1400
Monochlorobenzene	170,000
Methanol	630,000
Phosgene	840,000
1,1,1 TCA	2575
TOTAL	**1,665,775**

Pollution prevention assessments begin with an inventory of all wastes generated at a plant before treatment, recycling, or emission. They also require a chemical-use inventory for the site, which includes consumed chemicals that do not contribute to waste and emissions.

The inventory at La Porte required considerable work, as is common in a pollution prevention assessment. Even though major waste streams had been identified and tracked at the facility for pollution control (regulatory) purposes, data on their chemical composition were lacking. Such information is necessary for pollution prevention. For example, to continue the pollution prevention potential for one waste stream, we needed to know what components were present in the waste stream in less than 5% concentration. In another waste stream, a high degree of confidence in the distribution of minor constituents was required. Because this sort of information often is of no regulatory or process significance, it is not gathered. Much of the data could have been easily collected, however, if made a priority at the plant.

Following the waste and chemical-use inventory, we assessed opportunities for reduction through substitution, efficiency improvements in the process, recycling, and other options. Priorities can be set on the basis of financial considerations by working first on those projects that would deliver the highest rates of return. At La Porte, we set priorities primarily according to potential human health and environmental impact, which translated into high interest in MCB emissions to the air and the wastes burned in the hazardous waste incinerator.

For each chemical or waste being assessed for reduction opportunity, the team had to identify the reason the waste was generated. Answering this question required in-depth knowledge of the manufacturing process and basic plant chemistry as well as the conditions under which waste was generated (e.g., was it generated continuously or intermittently, under upset or normal conditions, etc.).

Because these critical pollution prevention issues are considered of little or no relevance in conventional assessments of a plant's environmental issues, it becomes critical to engage the production engineering personnel, especially the process chemistry experts, in pollution prevention planning. At La Porte, meetings with process engineers were key to developing a process flow diagram that showed material flow throughout the plant and indicated where each priority waste was generated.

Pollution Prevention Opportunities

After examining the process flow and waste information, project participants concluded that the site's single largest environmental opportunity would involve capturing and recycling MCB and ending the incineration of 500,000 lb of chlorinated hydrocarbons. If all of the other waste streams to the on-site RCRA incinerator (called a thermal oxidizer, or TOX) could also be reduced, recycled, or otherwise managed, the TOX could be closed altogether, a broader prospect with superb environmental and cost-savings benefits. (The cost saving from eliminating the TOX was determined to be substantial, in light of upcoming re-permitting requirements, soon-to-be-required upgrades in the unit, and operation costs.) Thus, we had found an option that would be good for the environment and save the company a lot of money.

To proceed on the TOX closure opportunity, six major waste streams that entered the TOX had to be examined: methanol contaminated with amines and ammonia, MCB air emissions captured by the pressure swing absorption/carbon adsorption system, other organic compounds captured on carbon, a phenyl isocyanate/MCB waste, phosgene manufacture vent gases (phosgene and carbon monoxide), and MCB from a groundwater pump-and-treat system that processed historical contamination from a previous owner. A pollution prevention assessment was undertaken for each of these waste streams. What follows is a consideration of options for the first two wastes, methanol and MCB, which offered the most interesting opportunities for reduction. We determined that the others were best addressed by using alternative treatment options.

Dealing With Waste Streams

At La Porte, options for conventional end-of-the-pipe alternative treatment of the methanol waste stream include processing in the wastewater treatment plant or sale as a product. Incineration, however, had been considered the best option because methanol is essentially a clean fuel, and its use reduced the need for TOX operations' supplementary fuels. The pollution prevention assessment started with a different set of questions about methanol, and it presented some interesting options.

First we asked, Where does the methanol originate? Although methanol is a major waste stream generated at the plant, a flow chart of the basic process chemistry does not show an obvious source. Interviews with plant personnel revealed that methanol enters the process in the formaldehyde-water solution (formalin) used to manufacture MDA. Formaldehyde is manufactured from methanol; and some residual methanol, in this case about 0.5%, is kept in the commercial product as a stabilizer. This methanol must be removed by La Porte before the phosgenation step. Because La Porte uses millions of pounds of formaldehyde each year, the amount of methanol waste being burned in the TOX reaches hundreds of thousands of pounds annually.

The next obvious question was, Can we substitute formaldehyde with a less toxic chemical that does not generate a waste stream? Formaldehyde is used in the plant to provide a carbon atom to connect two aniline molecules and form the intermediate MDA. Perhaps carbon dioxide (CO_2) could be used to achieve the same end. The assessor researched the use of CO_2 and found that, although there was a patent on the use of CO_2 in the manufacture of toluene diisocyanate (TDI), the process had never been commercialized and was not applicable to the manufacture of MDI. Other alternative sources of the carbon bridge atom were discussed with Dow research and development personnel, but we found no other good options and stopped this line of inquiry.

The next option was to look for ways to reduce or eliminate the methanol in the formalin. But conversations with a major formaldehyde manufacturer indicated that, under the temperature, humidity, and transit time conditions common in the Gulf Coast, at least 0.3% methanol is necessary to prevent the in-transit polymerization of formaldehyde. Thus, we could not decrease methanol waste to insignificant quantities by using this approach at this plant.

The identification of an alternative stabilizer, one that might be effective at part-per-million (ppm) levels, was not explored for La Porte; the driving force for developing an alternative would have to come more broadly from other formaldehyde users across the country. However, the assessment did raise interesting questions about the treatment and disposal costs incurred nationwide for the management of waste methanol at plants that use formaldehyde as a raw material. Approximately 8 billion lb of formaldehyde are used annually in U.S. manufacturing, and calculations show that 40 million lb of methanol are being handled or disposed of by the plants purchasing this chemical. Full-cost accounting of the cost per pound of formaldehyde purchased as a raw material could reveal that the cost for residual waste methanol management is as high as or higher than the raw material cost—and open a market opportunity for higher priced, methanol-free formaldehyde.

Next we asked, If methanol is needed to stabilize the formaldehyde, might there be a way to remove the methanol as a clean waste stream before the formaldehyde enters production and comes in contact with aniline? Interest in this option was depressed by the low value of recovered methanol and the high cost of constructing and operating separation equipment to process millions of pounds of formaldehyde each year.

Because the trace quantities of aniline and other nitrogenous compounds make the waste methanol difficult to sell, we then asked whether the aniline could be removed from the methanol. Even though trace quantities of aniline could be removed and yield a salable methanol–water mix, the assessment team did not believe that aniline could be eliminated to levels considered safe for unrestricted commercial use of the methanol waste.

Project organizers then asked whether a customer could be found for methanol containing aniline. A cursory review of TRI data revealed no facility close to La Porte that had methanol and aniline emissions, and we did not pursue an in-depth review.

The final option we evaluated was returning the waste methanol to the formalin manufacturer for future processing into the formalin product Dow was interested in this option and would consider a "take back" provision in its purchase contract quite favorably, although this option has not been issued to date.

MCB Waste Stream

Analysis of MCB took the same initial path: identifying the use of MCB in processing (solvent carrier in the phosgenation step of the process) and seeking less toxic–chlorinated alternatives for this purpose. Finding no alternatives, we shifted our focus to recycling MCB instead of incinerating it. Plant personnel reported they had briefly considered this alternative when they decided to capture the MCB air emissions to reduce their air pollution, but they decided to incinerate the solvent because it was convenient and legal to do so, and they did not believe this practice would pose significant risk.

The plant personnel also believed that the presence of water in the used MCB would preclude recycling. The assessor researched this and found that water was not actually the principal barrier to recycling. To the contrary, the virgin MCB purchased by the plant is routinely treated on site to remove water introduction in a molecular sieve bed before introduction into the process. Production staff then raised a more important concern: If the waste MCB contained any impurities, they could build up in the system if MCB were recycled. Impurities were possible, but sufficient information was not available about the time course of their buildup. The uncertainty raised by this issue could be easily resolved by analysis of the actual level and identity of trace contaminants in this stream, however.

At the end, the assessment team's short-term recommendations included recycling the MCB waste stream that is currently incinerated; selling the methanol or burning it at an alternative, off-site incinerator; reducing levels of phosgene sent for treatment; and scrubbing the remaining waste phosgene instead of incinerating it.

If all these waste streams could thus be addressed, and several very minor waste streams were sent off site, Dow La Porte could conceivably shut down its on-site hazardous waste incinerator and avoid the cost of RCRA re-permitting. Dow personnel estimated the rate of return on investment for this project at 20–70%. The investment would pay for itself somewhere between 15 months and 5 years, depending on how the various projects were configured. Estimated savings are $1 million a year, derived from reduced raw material costs, incinerator operating costs, and re-permitting costs with regulatory authorities. Virgin solvent purchases alone might drop by 90%.

Barriers to Implementation

There were no significant regulatory barriers to adopting this plan. The problem lay within the company. Specifically, these opportunities were weak candidates in the capital investment process at Dow. The project was considered for implementation twice by the urethane business group within Dow Chemical and was put off both times because other, more financially attractive business opportunities were given higher priority.

Had EPA required that Dow reduce these waste streams, the 4P projects would have been mandatory, and the rate of return of the project would have been irrelevant to Dow's decision making. However, because these were voluntary opportunities, they were considered in the same way as other business opportunities would be. To succeed, these opportunities needed to do more than reduce waste and save money; they needed to be superior to other options for capital investment.

Although Dow hopes to implement the pollution prevention plan in the future, the La Porte pollution prevention project rests in an odd position: It is not required for the purposes of environmental compliance, and it is not of central interest to production engineers whose main priorities are in capacity building. Nor is the project highly compelling to business line personnel with profit-and-loss authority. They are more concerned with maximizing profit for their business among various Dow plant locations around the world.

Conventional wisdom says that most good opportunities to reduce waste and emissions have already been identified, but that belief was incorrect in this case: The 4P project found very promising opportunities that had not been identified by Dow. More significantly once pollution prevention opportunities were found, corporate business priorities and decision-making structures posed formidable barriers to implementing those opportunities.

Most environmental professionals outside of industry incorrectly assume that a pollution prevention plan that actually saves money and is good for the environment will be quickly seized upon by U.S. businesses. This work shows that at least in one firm, such opportunities may not be sufficiently compelling as a business matter to ensure their voluntary implementation.

References

1. Porter, M. E.; van der Linde, C. *Harvard Business Review* September-October 1995, 120–34.
2. New Jersey Department of Environmental Protection and Energy. *Industrial Pollution Prevention Planning: Meeting Requirements Under the New Jersey Pollution Prevention Act;* Office of Pollution Prevention, State of New Jersey: Trenton, July 1993.
3. Little, A. D. "Hitting the Green Wall," *Perspectives;* Arthur D. Little: Cambridge, MA, 1995.
4. Schmitt, R. E. *Natural Resources and Environment,* 1994, 9, 11–13, 51.

POSTSCRIPT

Will Voluntary Action by Industry Reduce the Need for Future Environmental Regulation?

Most assessments of the environmental achievements of the past three decades agree with Patchak and Smith that regulations have achieved a great deal. At the same time, however, they also note that degradation of the world's ecosystems has only been slowed, not halted or reversed. It is true that governmental environmental agencies have encouraged voluntary efforts like Responsible Care and ISO 14000, but there is little evidence that the idea that these initiatives can replace or reduce the need for "command and control" regulation and enforcement is being broadly embraced. At best, as Paulette L. Stenzel writes in "Can the ISO 14000 Series Environmental Management Standards Provide a Viable Alternative to Government Regulation?" *American Business Law Journal* (Winter 2000), the ISO 14000 standards "provide a useful supplement to environmental regulation. They can facilitate the work of the U.S. Environmental Protection Agency and promote worldwide pursuit of sustainable development."

In "Barriers and Incentives to the Adoption of ISO 14001 by Firms in the United States," *Duke Environmental Law & Policy Forum* (Fall 2000), Magali A. Delmas says that in the 1990s "environmental regulation entered its third generation placing great emphasis on voluntary environmental initiatives" such as the ISO 14001 standard. However, "since the standard does not present tangible results of the actual improvement of environmental performance to a firm's stakeholders, it becomes vital that all stakeholders believe in [and promote] the underlying benefits" of the standard. For an industry assessment of the potential of ISO 14000, see "Environmental Management With ISO 14000," by Steven Voien, *EPRI Journal* (March/April 1998).

As Greer and van Löben Sels argue, many environmental advocates assume that all that is necessary to get industry to implement more environmentally appropriate practices is to demonstrate that the changes will have a short-term economic payback. The discovery that this is often untrue has soured several cooperative efforts between corporations and environmental organizations. The problem is that implementing most production changes requires an up-front investment, and management will usually look for the most profitable investment of available revenues. This means that many opportunities that can be proven to have less potential profitability, even if they may appear to be cost-effective, will be rejected. Of course, environmental protection also often requires industries to make changes that are costly and that will not improve the bottom line!

ISSUE 19

Is Sustainable Development Compatible With Human Welfare?

YES: Dinah M. Payne and Cecily A. Raiborn, from "Sustainable Development: The Ethics Support the Economics," *Journal of Business Ethics* (July 2001)

NO: Jacqueline R. Kasun, from "Doomsday Every Day: Sustainable Economics, Sustainable Tyranny," *The Independent Review* (Summer 1999)

ISSUE SUMMARY

YES: Professor of management Dinah M. Payne and professor of accounting Cecily A. Raiborn argue that environmental responsibility and sustainable development are essential parts of modern business ethics and that only through them can both business and humans thrive.

NO: Professor of economics Jacqueline R. Kasun argues that sustainable development poses threats to human freedom, dignity, and material welfare.

O ver the last 30 years many people have expressed concerns that humanity cannot continue to increase population, industrial development, and consumption indefinitely. The trends and their impacts on the environment are amply described in numerous books, including historian J. R. McNeill's *Something New Under the Sun: An Environmental History of the Twentieth-Century World* (W. W. Norton, 2000).

Can we keep it up? is the basic question behind the issue of sustainability. In the 1960s and 1970s sustainable development was expressed as the "Spaceship Earth" metaphor, which said that since we have limited supplies of energy, resources, and room, we must conserve and recycle if we are to avoid crucial shortages. Sustainability entered the global debate in the early 1980s, when the United Nations' secretary general, Javier Perez de Cueller, asked Gro Harlem Brundtland, a former prime minister and minister of environment in Norway, to organize and chair a World Commission on Environment and Development and to produce a "global agenda for change." The resulting report, *Our Common Future* (Oxford University Press, 1987), defines "sustainable development" as "development that meets the needs of the present without compromising the ability of future generations to meet their own needs." It acknowledges that limits on population size and resource use cannot be known precisely, that problems may arise gradually rather than suddenly and will be marked by rising costs, and that limits may be redefined by changes in technology. But the report also recognizes that limits exist and must be taken into account when governments, corporations, and individuals plan for the future.

The Brundtland report led to the United Nations Conference on Environment and Development held in Rio de Janeiro in 1992. This conference set sustainability firmly on the global agenda and made it an essential part of efforts to deal with global environmental issues and to promote equitable economic development. In brief, sustainability means such things as cutting forests no faster than they can grow back, using ground water no faster than it is recharged by precipitation, stressing renewable energy sources rather than exhaustible fossil fuels, and farming in such a way that soil fertility does not decline. In addition, economics must be revamped to take account of environmental costs as well as capital, labor, raw materials, and energy costs. Many add that the distribution of the earth's wealth must be made more equitable as well.

Given growth in population and demand for resources, sustainable development is a difficult proposition. Some think that it can be done, but others maintain that for sustainability to work, either population or resource demand must be reduced. Not surprisingly, many people consider sustainable development to be in conflict with business and industrial activities, private property rights, and such human freedoms as the freedoms to have many children, to accumulate wealth, and to use the environment as one wishes.

In the following selections, Dinah M. Payne and Cecily A. Raiborn argue that environmental responsibility and sustainable development are essential parts of modern business ethics. They maintain that the consequence of the activities of all members of society, including businesses, "is an environment that is either habitable or one that is not" and that all "have a responsibility towards the environment and each other." Jacqueline R. Kasun argues that traditional economics is a truer way to look at human activities in the world and that sustainable development will require sacrificing human freedom, dignity, and material welfare.

**Dinah M. Payne and
Cecily A. Raiborn**

Sustainable Development: The Ethics
Support the Economics

Introduction

The field of business ethics is rampant with diverse issues and dilemmas. One critical ethical issue has, for many years, received significantly less attention than it merited: the responsibility of business organizations to their environments. Organizations world-wide have created and have faced resource depletion and pollution. However, there now seems to be a distinct and overt embracing of environmental social responsibility by many companies. This new-found interest may have been generated, in part, by gatherings such as the Rio de Janeiro (Earth) Summit and Kyoto Protocol. But, more importantly, these gatherings have spawned a plethora of groups focused on the issue of environmental social responsibility and, specifically, the issue of sustainable development. What is this concept and why should it concern businesses and their managers? Why should sustainable development be viewed as an ethical responsibility of businesses? To what extent should businesses attempt to engage in sustainable development activities? And what actions, beyond legal requirements, can be and are being taken by businesses to promote this concept with its resultant benefit to all business stakeholders?

Issue Definition and Identification

The term *sustainable development* was introduced in the 1970s, but actually became part of mainstream vocabulary during and after the 1987 World Commission on Environment and Development (also known as the Brundtland Commission). The Commission defined sustainable development as "development that meets the needs of the present without compromising the ability of future generations to meet their own needs." On the surface, this definition seems to be fairly simplistic, but the issue's breadth and depth create complexities.

To more fully and meaningfully refine the concept, the Earth Council indicated that such development should be economically viable, socially just, and environmentally appropriate. An additional expansion suggested that sustainable development should mean that the basic needs of all are met and that all should have the opportunity to fulfill their aspirations for a better life. The definition postulated by the World Business Council for Sustainable Development is that sustainable development is "the integration of economic development with environmental protection and social equity." Several complicated and sensitive issues are inherent in these definitions.

First, how can the "needs" of the present be differentiated from the "wants" of the present as well as how can the needs of the future be ascertained currently? Into this debate fall questions such as how can nonexistent future generations be protected and to what extent should today's civilization be sacrificed to protect future generations? Although the answers to these questions are arguably unanswerable, it is apparent that business have some responsibility to provide goods and services to the world. The free market helps "push" businesses to produce the goods and services currently desired for purchase (whether these goods are needed or simply wanted). Additionally, businesses partially establish future needs and wants of consumers through product development in response to current external pressures (desires communicated from consumers) as well as current internal abilities (research and scientific discoveries). In responding to these current pressures or abilities, many businesses utilize life-cycle analysis to assess potential future environmental impacts of product design, manufacturability, and recyclability.

Second, relative to what context or benchmark should "economically viable" be determined? This term could mean radically different things between businesses in developing and in developed nations, between start-up and long-standing businesses, or between business having significant environmental impacts and those having minimal environmental impacts. In each of these three scenarios, the cost of sustainable development would generally be more expensive (in relative cost to revenue proportions) to the former companies than to the latter. Thus, what might be deemed economically viable for a large retailer in England might mean financial ruin for a small mining company in Haiti.

There is a clear trend in the developing world towards better environmental policies that include the pursuit of economic development alternatives that minimize negative environmental impacts. Evidence also exists to indicate that "through technological change, substitution between resources, and higher prices for goods that pollute, environmental objectives and economic growth can be made more compatible." In regard to technological change, it is generally true that as technology advances, it becomes more efficient. Thus, because the industrialization process in developing countries often begins with the use of outdated technology, production may be environmentally expensive (the lower efficiency contributes to increased resource depletion and less emphasis on pollution control). As technology becomes more sophisticated, efficiency increases causing an increase in productive activity with fewer defects and spoilage, and thus a decline in the rate at which resource depletion occurs. Additionally, as the country advances, less environmental pollution may be tolerated. In the last stage of industrialization, organizations use advanced (more efficient and cleaner) technology, causing a net decline in resource depletion and pollution. Per capita income and social and governmental consciousness about the environment also rise; more "green" laws are written and enforced. Thus, an inverted U-shaped curve can be used to represent the changes in a society that starts at a point without environmental quality, rapidly advances, and then slows and turns around when that society has the time and/or money to spend to protect the environment.

Third, how and by what party should "socially just" development or "social equity" be determined? These factors would depend on who was obtaining the benefit from the development, what form that benefit took, what level of economic development existed in the area, whether resources consumed were replenishable, and what political and social issues were being faced or remedied.

Last, how and by what party is "environmentally appropriate" development to be judged? This judgment must reflect the answer to whether the environment should be protected for its own sake and/or for the sake of human inhabitants. Ecological ethicists argue that non-human inhabitants are intrinsically valuable and, thus, deserve respect and that humans have duties of preservation towards them. Alternatively, even if the intrinsic value of the environment and its non-human inhabitants is refuted, a livable environment is owed to all humans so that they may be permitted to fulfill their capacities as rational and free beings. Healy asserts that future sustainability will require a reorientation away from the human-centered (or anthropocentric) anthropological view towards more nature-centered (or ecocentric) view. Thus, determination of "environmentally appropriate" would commonly be more an issue of perspective than one of specific activity. Some individuals and businesses will take a broad perspective and assess the impact of an activity on the overall current and future physical environment (not just that part inhabited or used by humans). Other individuals and businesses will take a narrow perspective and assess the impact of the activity on the surrounding environment in the here-and-now.

Regardless of the definition or the diverse possible answers to definitional issues, it is clear that all publics (businesses, consumers, regulatory agencies, scientists, communities, and governments) are touched by the concept of sustainable development. All of these publics interact, directly or indirectly, and face the same outcome, which will not be locale-by-locale, industry-by-industry, or political party-by-political party based. The long-term consequence of the activities of all publics is an environment that is either habitable or one that is not. That being the case, each public separately and all publics collectively have a responsibility towards the environment and each other to better understand sustainable development and to strive to achieve meaningful progress towards its attainment. Thus, the ethical issue in sustainable development is the basic issue of life versus death; if business and all other publics do not begin practicing the tenets of sustainable development, life as it currently exists will be extinct.

Businesses and their managers should be concerned about sustainable development for many reasons. Economic pragmatists would base their arguments on the simple fact that, without sustainable development, neither businesses nor the societies in which they exist will have a long-run future. Others believe that engaging in sustainable development will be a megatrend that will enhance organizational reputations. Others believe that sustainable development can be used by businesses as a unique core competency to obtain a strategic competitive advantage. All three rationales are valid and serve to stress the need for responsible business to pursue sustainable development in the current competitive reality.

Sustainable Development as an Ethical Issue

A 1996 survey of American and Canadian corporate executives included the question, "Why does, or will, your company practice sustainable development?" On a 10-point scale of level of importance, the responses of (1) promoting good relations and (2) creating shareholder value scored, respectively, 8.1 and 7.3. However, more importantly, the two most highly ranked responses were (1) to comply with legal regulations and (2) a moral commitment to environmental stewardship (8.8 and 8.5, respectively). Thus, there is evidence that business executives recognize that sustainable development can and should be viewed as part of the interwoven framework of business ethics. Ethicists would applaud such a view and could use the theories of utilitarianism, rights/duties, and the categorical imperative to provide the underlying support.

In making a utilitarian analysis of businesses' implementation of sustainable development concepts, the "greatest god or least harm for the greatest number" principle can be easily envisioned. The stakeholders involved are all the earth's inhabitants, both human and non-human. Sustainable development would create the greatest good or least harm by allowing those inhabitants (and potential offspring) to exist in a world where the air is breathable, the water is drinkable, the soil is fertile, and renewable resources thrive. It is difficult to use traditional monetary cost-benefit analysis to determine whether sustainable development is worthwhile. First, although many current and future costs could be estimated and discounted back to present values, it is probably impossible to even comprehend what types and amounts of costs might be necessary in the future. Second, the benefits of sustainable development are significantly more qualitative than monetarily quantitative; for example, how can the value of a living species be estimated? But, even without finances attached, the result would be undeniably conclusive: no matter how high the costs of sustainable development are, the benefits of current and continued existence by the earth's species *must* exceed that cost. Ethically, the benefits of life outweigh the costs to obtain it.

Analyzing sustainable development activities by business entities using the theory of right/duties addresses the issue of whether an inhabitable environment is a moral right.... Blackstone postulated that access to livable environment is a human right because such an environment is essential for humans to fulfill their capacities. Thus, everyone has the correlative moral obligation to respect that right. Rawls and Kant would support this concept because of the rationality of people being entitled to rights that do not infringe upon others' rights. A human's inhabitable environment includes other living creatures, flora, fauna, and resources (e.g., air, water, and minerals). These non-human elements of the planet are not responsible for, nor can they correct, the ecologically damaging discharges of pollution or disproportionate use of resources created by humans. Thus, businesses, as collections of human beings, have the duty to engage in sustainable development activities so as to mitigate their environmental impacts and help in providing, protecting, and preserving a livable environment.

In its determination of morality as objectively and universally binding, Kant's categorical imperative would support businesses' sustainable development actions. Proponents of Kantianism, however, would be quick to point out that sustainable development activities should be performed from duty, not simply from inclination or self-interest. In other words, businesses should not engage in sustainable development because such activities will reduce costs, increase revenues, or provide an advantageous reputation. Businesses should engage in sustainable development because, in the minds of all rational people, reclaiming and preserving the earth's environment as well as limiting pollution and resource depletion is the "right" thing to do. In the final analysis, sustainable development represents an action that would be right and valid "even if everyone were to violate it in actual conduct."

Sustainable development is, then, an important and ethical value to be upheld by businesses. But some aspects of sustainable development are more clearly pursued, or pursued to different degrees, by some publics than by others.

Level of Sustainable Development Efforts for Businesses

From the standpoint of businesses, it is important to ascertain which sustainable development issues can and cannot be addressed. Businesses cannot pass laws or treaties to protect the environment, enact land reforms, or control populations. Businesses cannot force consumers to recycle, reuse, or slow consumption. Businesses, in general, cannot produce the scientific knowledge that will end global warming, save the rain forests, or eliminate pollution. Businesses cannot stop societal development. And businesses cannot decide to pursue totally altruistic environmental goals without any concern for profitability or longevity. (To do so would be to guarantee organizational failure: owners would remove financial backing because they could not achieve a reasonable return on investment; employees would look elsewhere for jobs because they could not rely on continued employment; and suppliers would limit or revoke credit because they could not be assured of payment.)

Although businesses cannot do any of the things mentioned above unilaterally, there are many things that they can do. Businesses can influence passage of laws through lobbying and other efforts. They can influence consumer behavior (through product development and packaging, encouraging consumer recycling and reuse, and community awareness activities). Businesses can (through research agendas and new product discovery and development) help reduce or eliminate pollution causes. Businesses can also influence how societal development will occur and what the impact of that development will be through their location and technological investment choices. And businesses can undertake a strategy of pursuing sustainable development in conjunction with profitability and longevity to the benefit of all organizational stakeholders. Such a strategy would focus on both current and future eco-efficiencies.

Given the myriad of opportunities for engaging in environmentally "correct" or, at a higher level, sustainable development activities, how should a business determine its participation? One possible technique would be the use of the hierarchy of ethical behavior suggested by Raiborn and Payne. The hierarchy consists of four degrees of achievement:

- basic (reflects minimally acceptable behavior that complies with the letter, but not the spirit, of the law);
- currently attainable (reflects behavior deemed moral, but not laudable, by society);
- practical (reflects extreme diligence toward moral behavior; achievable but difficult); and
- theoretical (reflects the highest potential for good or the spirit of morality).

Basic Level of Behavior

A business operating at the basic level of behavior would merely comply with the laws of the jurisdictions in which it operates. Such an organization would make no sustainable development efforts because the concept is not embedded into the law in any country in the world. This organization would remain within legally acceptable pollution levels, although it would possibly view those levels as hindrances to productive activities. Such organizations... would more than likely espouse (although quietly) the following beliefs: *We recognize that the environment is not a "free and unlimited" good. However, environmental laws cost money that could be going to support the economic goal of increased shareholder value. We will operate within the law, but will not seek environmental improvements beyond the law.* Thus, these companies' behaviors would be deemed legal, but not necessarily ethical.

Currently Attainable Level of Behavior

A business operating at the currently attainable level of behavior would acknowledge that some benefits do arise from engaging in environmentally-friendly activities that are not legally mandated. These organizations, however, probably engage in such activities for the "wrong" reasons (according to the categorical imperative): cost reduction, revenue enhancement, or reputation improvement. In other words, the activities are likely to provide short-term monetary benefits greater than their costs.... These companies would more than likely espouse the following belief: *We recognize that the environment is not a "free and unlimited" good. Environmental laws are necessary because business should be held responsible to remove the damaging effects they have had and to reduce or limit the future impacts they will have on the earth's ecosystems in their role as society's major tangible goods producers. We will operate within the law and will seek to find environmental improvements that reduce costs or improve productive activities so that short-term profits are enhanced and shareholder value is increased.* These organizations may be viewed by society as environmentally-conscious companies that are operating for the greater good . . . but, in reality, the greater good is primarily that of the organization.

Practical Level of Behavior

A business operating at the practical level of behavior would also acknowledge that benefits arise from engaging in environmentally-friendly activities. These organizations, however, would strive to do the "right" thing relative to the environment because it is "right" rather than because of short-term profits or reputation. These businesses and their managers recognize the need for, and worth of, environmentally sound production and marketing practices. These organizations would attempt, in their varying activities, to engage in environmental innovations that might be expensive but that would provide the most beneficial future outcomes. In doing so, the businesses would hope that consumers would recognize the benefits of such innovative practices are worth purchasing at a higher cost than those of less environmentally sensitive competitors. There should be no question that these businesses are profit motivated: management has a fiduciary duty towards a number of groups (among which are shareholders, creditors, employees, and consumers) to maximize profits and, therefore, efficiency. "For both infrastructure and services, it has to be recognized that private sector participation will be achieved only on the basis of an acceptable expected revenue scheme."

. . . These companies would more than likely espouse the following belief: *We recognize that the environment must be protected, not only through laws but also through our own proactive involvement. We will find and implement environmental improvements and innovations for our products and processes, knowing that consumers will recognize the long-run benefits of our actions and be willing to support those actions with their purchasing decisions. Through this strategy, we believe that we will provide high quality products that have the least detrimental environmental impact on our local and global community.* Thus, these organizations view themselves as forerunners in the area of environmental protection, for the sake of all stakeholders. But these companies have not crossed the line from overt environmental concern to cutting edge, world-class leadership in sustainability.

Theoretical Level of Behavior

A business operating at the theoretical level of behavior would have incorporated the idea of sustainable development into its organizational strategy. There would be no "piecemeal projects aimed at controlling or preventing pollution. Focusing on sustainability requires putting business strategies to a new test. Taking the entire planet as the context in which they do business, companies must ask whether they are part of the solution to social and environmental problems or part of the problem." These organizations... would more than likely espouse the following belief: *The new paradigm must view the environment as fundamental to the business', society's, and the earth's continued existence. It is to be protected and replenished through all human and machine investments that are necessary to secure our place and the place of others (both human and nonhuman) on this planet. In doing so, our organization will be cost efficient from waste reduction and resource productivity maximization. Our business will be respected by our stakeholders; our products and services will be desired and recognized as value-added; and our eco-efficiency will enhance organizational profitability and promote organizational longevity.* These organizations take the concept of "walking the talk" completely literally.

What Actions Can and Are Being Taken by Businesses?

One statistic starkly exhibits the crisis that looms: "By the year 2030, world population will double from 5.5 billion to 11 billion.... To provide basic amenities to all people, it is estimated that production of goods and energy will need to increase 5 to 35 times today's levels." Such changes will cause further environmental strain and perhaps irreparable damage. Can the earth assimilate the massive pollution and resource depletion inherent in such growth? Should economic growth be pitted against environmental and human health? Will implementation of sustainable development activities require a change in consumption habits and, if so, what habits of whom should be altered? Will technological innovation arise as the hoped-for panacea, such that consumption habits may remain unchanged? Can stakeholders accept, encourage, and reward through product/service purchases and organizational investment business actions toward sustainable development? Answers to these questions would obviously ameliorate the chance for efficacious solutions. Unfortunately, only simple answers can be provided for these complex questions at this time. Significant research needs to be performed to ascertain the answers that are the most ethical and the most eco-efficient. But one thing is clear: if businesses, as the manufacturers and providers of the world's products and services, do not begin individually and collectively to immediately work toward a solution, after some point there will be no solution to achieve.

Businesses should not be considered as irresponsible entities that must be forced into doing the ethical thing with regard to environmental protection or sustainable development. Businesses recognize the symbiotic relationship between the environment, consumers' demands, and the provision of goods and services to the world's communities. Businesses also recognize the synergistic relationship between them and the environment/society in which they operate. It would be irrational to suggest that business could exist without society and equally irrational to suggest that society could exist as well as, better, or at all in the absence of business. In other words, business and society need each other for practical reasons: businesses want to provide goods and services that society needs and/or wants....

Shrivastava has suggested that, as a beginning, businesses strive to attain various goals that are commensurate with the goals of sustainable development. He suggests that energy conservation techniques could be employed that would have a positive impact on pollution and resource depletion. Businesses could also engage in resource regeneration aimed specifically at the reduction of resource depletion. Additionally, he promotes environmental preservation, which strengthens and is strengthened by arguments that the environment itself is worthy of care and protection, aside from its human-associated values. To implement these three goals, businesses can improve processes, educate employees, provide consumer advice, perform research, be prepared for emergencies, and listen openly to concerns.

A final, but very important method by which businesses can strive toward sustainable development is to join with others to form organizations focused on this goal. Some of these organizations include the World Trade Organization's Committee on Trade and Environment, the World Business Council for Sustainable Development, the International Chamber of Commerce's Commission on Environment, and the United Nations Environment Program. As aptly stated in the International Chamber of Commerce Commitment to Sustainable Development.

> all sectors of society, including government, business, public interest groups and consumers, have a role to play in contributing to sustainable development, and they must work in partnership, bringing their values and experience to bear on the challenge. Sustainable development will only be achieved if each one plays its part. Each sector should focus on what it can do best, but, through partnerships, local, national or even global, we can build on the strengths of each group.... Business is best suited to contributing to sustainable development in the economic sphere—through the creation of wealth in an environmentally sound manner.

Conclusions

Businesses need to assert their commitment to sustainable development over and above environmental legalities. As indicated by Porter and van der Linde, "Regulators tend to set regulations in ways that deter innovation. Companies, in turn, oppose and delay regulations instead of innovating to address them. The whole process has spawned an industry of litigators and consultants that drains resources away from real solutions."

Who, in business, should lead the way in the pursuit of sustainable development goals? The easiest answer is that global, multinationals based in highly developed countries should be the leaders; some of these entities have already begun the journey. Another answer is that those entities creating the biggest environmental problems should lead the way. The most appropriate answer, however, is that organizations whose stakeholders recognize the necessity of sustainable development as part and parcel of the company's need to act ethically should be the role models.

Businesses, acting alone, cannot create sustainability. If the internal and external stakeholders are not willing to adopt the concept of sustainability as a long term necessity, then should businesses view the idea as not worthy and expunge it from the organizational strategy? Absolutely not! As indicated within the paper, there is significant interaction between and among all value chain constituents. And, similar to the spread of high product and service quality as a priority among value chain members, as one member of the value chain demands a view of sustainable development, so will others. In some cases, there will be a trickle-down effect; in others, there will be a waterfall.

It is time that businesses realized that environmental responsibility and sustainable development are part and parcel of business ethics. Rules can be written and laws can be passed about pollution control or environmental degradation, but the framework to which these are bound is the minimum or basic level of acceptable behavior. Like a corporate code of ethics, an environmental policy will reflect the corporate culture from which it stems. The companies that move in a continuous path up the hierarchy of ethical behavior from merely complying with legalities to integrating sustainable development concepts into strategic initiatives and mission statements are companies whose managers understand, espouse, advocate, and uphold the fundamentals of business ethics. These are also the companies and managers that are well aware that ethical business is good business. These are the long term survivors.

Jacqueline R. Kasun

 **NO**

Doomsday Every Day

Sustainable development" was the galvanizing theme of the 1992 Earth Summit in Rio de Janeiro. Based on the work of the Brundtland Commission in 1987, the goal of sustainable development has been enthusiastically promoted by the World Bank, the U.N. Development Fund, the U.N. Environment Programme, and the United Nations agencies promoting "world governance." It inspires President Clinton's Council on Sustainable Development. It has precipitated an avalanche of World Bank publications, such as the fourteen volumes of the *Environmentally Sustainable Development Proceedings* series of the 1990s, transforming untold acreages of forest into official paper. The phrase occurs frequently in the Chinese Communist press, usually in conjunction with news about the progress being made in the family planning program (Hong 1998). The two topics—sustainable development and "family planning"—are linked throughout the literature.

Economists have struggled, without much success, to reconcile the various definitions that have been offered for "sustainable development." Herman Daly, an economist who has been involved since the beginning, says not to worry —lots of good ideas can't be defined (1996,2). Daly, long associated with the World Bank, has written the seminal works in the field and is now joined by a host of authors producing textbooks for the college generation. Instruction in "sustainable economics" suffuses or replaces introductory economics courses at a number of institutions.

Whatever it is, sustainable development promises to transform life on this planet. The Rio conference produced agreements on everything from land-use planning (including "sustainable mountain development") and greenhouse gases to, or course, birth control. There were agreements on "human settlements," "sustainable agriculture," "biodiversity," and on and on in its "Agenda 21" and its Climate Convention and its Convention on Biological Diversity (*Agenda 21* 1992). Though Congress did not adopt the program, the Clinton administration proceeded as if it had, adopting new federal regulations and appointing a President's Council on Sustainable Development, made up of federal officials and prominent environmentalists, to pursue the agenda with vigor.

From Jacqueline R. Kasun, "Doomsday Every Day: Sustainable Economics, Sustainable Tyranny," *The Independent Review*, vol. 4, no. 1 (Summer 1999). Copyright © 1999 by The Independent Institute, Oakland, CA. Reprinted by permission of *The Independent Review*, a journal of political economy. References omitted.

The Clinton Council on Sustainable Development has issued its own version of Agenda 21, declaring that we must "change consumption patterns," "restructure" education, "conduct a high-visibility public awareness campaign... to adopt sustainable practices," "create a network of conservation areas for each bioregion... based on public/private partnerships" (so much for private property), "realign social, economic and market forces... to embrace conservation," "use building codes [to secure]... environmental benefits," have "local... community planning... to develop a common vision," create "a council of... key stakeholders to... achieve sustainable management of forests," and "promote development of compact... neighborhoods" (good-bye, suburbs) (President's Council 1995).

Moreover, it decreed that "population must be stabilized at a level consistent with the capacity of the earth to support its inhabitants," whatever that capacity might be (President's Council 1995). The definitions may be elusive, but the program is uniform throughout the literature. It is to create massive, new bioregional conservation areas; control land use, consumption, and markets; re-educate the masses; and control population.

The Sierra Club announced at the U.N. Population Conference in Cairo in 1994 that "local activists" of the club in the United States were working "in a consensus-based... process to establish... thresholds for... population and consumption impact on the local ecoregion.... Addressing local carrying capacities will improve the quality of life for all and help develop sustainable communities" (Sierra Club 1994). The club didn't specify what action those local activists would take if it turns out that local populations exceed carrying capacity, but, as will be shown, other devotees of sustainability have done so.

Since the Rio conference, more than 130 countries have created new bureaucracies to implement Agenda 21 and its requirements for sustainable development, according to the Earth Council, whose head is Maurice Strong, director of the Rio conference and now assistant secretary general of the United Nations (Earth Council 1997). Many local and regional compacts for sustainable development exist in the United States, stretching from Florida through Missouri to Santa Cruz and Humboldt County, California. Henry Lamb of the Environmental Conservation Organization has described some of them, including the statewide plans for Florida and Missouri (1998).

Sustained by foundation money and federal grants, rarely mentioning Agenda 21, salaried environmental activists are convening unsuspecting local citizens to engage in the "visioning" process to plan for the sustainable community in their future. Vice President Gore's Clean Water Initiative and the administration's American Heritage Rivers Initiative are nurturing the process by encouraging local "watershed councils" to make comprehensive plans for their regions.

Herman Daly's Apocalyptic Vision

Probably not many of these souls have read the works of Herman Daly or Maurice Strong, the Rio documents, or the modern college textbooks in sustainable economics. If they had, they might be less eager to help. Daly, an economist, first came to national attention during the 1970s when the Joint Economic Committee of Congress published his plan for reducing births by government licensing. As in China, the government would issue the licenses in the restricted numbers requisite for achieving its population targets, and persons attempting to give birth without licenses would be punished. Unlike the Chinese system, the licenses could be bought or sold, as in the modern schemes for emissions control (Daly 1976).

People of common sense hearing such schemes tend to find them fantastic and amusing. But the World Bank was so enchanted by Daly's notions that it gave him a job as a senior economist in the Environment Department. In 1990 he and a theologian co-author, John B. Cobb, Jr., published their comprehensive plan for the salvation of the world, *For the Common Good: Redirecting the Economy towards Community, the Environment and a Sustainable Future.* Disputing major teachings of economics, the authors called for university "reform" to reduce the influence of economics and increase attention to the "social and global crisis" (357–60). That reform, of course, is now going forward. Like other leaders of mass movements, they argued that logical reasoning is greatly overdone and called for "a conscious shift toward... relativisation" (359). Such a shift also is rapidly occurring. Daly's hostility toward economics is not unique; many aspiring world-changers have seen economics, with its emphasis on logical reasoning based on fact, as the enemy of their plans.

Daly and Cobb called for the conversion of "half or more" of the land area of the United States to unsettled wilderness inhabited by wild animals (255), the abolition of private land ownership (256–59), a giant forced reduction in trade and a change to self-sufficiency at not only the national level but at local levels also (229–35, 269–72), government controls to reduce output to "sustainable biophysical limits" (whatever those might be) (143), and the resettlement of a large portion of the population to rural areas (264, 311)—remember Cambodia and Pol Pot, who has been called "the ultimate deep ecologist."

Moreover, they wanted a prohibition of the movement of private wealth (221, 233)—so much for any escape from the sustainable paradise—the abolition of direct elections, except for local officials who would in turn elect higher officers of the government (177), and, of course, complete population control by means of birth licenses. The intent was to promote the "biospheric vision" in the spirit of "deep ecology," which sees the need for a "substantial decrease in the human population" to promote "the flourishing of nonhuman life" (377). They added that this necessary reduction in the "human niche," a phrase echoed in subsequent United Nations documents, might be achieved either by a fall in population or by a decline in resource consumption (378).

Daly and Cobb understood that these vast changes would require some readjustments in attitudes, to say the least, and saw hope in the "influence of ecological and feminist sensitivities" (377). Not only have those attitude adjustments materialized, but academic economics, identified by Daly as the enemy, has also been remarkably helpful, producing quantities of new books and courses on sustainable development and related topics. Generous grants from government, foundations, and international agencies have encouraged this outpouring.

The justification for these massive changes in human life on the planet lay in what Daly and Cobb called "the wild facts"—that is, the alleged extinction of species, the ozone hole, the greenhouse effect, acid rain, and the imminent exhaustion of oil supplies. The last, of course, has disappeared from the current list of portending calamities; but never mind, we now have deforestation and the methane crisis. In any event, the bottom line was that we suffer from an excessively human-centured point of view, and people should be taught to adopt the "biospheric vision" (376) in recognition of our "community with other living things" in the spirit of "deep ecology."

Daly and Cobb provided no evidence of any of the catastrophes they listed and even acknowledged some uncertainty about the "precise physical effects" (416). Nevertheless, they insisted that the impending crises were "facts" that could not be denied. Scientific disputes over these matters have expanded since then, prompting the True Believers to develop new arguments.

Some of us may wonder whether the work we do makes any difference in the scheme of things. Daly and Cobb need have no such concerns. Their words, phrases, and arguments now appear throughout the United Nations documents on the sustainable society and the literature of sustainable economics. And Daly, now at the University of Maryland, has reiterated his vision in a 1996 book, *Beyond Growth: The Economics of Sustainable Development*. Together with Robert Costanza, Daly now directs the International Society for Ecological Economics, based in Solomons, Maryland.

Steven Hackett's Contribution

The nature of current college instruction in the field can be seen in a new textbook, *Environmental and Natural Resources Economics: Theory, Policy, and the Sustainable Society* (1998), by Steven C. Hackett, who teaches economics at Humboldt State University. As in Daly's case, Hackett's justifications for proposing fundamental social change are the imperiled biosphere and "the continued growth of human population," which causes "loss of biodiversity" and "deteriorating... wilderness areas" (12, 13), and many other ills.

On these points, there is serious debate, as the author admits. He insists nevertheless on "the potential for catastrophic change in the global climate... rising sea levels... inundation of... low-lying areas... desertification of... grain-producing areas... mass hunger... and... rapid loss of biodiversity" (12). These dire forecasts, of course, have been featured on television for a generation and will probably not unduly alarm modern students. Nor will these hardened young consumers of doomsday prophecies be surprised to learn that population growth threatens the "habitats of many of the world's species of animals and plants... the integrity of the world's remaining temperate zone wilderness areas, coral reefs and other marine ecosystems, and tropical rainforests" (12, 13)....

Is the Earth Overpopulated?

Overpopulation, according to Hackett, is a major cause of our doleful condition. Having softened the obviously elitist implications of the diagnosis by professing his concern for injustice, he can get on with the real message. The prolific people of the less developed countries are wreaking havoc on their "fragile environments," engaging in "deforestation... migration to... polluted urban areas... massive environmental degradation" (13), and so forth. Unmentioned are the government policies that create these disasters, such as the destructive taxation of farmers' productivity, the government monopolies that underpay and overcharge the people, the confiscation of traders' stocks and pack animals, the endless wars financed by foreign aid.

Hackett doesn't mention the large current declines in fertility and population growth rates throughout the world. United Nations figures show that seventy-nine countries with 40 percent of the world's population now have fertility rates too low to prevent ultimate population decline in those countries (U.N. Population Division 1996). But this evidence gives little comfort to Hackett, who quotes estimates showing that "2 to 5 hectares of productive land are needed to support... the average person... in an industrialized country [whereas]... the world has only 1.5 hectares per capita of ecologically productive land... and... only 0.3 hectare per capita are suitable for agricultural production" (263). In other words, not only does the less developed world have far too many rapidly multiplying people, but *population in the industrialized countries is several times too large.*

As he does throughout the book, Hackett hedges by saying that we don't really know our "carrying capacity," but the undergraduate reader is going to learn that, whatever that capacity may be, there are already far, far too many people on the earth. In a like vein, Paul Ehrlich, famous for his unblemished record of wrong forecasts, has said the world has "perhaps" five times as many people as it can tolerate (Ehrlich 1989).

Let us not imagine, therefore, that the advocates of the sustainable society are merely talking about cleaning up pollution and giving birth control pills to people in Africa, Asia, and Latin America. Although present State Department and U.N. efforts to restrain the increase of dark-skinned people are very strenuous indeed, they are seen as not nearly enough. Hackett quotes Devall on the desirability of "a substantial decrease of the human population" (20). And he describes the "coercive fertility-control" in China (234) and the proposals of Daly and Cobb and Kenneth Boulding for birth quotas. Spokesmen for the Clinton administration, such as Timothy Wirth, have specified that world population control must include the United States (Wirth 1996). Notice, too, that all of the sustainable society documents call for "population stabilization," without saying whether that is to occur at a population size larger or smaller than the present population.

We hope no guilt-ridden students rush to jump out of our overladen lifeboat before, first, asking why Hackett, Daly, Cobb, and Ehrlich have not done so already and, second, hearing some other information. Again according to Ehrlich and other, more reliable, sources, human beings actually occupy between 1 and 3 percent of the world's land area (Vitousek et al. 1986). The entire world population could be put into the state of Texas, leaving the rest of the world devoid of people. The population density of that giant city of Texas would be about 20,000 persons per square mile, which is somewhat higher than in San Francisco but lower than in Brooklyn (5.9 billion world population divided by 262,000 square miles of land in Texas implies 22,500 persons per square mile, or 1,200 square feet per person).

Farmers use less than half of the world's arable land (Revelle 1984). The world food supply has increased a great deal faster than population since 1950, according to the Food and Agriculture Organization (U.N. Food and Agriculture Organization 1996)....

Market Failure?

Hackett has little hope that existing institutions can steer the earth away from the looming catastrophes. As for markets, they *"reinforce* self-interested behavior" (29). One searches Hackett's book in vain for any sign of understanding Adam Smith's "invisible hand" that leads men to serve one another and to economize in their use of resources as they pursue their own self-interest. There is no sign that Hackett has ever read the great economist John Maurice Clark, who called the market "our main safeguard against exploitation" because it performs "the simple miracle whereby each one increases his gains by increasing his services rather than by reducing them" (1948). He seems unaware of Walter Eucken's perception that markets break up the great concentrations of economic power (1950) or F. A. Hayek's (1948) and Ludwig von Mises's (1949) realization that markets provide otherwise unavailable information about the scarcity of the resources that are the focus of his concerns.

This is not to argue that markets will solve all economic problems. Well-known and much-discussed problems of externalities, public goods, and common pool resources, sometimes arise, as Hackett notes. But the nonmarket economies of this century have provided vivid object lessons in the pitfalls of "communitarian" planning, and the work of James Buchanan, Gordon Tullock, and others has pointed up the perverse incentives that infest the public sector as it goes about trying to correct "market failure."

At times Hackett acknowledges that public ownership and management do not always produce ideal results, but for the most part he sees the market as the villain and concludes that our best hope lies in "cooperative rather than noncooperative decision making" (91). It is a conclusion he draws from game theory, and it leads to his hopes for "sustainable development" through small-group negotiations. On this issue, more later....

Not surprisingly, Hackett finds private property highly suspect: "It is clear that systems centered around private property ... can conflict with the common good" (26). After a brief discussion of John Locke and proposals for protecting natural resources by assigning private property rights to them, Hackett points students to a patron saint of the French Revolution: "From Rousseau's perspective ... private property rights ... alienate people from nature ... [and] lead to inequality ... and wars." He quotes the great man: "Competition and rivalry ...

opposition of interests . . . and always the hidden desire to profit at the expense of others. All these evils were the first effect of property" (25–26).

Such an indictment demands a response. Private owners did not hunt the buffalo almost to extinction. And it was not a private property system that sent millions to the gulag. When the Ethiopian government socialized the privately owned donkeys, most of them perished (Deressa 1985). I keep the off-road vehicles out of my private forest. And the biblical good shepherd was not the government or the assembly of "stakeholders" in the "sustainable community"; he was the *owner* of the sheep. Where does the common good lie in these decisions? And, most important, Who decides what the common good is? In fairness, also, Hackett might have mentioned the bloodbath that Rousseau's ideas encouraged. Like Devall, Daly, and other environmental utopians of our own time, Rousseau distrusted reason and argued for going "back to Nature." Ever the romantic, he sent his five children to a foundling home (Gauss 1972).

Economists have long noted that voluntary trade must make its participants better off or they wouldn't engage in it, whether they are children trading the contents of their trick-or-treat bags or Mexicans buying used bottles from California to turn them into gravel. Adam Smith and David Ricardo, and even Sir Dudley North before them, saw it as the solution to the uneven distribution of resources. Hackett, however, like Daly and Cobb, whom he quotes at length, lists many objections to trade. It "may . . . allow rich countries to import pollution-intensive, resource-intensive, and endangered-species products they do not wish to produce themselves and to export their toxics and trash" (225). It "tends to erode livable wages, the bargaining power of unions, and environmental and other standards of communities" (226). It "undermines sustainability" (227) and "has put great pressure on . . . endangered wildlife" (229).

Nevertheless, Hackett concludes that although there are "important questions" about how much and what kind of trade to allow, "it is neither practical nor desirable to eliminate trade completely" (230). What a relief. Clearly, however, what is left will be a far cry from free trade, just as all other human activity will be far from free in the "sustainable society."

Throughout the world, controllers and would-be controllers have seen, to use Smith's phrase, the human "propensity to truck, barter, and exchange" as a resource to be exploited or suppressed for the benefit of those in power. From mercantilist England, France, and Spain to the recent Soviet Union and modern Ethiopia, governments have sought to channel this propensity, always with the result of impoverishing their subjects. To illuminate the ill effects of trade controls was the main task of Smith's *Wealth of Nations*. That modern proponents of the "sustainable society" should be so eager to revive such controls should give us pause—doubly so because these people *intend* to reduce human consumption, and they understand very well that trade restrictions do impoverish people.

Like Daly, Hackett takes a dark view of what he calls "mainstream economics." Students who have studied economics, according to Hackett, are less altruistic than other students (28). Economics itself, he maintains, tends to reduce everything to a monetary cost-benefit comparison without recognizing "intrinsic" values. In his view, however, not all intrinsic values are equally worthy of recognition. Individual rights are especially suspect. By contrast, the "sustainability ethic holds the interdependent health and well-being of human communities and earth's ecology over time as the basis of value" (209), and is therefore clearly superior to the viewpoint of mainstream economics.

Economics and Ethics

Private property, the market, and economics itself, it would seem, are the bad fruit of a bad tree, the disordered ethical system of contemporary society. Hackett blames the shortcomings of economics on its "teleological ethics"—that is, the end justifies the means—attributing the idea to "religious philosophers" (21). This reference enables him to take a swipe at both religion and economics. Evidently, Hackett either never had catechism or was inattentive when Sister told him the end does not justify the means. His example is "utilitarianism," which he describes as the "normative base" for "much of the traditional economic perspective" (21). His straw man is Jeremy Bentham, a nineteenth-century eccentric who had his body stuffed and put in a glass case after he died so it could be on view for University College, London, undergraduates for all time (Mack 1972).

Bentham's mechanical pleasure-pain calculus has amused students for generations, but other men—Smith, Jean Baptiste Say, Ricardo, Carl Menger, Alfred Marshall, and others—did the serious work of showing how the market reveals and reconciles the varied and conflicting desires of multitudes of individuals, channeling their self-interest to the service of others in their pursuit of individual gain.

These monumental themes receive barely a glance from Hackett, who remains intent on showing the failures of market calculations and the need for more sublime direction by persons imbued with the spirit of the sustainable community and tutored in sustainable economics. To illustrate, Hackett poses the "question of whether an action (for example, policy protecting old-growth forest) is to be judged on its intrinsic rightness or based on the measurable benefits and costs that might result" and "the proper balance between individual self-interest and the common good," again undefined (17–18).

There ensues a discussion of the "fundamentals of ethical systems," beginning with "deontological ethics," which judges an action by "its intrinsic rightness" (19). As an example, Hackett quotes at length from the "ecosophy," or "earth wisdom," of Bill Devall, George Sessions, and Arne Naess:

> The well-being and flourishing of human and non-human life on Earth have value in themselves....
>
> The flourishing of human life and cultures is compatible with a substantial decrease of the human population. The flourishing of non-human life requires such a decrease.
>
> Those who subscribe to the foregoing points have an obligation... to... implement the necessary changes. (20, quoting Devall 1988)

Clearly, this call is not for minor adjustments in lifestyle. A "substantial decrease of the human population" is no small thing. Our "obligation ... to ... implement the necessary changes" is a profoundly serious matter. This proposal is not a nickel-and-dime deal. True, Hackett is only quoting Devall at this point, but his discussion makes it clear that Devall's insistence on "intrinsic rightness" is a far more beautiful thing than the crass monetary valuations of "utilitarian" economics.

To make the issue perfectly clear, Hackett offers an example. Suppose an endangered species is threatened by development. Guess what will happen in a "society that views the existence of a species as being of intrinsic value" (... la Bill Devall). Then guess what will happen if a monetary cost-benefit comparison determines the outcome. Obviously, all economists, except an enlightened few, should be taken out and shot.

Nowhere in Hackett's discussion of ethics does he refer to the Judeo-Christian tradition of stewardship—the admonition to "keep" the earth (Gen. 2:15), the prescribed days of rest for men and beasts (Deut. 5:14), the prescribed years of rest for the land (Lev. 25:4), the love of nature with its "Leviathan" taking its sport in the sea and its "coneys" among the rocks (Ps. 104), its cedars of Lebanon (Ps. 92), its hills that "rejoice on every side" and its valleys that "laugh and sing" (Ps. 65), and the strict injunctions against the worship of nature and the human sacrifice that often accompanied it (Deut. 17:3, 20:2–6; 2 Kings 17; Job 31:26).

Modern economic reasoning does not destroy these values any more than modern atmospheric science destroys the beauty of a sunset. Certainly, the sin of greed has always beset the race, as has idolatry. Just as certainly, modern economics has its idolaters as well as its Midases, but such corruption is nothing new on earth. Economic reasoning enables us to compare alternatives. It enables us to see that a society following the romanticism of Devall or Daly would probably be no more attractive or healthful than the one we have. One of the greatest tragedies of our time is not that undergraduates study economics but that they study so little of the great civilizing themes of our heritage—our great literature, art, and music, our legacies from the ancient Greeks, our tradition of human rights and our history of the struggle for liberty—and that they know so little about Christianity or Judaism. Thus deprived, they are left vulnerable, not so much to "utilitarianism" as to environmental lunacy.

Worse yet, as John Grobey, professor of economics and a senior colleague of Hackett at Humboldt State University, has noted, the result must be to deprive young people of the traditional birthright of youth—hope for the future. Taught from their earliest years that their own burgeoning humanity is destroying the earth and all of nature, the youth of today face a more depressing prospect than perhaps any previous generation. No wonder the doubling of the suicide rate among children aged ten to fourteen since 1980 (U.S. Bureau of the Census 1997). No wonder the epidemic of school shootings. No wonder the recent case in Humboldt County in which a young man on trial for attempted murder gave as his defense "overpopulation, dwindling resources and the certain doom of the planet" (Parker 1998).

The changes in "basic economic, technological, and ideological structures" called for by Devall obviously threaten traditional views of individual rights to life, liberty, and property. The question that occurs to a mainstream economist at this point is, Just which individuals will be given the awesome responsibility of determining the "common good" and the best interests of the community and the ecology? And what will happen to human beings, stripped of individual rights, who get in the way of the grand march to the sustainable community? Hackett gives hints but no answers. He acknowledges the seminal work of Daly, but without mentioning Daly's call for massive resettlement of populations. The question remains: Is the centuries-long pilgrimage from Magna Carta through *Areopagitica* and the Bill of Rights to Selma to be renounced now in the name of the environment? Will this denouement be the Clinton legacy?

No Price Is Too Great

. . . Hackett makes it sound as if the sustainable society will be brought about by local meetings of "stakeholders" negotiating over local issues. But undergirding these cozy negotiations will be "regulations, taxes, subsidies, and direct finding of clean technology" (277). Of course, the Sierra Club will be there to help.

Here is the rub. To avert a highly problematic future disaster, much disputed by competent scientists, Hackett and his soul-mates in the United Nations and the Clinton Council on Sustainable Development would require human beings to submit to a gigantic present sacrifice of freedom, human dignity, and material welfare in a regime controlled by unelected officials of a global eco-bureaucracy. Have we learned nothing from the utopian horrors devised for us during the past century?

People do love nature. The tremendous expansion of national parks and conservation areas during this century testifies to that love. The environmental movement itself is an expression of our determination not to let the industrial age destroy the oceanic Leviathan and the cedars of Lebanon. The real danger now, however, is not that we stand on the verge of destroying nature but that, stampeded by environmental terrors on every hand, we are plunging over the cliff into totalitarianism.

POSTSCRIPT

Is Sustainable Development Compatible With Human Welfare?

It is a truism that human impact on the environment depends on both human numbers and human activities. Those who contend that sustainability is not an issue often point out that all six-billion-plus living humans could be moved into a relatively small area such as Texas, leaving all the rest of the planet empty. However, it must be remembered that each human requires space on which to grow food and fiber and from which to extract energy and mineral resources. Also, the majority of the world's people do not share in the standard of living typical of the developed nations and are trying very hard to change that. If they succeed, the human impact on the environment will increase tremendously. Most projections say that demand for land, energy, and resources would then exceed supply. The world would be raped to meet the needs of the present generation, and nothing would be left for future generations. In "Windows on the Future: Global Scenarios and Sustainability," *Environment* (April 1998), Gilberto C. Gallopin and Paul Raskin assert that because the world is so interconnected today, there can be no separate solutions for the rich and the poor. They state, "There is no question that the contradiction between the modern world's imperative toward growth and the Earth's finite resources will ultimately be resolved in some way. The only question is how that will come about—whether through enlightened management, economic and environmental catastrophe, or some other means." See also Garrett Hardin's *Living Within Limits: Ecology, Economics, and Population Taboos* (Oxford University Press, 1993).

Sustainable economics is addressed in Louis P. Pojman's *Global Environmental Ethics* (Mayfield, 2000). In it, Pojman reviews the difficulties posed by global inequities.

The first of the Rio Declaration on Environment and Development's 27 principles says, "Human beings are at the centre of concerns for sustainable development. They are entitled to a healthy and productive life in harmony with nature." Therefore, any solution to the sustainability problem should not infringe on human welfare. This makes any solution that involves limiting or reducing human population or blocking improvements in standards of living very difficult to sell. Yet solutions may be possible. In *The Natural Wealth of Nations: Harnessing the Market for the Environment* (W. W. Norton, 1998), David Malin Roodman suggests that taxing polluting activities instead of profit or income would stimulate corporations and individuals to reduce such activities or to discover nonpolluting alternatives. In "Building a Sustainable Society," in *State of the World 1999* (W. W. Norton, 1999), Roodman adds recommendations for citizen participation in decision making, education efforts, and global cooperation, without which we are heading for "a world order [that] almost no one wants." (He is referring to a future of environmental crises, not the "new world order" feared by many conservatives, in which national policies are dictated by international [UN] regulators—the "unelected officials of a global eco-bureaucracy" decried by Kasun.)

Julie L. Davidson, in "Sustainable Development: Business as Usual or a New Way of Living?" *Environmental Ethics* (Spring 2000), states that efforts to achieve sustainability cannot by themselves save the world but that such efforts may give us time to achieve new and more suitable values. Davidson would likely be heartened by the United Nations World Summit on Sustainable Development, which was held in August 2002 in Johannesburg, South Africa. Its aim was to strengthen partnerships between governments, businesses, nongovernmental organizations, and other stakeholders and to eradicate poverty and make the distribution of the benefits of globalization more equal. See Gary Gardner, "The Challenge for Johannesburg: Creating a More Secure World," in *State of the World 2002* (W. W. Norton, 2002) and the United Nations Environmental Programme's *Global Environmental Outlook 3* (Earthscan, 2002), produced as a "global state of the environment report" in preparation for the Johannesburg summit.

Contributors to This Volume

EDITORS

THOMAS A. EASTON is a professor of life sciences at Thomas College in Waterville, Maine, where he has been teaching since 1983. He received a B.A. in biology from Colby College in 1966 and a Ph.D. in theoretical biology from the University of Chicago in 1971. He has also taught at Unity College, Husson College, and the University of Maine. He is a prolific writer, and his articles on scientific and futuristic issues have appeared in the scholarly journals *Experimental Neurology* and *American Scientist,* as well as in such popular magazines as *Astronomy, Consumer Reports,* and *Robotics Age.* He is also the science columnist for the online magazine *Tomorrowsf* (http://www.tomorrowsf.com). His publications include *Focus on Human Biology,* 2d ed., coauthored with Carl E. Rischer (HarperCollins, 1995) and *Careers in Science,* 3rd ed. (National Textbook, 1996). Dr. Easton is also a well-known writer and critic of science fiction.

THEODORE D. GOLDFARB (d. 2002) was a professor of chemistry at the State University of New York at Stony Brook. He earned a B.A. from Cornell University and a Ph.D. from the University of California, Berkeley. He was the author of over 35 research papers and articles on molecular structure, environmental chemistry, and science policy, as well as the book *A Search for Order in the Physical Universe* (W. H. Freeman, 1974). He was also the editor of *Sources: Notable Selections in Environmental Studies* (Dushkin/McGraw-Hill). Dr. Goldfarb was a recipient of the State University of New York's Chancellor's Award for Excellence in Teaching. He was an active member of several professional organizations, including the American Chemical Society, the American Association for the Advancement of Science, and the New York Academy of Sciences.

STAFF

Theodore Knight   List Manager
David Brackley   Senior Developmental Editor
Juliana Gribbins   Developmental Editor
Rose Gleich   Administrative Assistant
Brenda S. Filley   Director of Production/Design
Juliana Arbo   Typesetting Supervisor
Diane Barker   Proofreader
Richard Tietjen   Publishing Systems Manager
Larry Killian   Copier Coordinator

AUTHORS

SPENCER ABRAHAM was sworn in as the tenth secretary of energy on January 20, 2001. Before that he had been a Senator (R-Michigan) and cochairman of the National Republican Congressional Committee.

JANET N. ABRAMOVITZ is a senior researcher at the Worldwatch Institute.

RONALD BAILEY is science correspondent for *Reason* magazine. A member of the Society of Environmental Journalists, his articles have appeared in many popular publications, including the *Wall Street Journal*, *The Public Interest*, and *National Review*. He has also produced several series and documentaries for PBS and ABC News. Bailey was the Warren T. Brookes Fellow in Environmental Journalism at the Competitive Enterprise Institute in 1993. He is the editor of *Earth Report 2000: Revisiting the True State of the Planet* (McGraw-Hill, 1999) and the author of *ECOSCAM: The False Prophets of Ecological Apocalypse* (St. Martin's Press, 1993).

ROGER BATE is a director of Africa Fighting Malaria, a South African humanitarian group. He is also an adjunct fellow at the Competitive Enterprise Institute.

CAROL M. BROWNER has been the administrator for the Environmental Protection Agency since 1993. She has also served as director of Florida's Department of Environmental Regulation, as legislative director for then-senator Al Gore, and as counsel for the U.S. Senate Committee on Energy and Natural Resources. She earned her bachelor's degree and her law degree from the University of Florida in 1977 and 1979, respectively.

ROBERT D. BULLARD is Ware Professor of Sociology and director of the Environmental Justice Resource Center at Clark Atlanta University. He is the author of numerous books, including *Dumping in Dixie: Race, Class and Environmental Quality* (Westview Press, 1990, 1994, 2000), *People of Color Environmental Groups Directory 2000* (Charles Stewart Mott Foundation, 2000), *Unequal Protection: Environmental Justice and Communities of Color* (Sierra Club Books, 1994), and *Confronting Environmental Racism: Voices From the Grassroots* (South End Press, 1993).

JON CHRISTENSEN is a freelance investigative reporter and science writer based in Carson City, Nevada. He has written for the *New York Times*, *High Country News*, *Mother Jones*, *Outside*, and others. He is the author of *Nevada* (Graphic Arts Center Publishing, 2001). Christensen also edits GreatBasinNews.com and produces features for Nevada Public Radio. He was awarded a Knight Journalism Fellowship at Stanford University for the 2002–2003 academic year and will focus on evaluating environmental conservation projects.

GREGORY CONKO is a policy analyst and director of Food Safety Policy at the Competitive Enterprise Institute. He specializes in issues of food and pharmaceutical drug safety regulation and on the general treatment of health risks in public policy. Conko's interests include the safety of genetically engineered foods and the application of the precautionary principle to domestic and international environmental and safety regulations.

GIULIO A. De LEO is an associate professor of applied ecology and environmental impact assessment in the Dipartimento di Scienze Ambientali at the Universit degli Studi di Parma in Parma, Italy.

RICHARD A. DENISON is a senior scientist with the Environmental Defense Fund (EDF) in Washington, D.C., and an expert on waste management and incineration. He headed the EDF's delegation on the EDF-McDonald's Waste Reduction Task Force, beginning in 1990. He earned his Ph.D. in molecular biophysics and biochemistry from Yale University. He is coeditor, with John Ruston, of *Recycling and Incineration: Evaluating the Choices* (Island Press, 1990).

ANNE H. EHRLICH is a senior research associate in biological sciences at Stanford University. She is coauthor, with Paul R. Ehrlich, of *Betrayal of Science and Reason: How Anti-Environmental Rhetoric Threatens Our Future* (Island Press, 1996).

PAUL R. EHRLICH is Bing Professor of Population Studies at Stanford University. He is the author of *The Population Bomb* (Ballentine Books, 1968). He is coauthor, with Anne H. Ehrlich, of *The Population Explosion* (Simon & Schuster, 1990).

DAVID FRIEDMAN is a writer, an international consultant, and a fellow in the MIT Japan program.

MARINO GATTO is a professor of applied ecology in the Dipartimento di Elettronica e Informazione at Politecnico di Milano in Milan, Italy. His main research interests include ecological models and the management of renewable resources. Gato is associate editor of *Theoretical Population Biology*.

LINDA GREER, a senior scientist with the Natural Resources Defense Council, has worked for more than 15 years on the regulation of toxic chemicals and hazardous waste.

BRIAN HALWEIL is a staff researcher at the Worldwatch Institute.

CHRIS HENDRICKSON is a professor of civil engineering in and associate dean of the College of Engineering at Carnegie Mellon University in Pittsburgh, Pennsylvania.

PETER W. HUBER is a senior fellow at the Manhattan Institute. He has authored or edited a number of books, including *Galileo's Revenge: Junk Science in the Courtroom* (Basic Books, 1993), *Law and Disorder in Cyberspace: Abolish the FCC and Let Common Law Rule the Telecosm* (Oxford University Press, 1997), and *Hard Green: Saving the Environment From the Environmentalists: A Conservative Manifesto* (Basic Books, 2000).

JACQUELINE R. KASUN is a professor emeritus of economics at Humboldt State University and editorial director of the Center for Economic Education in Bayside, California.

SHELDON KRIMSKY is a professor of urban and environmental policy at Tufts University and the author of *Hormonal Chaos: The Scientific and Social Origins of the Environmental Endocrine Hypothesis* (Johns Hopkins University Press, 2000).

DAVID N. LABAND is a professor of economics and policy at the Forest Policy Center in the School of Forestry and Wildlife Sciences at Auburn University in Alabama. He is the author, with George McClintock, of *The Transfer Society: Economic Expenditures on Transfer Activity* (National Book Network, 2001). Laband's research interests include forest economics, causes and consequences of environmental policy, and land use planning.

LESTER LAVE is the James H. Higgins Professor of Economics at Carnegie Mellon University in Pittsburgh, Pennsylvania, with appointments in the Graduate School of Industrial Administration, the School of Urban and Public Affairs, and the Department of Engineering and Public Policy. He received a Ph.D. in economics from Harvard University and was a senior fellow at the Brookings Institution from 1978 to 1982.

DWIGHT R. LEE is the Ramsey Professor of Economics and Private Enterprise in the Terry College of Business at the University of Georgia. He received his Ph.D. from the University of California at San Diego in 1972 and his research has covered a variety of areas, including personal finance, public finance, the economics of political decision making, and the economics of the environment and natural resources. Lee has published over 100 articles and commentaries in academic journals, magazines, and newspapers. He is coauthor, with Richard B. McKenzie, of *Getting Rich in America: Eight Simple Rules for Building a Fortune and a Satisfying Life* (HarperBusiness, 2000).

AMORY B. LOVINS is a physicist and cofounder and CEO of the Rocky Mountain Institute, a nonprofit applied research center that fosters the efficient and restorative use of resources to create a more secure, prosperous, and life-sustaining world. He has written numerous books and essays on energy policy.

L. HUNTER LOVINS is a lawyer, political scientist, and cofounder of the Rocky Mountain Institute. With Amory Lovins she was designated a 2000 *Time* magazine Hero for the Planet, and she shared the 1999 Lindbergh Award, the 1993 Nissan Award, the 1983 Right Livelihood Award, and the 1982 Mitchell Prize. She is coauthor, with Paul Hawken and Amory B. Lovins, of *Natural Capitalism: Creating the Next Industrial Revolution* (Back Bay Books, 2000).

ANNE PLATT McGINN is a senior researcher at the Worldwatch Institute and the author of "Why Poison Ourselves? A Precautionary Approach to Synthetic Chemicals," Worldwatch Paper 153 (November 2000).

FRANCIS McMICHAEL is the Blenko Professor of Environmental Engineering at Carnegie Mellon University in Pittsburgh, Pennsylvania.

DANIEL B. MENZEL is a professor in and chair of the Department of Community and Environmental Medicine at the University of California, Irvine. His current research interests are the effects of drinking water contaminants and the effects of air pollution on acute and chronic disease. He is coeditor, with Fred J. Miller, of *Fundamentals of Extrapolation Modeling of Inhaled Toxicants: Ozone and Nitrogen Dioxide* (Hemisphere, 1984).

HENRY I. MILLER is a research fellow at the Hoover Institution at Stanford University and the author of *Policy Controversy in Biotechnology: An Insider's View* (Academic Press, 1997). His research focuses on public policy toward science and technology, especially pharmaceutical development and the new biotechnology. His work often emphasizes the excessive costs of government regulation and models for regulatory reform.

STEPHEN MOORE is a director of the Cato Institute and a contributing editor to *National Review*. Moore is also the president of the Club for Growth.

RAYMOND J. PATCHAK, a certified hazardous materials manager, has over 10 years of environmental regulatory compliance experience, during which time he has been responsible for auditing and developing compliance programs at many different types of manufacturing and waste-handling facilities. An active member of the Academy of Certified Hazardous Materials Managers (ACHMM), he is a founding member of the ACHMM's ISO 14000 Committee as well as president of the Michigan chapter of the ACHMM.

DINAH M. PAYNE is a professor of management at the University of New Orleans in Louisiana. She teaches business law, business ethics, and international management.

DAVID PIMENTEL, a professor at Cornell University in Ithaca, New York, holds a joint appointment in the Department of Entomology and the Section of Ecology and Systematics. He has served as consultant to the Executive Office of the President, Office of Science and Technology, and as chairman of various panels, boards (including the Environmental Studies Board), and committees at the National Academy of Sciences, the United States Department of Energy, and the United States Congress.

CECILY A. RAIBORNE is a professor of accounting at Loyola University in New Orleans, Louisiana. She teaches financial and managerial accounting.

BERNARD J. REILLY is corporate counsel at DuPont, where he has been managing the legal aspects of the company's Superfund program since 1986.

JOHN F. RUSTON is an economic analyst with the Environmental Defense Fund in New York City. He earned his bachelor's degree in environmental policy and planning from the University of California at Davis and his master's degree in city planning from the Massachusetts Institute of Technology. He is coeditor, with Richard A. Denison, of *Recycling and Incineration: Evaluating the Choices* (Island Press, 1990).

STEPHEN H. SAFE is in the Department of Veterinary Physiology and Pharmacology at Texas A&M University in College Station, Texas. He is the founding principal of Wellington Environmental Consultants, now Wellington Laboratories, and he is a member of the American Association for Cancer Research and the American College of Toxicology.

CHARLES W. SCHMIDT is a freelance science writer based in Portland, Maine. He received an M.S. in public health from the University of Massachusetts at Amherst, with a concentration in toxicology. He has written for *Environmental Health Perspectives, Technology Review, New Scientist, Popular Science,* the *Washington Post, Child* magazine, *Environmental Science and Technology,* and more.

KEVIN A. SHAPIRO is a researcher in neuroscience at Harvard University.

WILLIAM R. SMITH, a certified hazardous materials manager and an ISO 14001 Environmental Management Systems (EMS) auditor, is a principal at Competitive Edge Environmental Management Systems, Inc. He has served on the Registrar Accreditation Board's Auditor Certification Board, which governs the ISO 14000 EMS auditor certification program in the United States, and he is a board member of the Academy of Certified Hazardous Materials Managers. He is also the founding chairman of the ISO 14000 Committee.

PAUL L. STEIN is a judge of the Court of Appeal in New South Wales, Australia.

BRIAN TOKAR is an associate faculty member at Goddard College in Plainfield, Vermont. A regular correspondent for *Z* magazine, he has been an activist for over 20 years in the peace, antinuclear, environmental, and green politics movements. He is the author of *The Green Alternative: Creating an Ecological Future,* 2d. ed. (R & E Miles, 1987).

CHRISTOPHER VAN LÖBEN SELS is a senior project analyst for the Natural Resources Defense Council in San Francisco, California.

TED WILLIAMS has been a regular contributor to *Audubon* for over 15 years, during which he has covered a variety of topics, including gold mining in Alaska and the Northern Forest of the Northeast.

E. O. WILSON is Pellegrino University Professor and curator of entomology at the Museum of Comparative Zoology at Harvard University. Two of his books, *On Human Nature* (Harvard University Press, 1978) and, with Bert Hölldobler, *The Ants* (Belknap Press, 1990), have been awarded the Pulitzer Prize. In 1998 he published *Consilience: The Unity of Knowledge* (Alfred A. Knopf).

Index